AutoDesk® VIZ in Manufacturing Design:

AutoDesk® VIZ / 3ds max™ for Engineering and Technology

AutoDesk® VIZ in Manufacturing Design:

AutoDesk® VIZ / 3ds max™ for Engineering and Technology

JON M. DUFF

autodesk Press

THOMSON
DELMAR LEARNING

Australia • Canada • Mexico • Singapore • Spain • United Kingdom • United States

autodesk Press

AutoDesk® VIZ in Manufacturing Design:
AutoDesk® VIZ / 3ds max™ for Engineering and Technology
Jon M. Duff

Executive Director:
Alar Elken

Executive Editor:
Sandy Clark

Acquisitions Editor:
James Devoe

Development Editor:
John Fisher

Executive Marketing Manager:
Maura Theriault

Channel Manager:
Fair Huntoon

Marketing Coordinator:
Sarena Douglass

Executive Production Manager:
Mary Ellen Black

Production Manager:
Larry Main

Production Editor:
Stacy Masucci

Art/Design Coordinator:
Mary Beth Vought

Editorial Assistant:
Mary Ellen Martino

Composition:
The West Highland Press

Cover Art:
Jon M. Duff

Printed in Canada

1 2 3 4 5 XX 06 05 04 03 02

For more information contact
Delmar Learning,
Executive Woods
5 Maxwell Drive, PO Box 8007,
Clifton Park, NY 12065-8007

Or find us on the World Wide Web at
http://www.delmar.com

Library of Congress
Cataloging-in-Publication Data:

ISBN 01-4018-8420-2

NOTICE TO THE READER

CONTENTS

CHAPTER 6—MODELING WITH NURBS

CHAPTER 7—MODELING WITH MODIFIERS

CHAPTER 8—MODELING ASSEMBLIES

CHAPTER 9—MATERIAL BASICS

CHAPTER 16—ANIMATION BASICS

CHAPTER 17—CAMERA ANIMATION/THE FLY-AROUND

CHAPTER 18—PROGRAMMING IN MAXSCRIPT

CHAPTER 19—STILL AND MOVIE OUTPUT

PREFACE

As a designer, you have found (or will soon find) that if two designs share equal merit, the one with the most effective presentation will normally get the nod. Design presentation graphics, whether they take the form of static renderings or animated assemblies, are an important part of the design process—from the very beginning where a designer is trying to explore design alternatives, to marketing and then maintaining the product.

The design process can be characterized simply in the following way:

- Identification and delimitation of the problem
- Design definition
- Research design team meetings
- Preliminary sketches
- Working drawings (data)
- Scale models
- Final presentation renderings

This process is the same whether or not computer tools are used. Nothing is magic. Nothing is automatic. The difference lies in the computer's ability to combine certain steps, and provide a rapid method of testing design alternatives.

Today, powerful tools are available to assist in this process. AutoDesk's 3D Studio VIZ (VIZ) and Discreet's MAX (MAX) represent state-of-the-art technology for creating technical visual presentations. For example, with these products the same geometry that defines a design can be used to create working drawings, a model, and any of an infinite number of rendered views, from simple to photorealistic. Engineers and technologists are primarily concerned with the functionality of their solutions and the subsequent description for manufacturing and construction. Until the advent of tools like these, mechanical designers were kept from making effective presentation graphics because these graphics were artistic, as opposed to technical. By understanding a handful of camera, lighting, and material concepts, engineers and technologists can now make extremely effective presentations.

This book is designed around 3D Studio MAX and the MAX interface in 3D Studio VIZ. Both interfaces are practically identical and as my students have found, moving between the two is natural. Where this book departs from traditional treatment is in its technical graphics orientation. This book stresses mechanical and manufacturing applications and the 3D Studio modeling techniques that parallel processes such as extrusion, turning, machining, metal spinning and shaping, molding, and assembly.

Because VIZ may be considered the entry–level 3D Studio tool, menu and command references in this book are referenced first to that product. This probably takes care of 95 percent of the activities in this book. Where MAX differs significantly, such differences are noted.

This book demonstrates how either VIZ or MAX can be used to complement the design and engineering drawing skills you already have. In teaching engineers, designers, and technologists for over 30 years, the author has been highly successful in promoting a smooth transition between traditional and computer graphics skills. The topics in *3D Studio in Manufacturing Design* are arranged in an order proven to result in extraordinary results. This text *is not* intended to replace 3D Studio documentation nor introductory material on engineering drawing and CAD. You know engineering drawing; you understand CAD; you have access to 3D Studio and are familiar with its design environment. Now, how do you use VIZ or MAX to solve typical problems encountered in industry?

Eventually you will take what you have learned here and develop your own techniques, your own shortcuts, and become more productive than you ever thought you could be. When this happens, step back and share what you have learned with a young, aspiring designer. You'll make one college professor proud.

INTRODUCTION

You probably can't wait to use 3D Studio and start making effective presentations—the kind of renderings and animations you see on the pages of this book, the kind you have seen on television, or displayed during product introductions. VIZ and MAX are such powerful tools that most users can produce phenomenal results in a very short time. But as a designer, you want to control the tool, rather than let the tool control you. You don't want your solutions to be constrained by the limitations of the software. You want your solutions to be constrained by money, time, materials, construction or manufacturing processes, or your own ability to synthesize these variables. But not the tool.

The best way to master a tool is use it to solve real design problems. So after participating in the structured activities in this book, why not apply the techniques you learned to a design you have already worked on, one with which you are intimately familiar? Just substitute your sketches and drawings for those described in the chapters and have at it!

WHAT YOU NEED TO KNOW TO USE THIS BOOK

Many readers will come to VIZ or MAX from a background in one or more releases of Autodesk's AutoCAD. Although this won't necessarily prepare you for 3D Studio, experience in AutoCAD will aid in creating geometry that can be used to create realistic presentations. You may want to continue modeling and drawing in AutoCAD, using 3D Studio for material mapping, rendering, and animation, or follow along with the constructions outlined in these chapters.

Or, you may come with a background in another modeling/animation program. If you have experience in 3D Studio Max, you'll find, with a few exceptions, the same approach in VIZ. Additionally, the models you created in 3D Studio MAX can be used directly in VIZ. If your experience is in another program such as Lightwave, SoftImage, StrataStudio, Maya (or a dozen others), rest assured that the concepts are the same, only the names and positions in the interface have been changed.

But more important than computer skills, you should have a keen interest in how things are made, how they work, how they look, and make you feel. You should be interested in materials and textures, color, light, and shadow. You should be unafraid to try alternatives and ask yourself, "What if I try it this way?"

And you should have sharp skills in spatial observation. Is this above or below that, in front or behind? Is this shape just like that shape if I just changed this feature? How is this object held, used, or experienced? After months of use, how will the surface finish change? With 3D Studio VIZ and MAX, you can ask these questions.

WHAT THIS BOOK CONTAINS

First, this book is not a reprint of 3D Studio VIZ or MAX manuals or tutorials. In these pages you will find a mechanical and manufacturing approach to presentation modeling and animation. You'll find no exploding monsters and you'll find no houses or architectural subjects. You'll find a technical orientation that requires models to be built to exacting engineering specifications, usually from engineering drawings.

This book is organized into nineteen chapters that will lead you through modeling tasks that mirror industrial design and manufacturing processes. Each chapter stresses the importance of modeling and sketching and uses an applied example to reinforce each concept. Not every software feature is explored. However, you will experience the majority of features directly applicable to industrial tasks. For practice, starting files are provided on the accompanying CD-ROM so that you can follow along with each chapter example.

One of the powerful features of VIZ is the ability to switch between default VIZ, architectural, mechanical, land development, and 3D Studio MAX interfaces. Because this book emphasizes industrial applications, the 3D Studio MAX interface is used throughout. This makes tools available that clearly follow manufacturing methods such as turning, stamping, milling, shaping, casting, and extruding. It also allows users of both products to see how mechanical and manufacturing topics can be handled.

Each chapter in *3D Studio in Manufacturing Design* begins by explaining constructs and commands used in the following pages. You should locate these commands as you read their descriptions before continuing with the instruction. The importance of each chapter topic is explained thoroughly, especially in relation to design and manufacturing practices.

The importance of *geometric analysis* is stressed in each chapter. By understanding the geometry, you can apply each modeling tool more effectively. We do this analysis by *sketching*. Through sketching you will "learn" the geometry; mistakes made on paper are much less expensive than mistakes made on the computer. And by working out your strategies on paper you'll be more productive once your hand hits the mouse.

The VIZ and MAX operations you need to understand are presented with specific industrial examples. To fully understand these operations you will want to set up the same design so you can follow the example. Only with practice will you gain the confidence necessary to begin using 3D Studio in your own design work. Finally, the step-by-step solution of an applied problem demonstrates each chapter topic. To master VIZ or MAX, you'll want to follow the steps on your own computer. Many times pieces and parts to use will be found on the accompanying CD-ROM. Problems complete the chapters. They are meant as extensions of the applied problem. For sources of additional manufacturing and construction problems, turn to References for Further Study at the end of this text.

USING THIS BOOK

The *3D Studio in Manufacturing Design* text is organized for two 15-week couses. Chapters 1 through 9 cover modeling. All effective technical presentations are based on accurate geometry, either dimensioned or scale views. These chapters would form the basis of a first course. Chapters 10 through 19 cover what you do, once you have accurate models, to make realistic presentations and animations. A second course covering this next group of chapters would expand on the modeling skills developed in Chapters 1 through 9 and cover materials and motion.

Or, one could use the nineteen chapters to form one course, albeit an overview of technical visualization and animation. In this way, all major topics could be covered but without the practice needed to develop measurable skills. If you are studying on your own, get on the "chapter-a-week" plan. In this way, you can follow the applied example and complete one or two of the problems found at the end of the chapters. Or you might want to substitute a problem of your own. It's your call.

CONVENTIONS USED IN THIS BOOK

Choosing refers to making a menu or tool choice. *Selecting* refers to clicking on an object in the scene to make it active.

When a VIZ menu option is presented, it is shown bold and in sans serif typeface. The passage:

Tools|Drafting Settings|Units Set Up: Millimeters

means "choose the **Tools** menu. From the **Tools** menu, choose **Drafting Settings**. From **Drafting Settings**, choose **Units Set Up**. On the **Units Set Up** dialogue box, choose **Millimeters**."

When a procedure requires clicking on an icon, the icon's name is also emphasized. Thus, the **Select and Move** command refers to the icon in the bar at the top of the interface representing that function.

The roll-out menu at the right of the interface is also referred to in capital letters. Reference to **Create**, **Modify**, **Hierarchy**, or **Display** menus refer to options there. Suboptions in the roll-out menu are also capitalized.

The passage:

Create|Compound Objects|Loft|Skin Parameters|Path Shapes:24

means "from the **Create** command choose **Compound Objects**. Then choose **Loft**. From the **Loft** roll-out menu, choose **Skin Parameters** and **Path Shapes**. Enter the number **24** into the **Path Shapes** field."

The first letter of the names of parts is capitalized. The phrase "merge the Cover Plate" refers to a part named "Cover Plate." The name of computer files or CD-ROM directory structure is *italicized*. The passage "from *chapter_4\problems\basefile.max*" means go to the CD-ROM and get the file *basefile.max* from the *problems* subdirectory inside the *chapter_4* directory.

TOPICAL ORGANIZATION

Unlike other VIZ and MAX books, *3D Studio in Manufacturing Design* is meant to be used front to back, in order. The pedagogy the author has successfully used with designers, technologists, and engineers is based on learning certain concepts, practicing them, and applying them to more sophisticated operations. It is tempting to jump right in with a powerful tool like VIZ or MAX and try to do everything at once. But if you approach modeling and animation as presented in these pages, you'll find yourself more productive, with greater knowledge and ability, and in the end, more satisfied with your results.

For these reasons, shortcuts, plug-ins, and tricks are avoided. Given time and resources, you will learn a number of methods that will make your modeling and animation more effective and efficient. As you work on the assignments contained in the following pages, someone will no doubt say: "You don't have to do that! You can do it much faster if you spin this and twist that, and see…there it is automatically!" The problem is, you haven't learned anything. Tricks from knowledge are effective. Tricks without knowledge are just tricks.

ON THE CD-ROM

The CD-ROM accompanying this text contains the problems featured in each of the chapters. In each case, a beginning file and a completed file are available. The completed file can be used to compare your results to that of the author. Additionally, the completed file can be used as the basis of subsequent operations should your solution be unusable.

The CD is organized to help you use this book. There is a directory for each chapter containing files you can use to follow the tutorials, as well as related resources so you can make full use of the methods and techniques discussed. You'll also find a *tools* directory containing planning sheets, grids, and MAX files to make you more productive. You'll also find Web links to the materials for two courses: Computer Graphics Modeling and a second course, Computer Animation. These 15-week courses are keyed to the text and provide a solid pedagogy for developing a manufacturing orientation to VIZ/MAX.

ACKNOWLEDGMENTS

There are a number of people without whose support and encouragement a project such as this would have been impossible. John Fisher at Delmar provided the developmental support and Jim DeVoe the publishing direction. Several reviewers provided invaluable input on this project: Bill Ross at Purdue University gave valuable suggestions from the vantage of someone who has been teaching 3D Studio for almost ten years; Tom Bledsaw from ITT Educational Services was involved from the very beginning; Tom Singer at Sinclair Community College and Dan Ward from Ivy Tech reviewed the manuscript from a community/technical college perspective.

And of course, I must thank my students in the College of Technology and Applied Sciences at Arizona State University East for their enthusiasm and hard work. Some of their efforts appear on the pages of this book and the assignments you find there, on the CD-ROM, and on the related Web site were shaped by their suggestions. Finally, thanks go to AutoDesk and their educational representatives. With their support, students around the world have access to tools that spark their imaginations, challenge their thinking, and prepare them for careers in Engineering and Technology.

CONTACTING THE AUTHOR

I enjoy hearing from readers of my books, both constructive criticism as well as "attaboys." Every textbook is a compromise but the author would be more than willing to consider adding or subtracting material in future editions based on what you, the reader, deems important.

Jon M. Duff (jmduff@asu.edu)
Arizona State University East
Technology Center, Building 50
7001 East Williams Field Road
Mesa, AZ 85212

Modeling with Primitives

CHAPTER OVERVIEW

Modeling with geometric primitives is the perfect place to start your experiences with 3D Studio. Both VIZ and MAX have a wide assortment of preprogrammed geometric objects that either by themselves, combined with other primitives, or edited into new shapes, can form the basis for many objects that you'll want to make.

In this chapter you will be introduced to many of 3D Studio's standard and extended primitives. You will learn how to precisely size and place these primitives in space to create a scene. Once familiar with these operations, you will be ready to plan your use of primitives by analyzing standard engineering drawings so you can choose primitives that best meet your needs. You'll get the chance to see the complete modeling-from-primitives process by analyzing and then modeling an object using only primitives. Finally, you'll learn how to establish a camera view, set standard lighting, and make your first print.

KEY COMMANDS AND TERMS

- **Arc Rotate Selected**—the command that allows you to rotate the viewpoint (called **Orbit** in this text).
- **Color Chip**—the visual indication in the **Command Panel** of the color currently assigned to the selected object.
- **Command Panel**—the area at the right of the interface that contains the majority of functions.
- **Create Menu**—the option in the **Command Panel** that provides the major functions in 3D Studio.

- **Draw|Drafting Tools|Units Setup Menu**—the menu selection that allows you to change the current units of measurement.
- **Materials|Environment Menu**—the menu that allows you to change the background color for rendered scenes.
- **Parameters**—characteristics of objects that can be numerically entered in the various roll-out menus in the **Command Panel**.
- **Quick Render**—the command that renders the active viewport using settings in the **Render Design** command.
- **Render Design**—the command that allows full control over the rendered scene including saving the rendered scene as a file.
- **Roll-out Menus**—menus in the **Command Panel** that may be collapsed (–) or expanded (+) to reveal additional options and controls.
- **Select**—the command that allows objects to be selected.
- **Select and Move**—the command that allows objects to be selected and then subsequently moved.
- **Select and Move (right-click)**—the command that brings up the **Move-Transform Type-In** dialogue box for a selected object.
- **Select and Move (shift key)**—the command that brings up the **Clone Options** dialogue box for making copies of selected objects.
- **Tools|Options|Colors Menu**—the menu that allows you to change colors in the VIZ interface. In MAX, **Customize|Preferences|Colors**.
- **View Label**—the label at the upper left of a viewport that identifies the type of view being displayed.
- **View Label (right-click)**—the action that brings up the **View Options** dialogue box for such choices as wireframe display and view choice.

SETTING UP YOUR WORKSPACE

Each person using 3D Studio VIZ or MAX will have personal preferences on how they would like their work environment to appear. You may like to draw white on black or black on white. But keep the following in mind:

- The color of highlighted objects. You should pick a color that is easily identifiable. I prefer magenta.
- The color of locked objects. If locked objects are black and nonlocked are black, you are in trouble. I prefer blue.
- The color of the viewport background. Because many industrial products are metallic, using a gray background may be confusing. By using a light blue, light

green, or light yellow background, holes, cavities, and unaligned geometry can be more easily detected. I prefer a slightly off-white green.

These preferences can be found in the **Modify|Options** menu and by selecting the **Colors** tab (Figure 1.1). If more than one designer will be using the software, it may be helpful to save the current environment in a *custom user interface* (CUI). By selecting **Tools|Customize|Save User Interface As...** each VIZ user can save a CUI and load it at the start of work with **Tools|Customize|Load User Interface**. In the same **Tools|Customize** menu is the option **Customize User Interface** where the fine-tuning of the environment is accomplished.

The other task you need to accomplish before modeling is to set the system of units you will be using. With computer graphics this isn't absolutely critical because geometry can be scaled later to any size, in any units. But starting with the units you will be using makes construction easier. Figure 1.2 shows the dialog box displayed with **Draw|Drafting Tools|Units Setup**. When using fractional feet and inches, it is important to set the division spinner to half the smallest possible division you anticipate. It is often convenient to halve a dimension to find its middle. For example, if the smallest dimension is $^3/_{32}$, set divisions to 64ths. That way, you can address $^3/_{64}$ (half of $^3/_{32}$) and 3D Studio won't try to round your numerical entries to the nearest division.

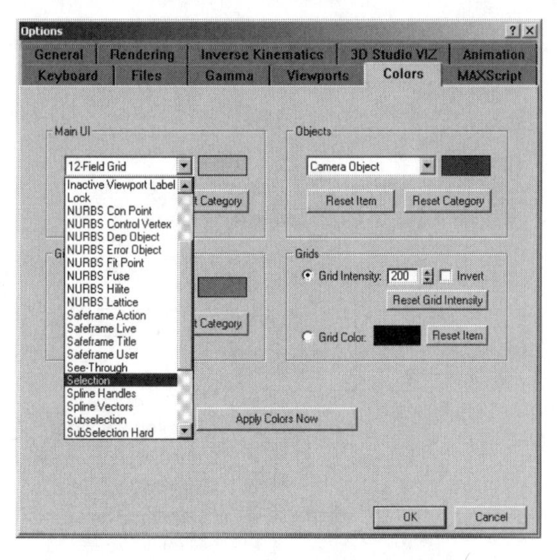

Figure 1.1 *Use **Tools|Options|Colors** to set up your work environment.*

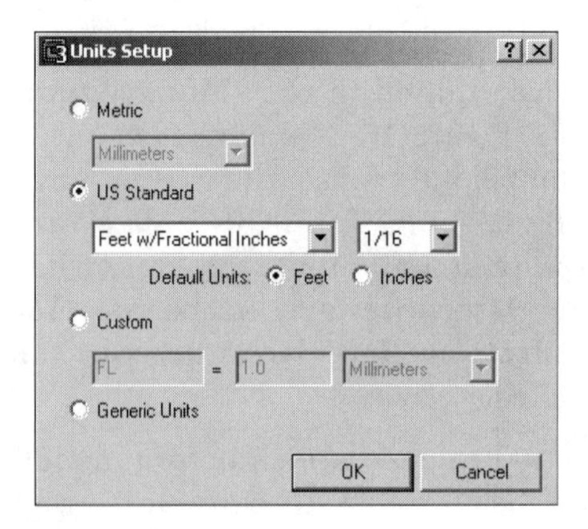

Figure 1.2 *Set **Units** so that dimensions can be entered directly.*

IMPORTANCE OF MODELING WITH PRIMITIVES

Industrial products are often constructed from standard material stock: cylinders, spheres, plates, bars, and so on. Shaping or machining with tool profiles modifies these standard shapes. 3D Studio provides a wealth of primitives from which more sophisticated shapes can be built (Figure 1.3).

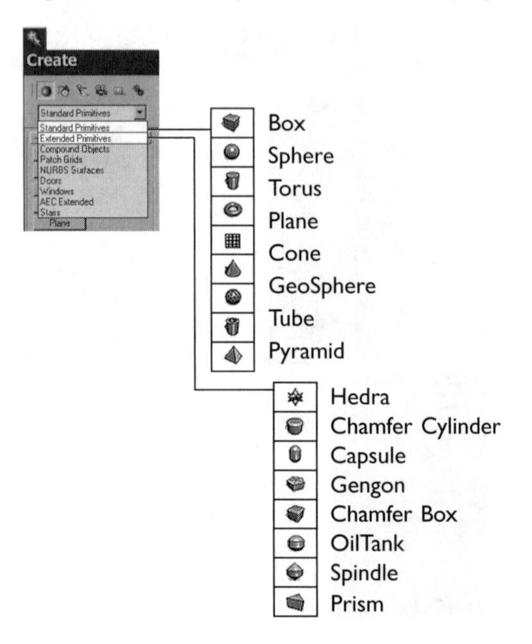

Figure 1.3 *Standard and extended primitives form the basis of many models.*

Tip: Always apply the "less is more" rule. Use the least sophisticated geometry to accomplish the task. You save time, computer resources, and money.

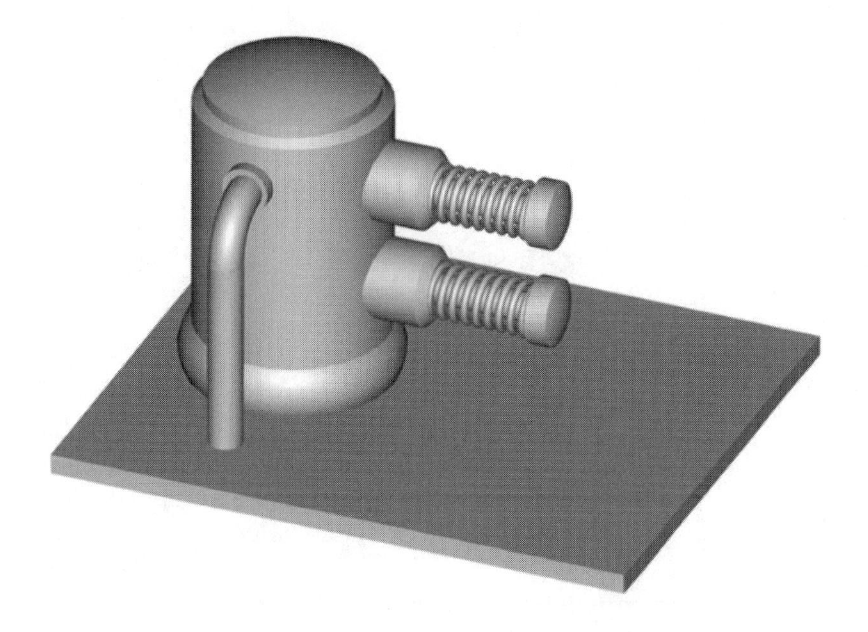

Figure 1.4 *Models with primitives allow spatial relationships to be quickly determined.*

For design visualization, you may want to start with primitives to get an idea of how various components interact with each other (Figure 1.4). Worry about general shapes and relationships first, details later. You wouldn't want to invest hundreds or thousands of hours in a project only to realize that you didn't do enough early planning. The greater your use of primitives, the less geometry you have to actually create yourself.

But to be able to intelligently use these shapes you have to be able to see "through" a design to the underlying geometry.

GEOMETRIC ANALYSIS

Figure 1.5 displays the views of a hand wheel. An experienced engineer or technologist can study these views and easily "visualize" the underlying geometry. One way to help do this is by making a geometric analysis sketch (Figure 1.6). Compare the sketch with the engineering drawing and you can see that the hand wheel is really composed of an outer ring, an inner hub, and spokes. The first spoke is rotated up and then arrayed 360 degrees with four additional copies. So the question is: "What primitives are available to quickly model the hand wheel?"

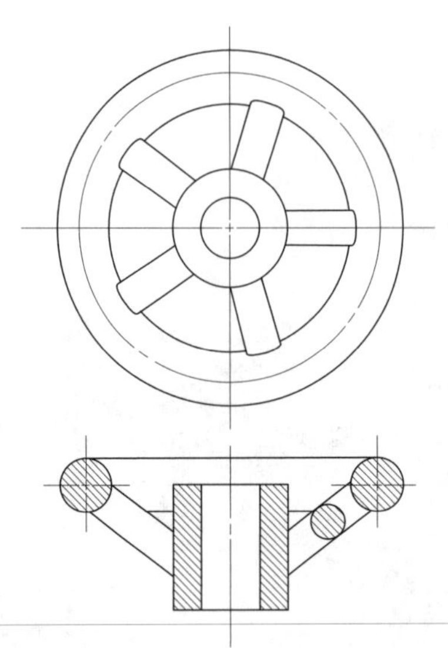

Figure 1.5 *Orthogonal views of a hand wheel.*

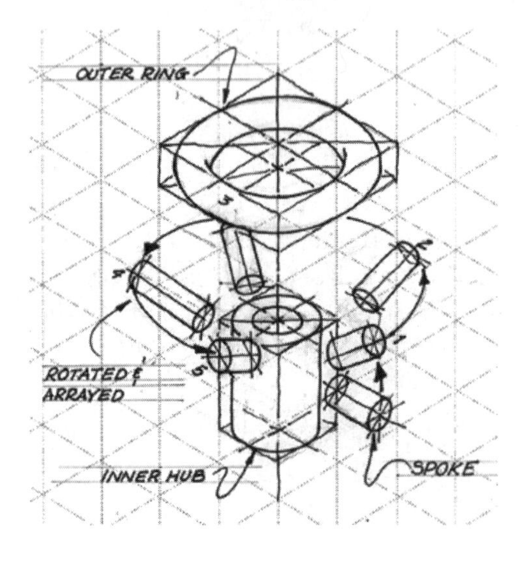

The Importance of Sketching

In industry, sketching is the way you talk about your designs.

Figure 1.6 *Geometric an analysis sketch breaks down the geometry into component parts.*

The outer ring is actually a torus; the inner hub is a tube; the spokes are identical cylinders. 3D Studio has these three primitives (Figure 1.7) and when assembled correctly in space, an accurate model of the hand wheel can be made (Figure 1.8).

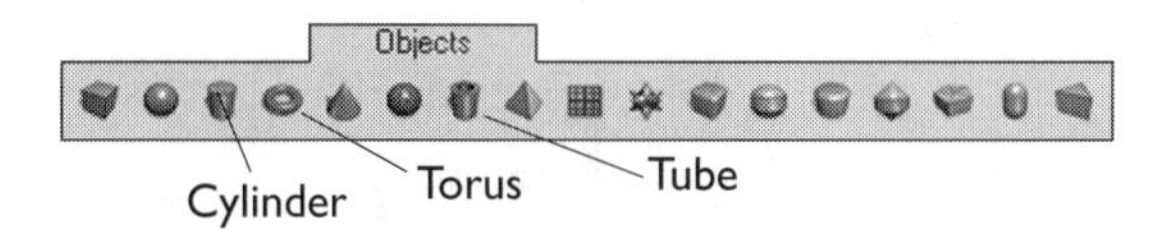

Figure 1.7 *Primitives correspond to geometric components of the hand wheel.*

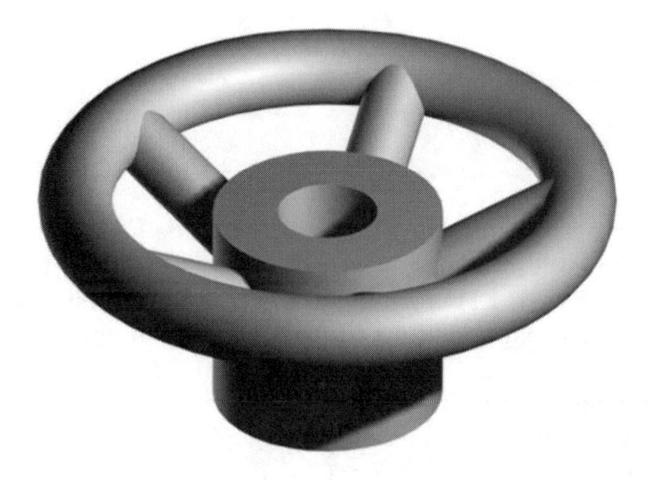

Figure 1.8 *Hand wheel modeled from primitives.*

EXAMPLE: MODELING THE HANDLE

For our first attempt at modeling with primitives, consider the views of the "Handle" displayed in Figure 1.9. This part may be cast, forged, or constructed from individual pieces welded together. Of course, we want to employ the most simple, accurate, and flexible modeling technique. A cursory inspection reveals that all shapes might be described by primitives. You'll also notice that the handle itself has been knurled in a diamond pattern and one portion has been threaded. We will not be modeling the knurls or the threads. Instead, in Chapter 12, Modeling with Bump Maps we will be adding these surface characteristics by using bump maps. This will yield correct representation without adding geometric complexity.

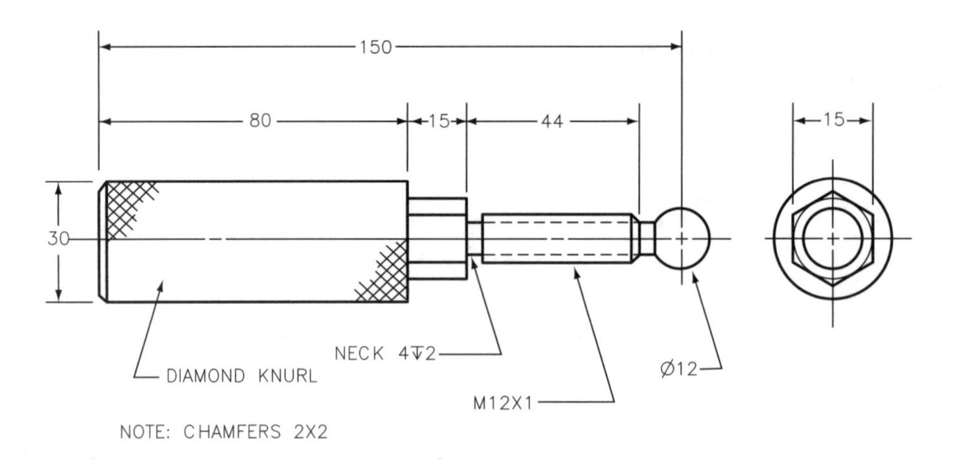

Figure 1.9 *Dimensioned views of the handle.*

STEP 1

Preform a geometric analysis of the part. Produce a sketch (Figure 1.10) describing the handle in terms of primitive objects. You'll find this sketch sheet (*iso_grid.eps*) in EPS format on the CD-ROM inside the *tools/* directory. Notice that each part on the sketch has a name that will be used in 3D Studio so you can keep the parts straight. An inspection of the handle reveals that the knurled part has a chamfer at one end only. Unfortunately 3D Studio produces only two-ended primitive chamfer cylinders. A chamfer cylinder must then form the left end while a regular cylinder forms the right end.

STEP 2

Set up your environment in millimeters (**Tools|Drafting Settings|Units Setup: Millimeters**). Display grid lines every 2 mm with major divisions every 5 grid divisions (**Tools|Drafting Settings|Grid and Snap Settings: Home Grid=2mm; Major Lines Every Nth=5**). This is **Customize|Units Setup** in MAX.

Tip: Save files often. The default path will put your scenes into the 3DSVIZ or MAX scenes directory. You will want to redirect the files to a more appropriate storage location.

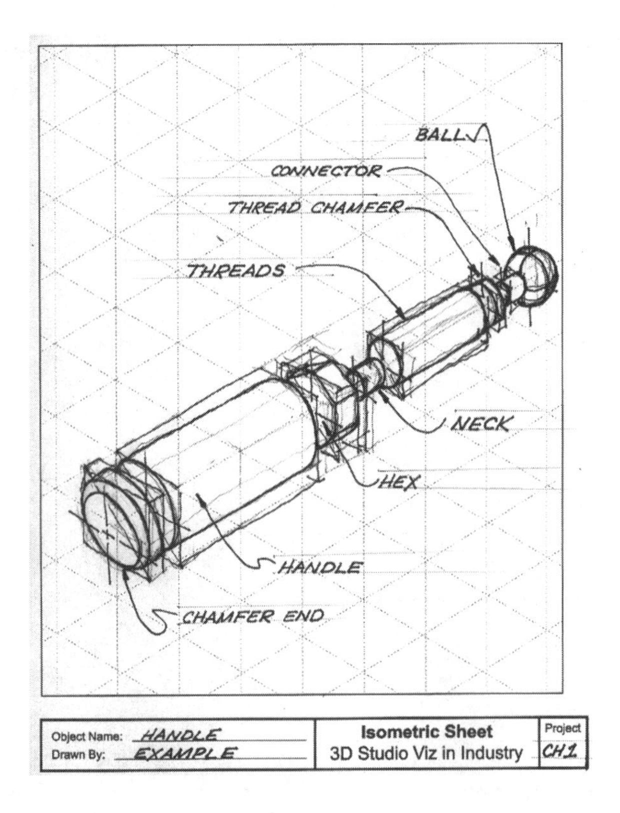

Figure 1.10 *Geometric analysis sketch of the handle.*

STEP 3

Locate the 0, 0 ,0 world origin at the left side, middle. The local origin (**Pivot**) of each part can then be determined relative to this point by using distance information from the engineering drawing. By choosing the appropriate view in which to start construction, you can place the **Pivot** on the left or right side as needed.

STEP 4

Create the objects to the correct sizes and location using the engineering drawing and your sketch as references.

Start by creating a chamfer cylinder in the Right Side view (either from the **Objects** tab or the **Extended Primitives** drop-down in the **Create** menu). This will place the cylinder's local origin on its base as the cylinder's height is drawn toward you. Enter 15 mm **Radius**, 30 mm

Height, 2 mm **Fillet** and 36 **Sides** in the create roll-out menu. Turn **Smooth** off for now (Figure 1.11).

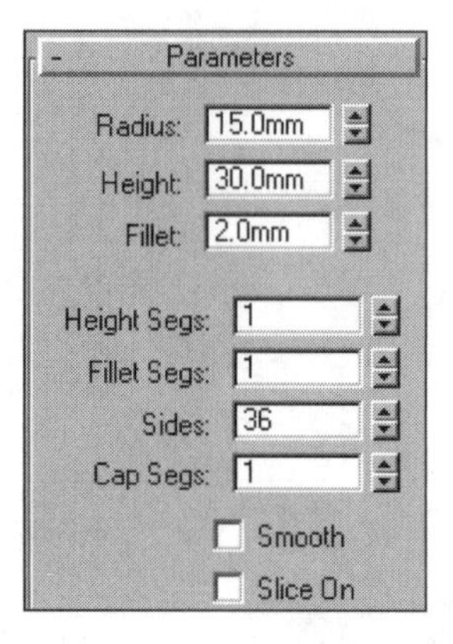

Figure 1.11 *Chamfer cylinder parameters.*

STEP 5

Switch to the front view. Select the **Chamfer Cylinder** object and right click on the **Select and Move** tool. Enter 0, 0, 0 in the world coordinate fields. The cylinder's left-middle-center is aligned to the world origin (Figure 1.12).

Tip: With a right-click inside a viewport you make that view active. By right-clicking, you don't run the risk of selecting an object and inadvertently moving it when trying to make a viewport active.

STEP 6

Create the Handle in the Left Side view. This will put its local origin on the right side where it can be located 80 mm from the world origin (refer back to Figure 1.8). Because the chamfer cylinder takes up some of the 80 mm, you can make the handle 15 mm **Radius**, 60 mm **Height,** 2 mm **Fillet** and 36 **Sides**. When positioned at X=80, Y=0, Z=0 the two ends form the complete part (Figure 1.13).

Note: If positioned accurately, 3D Studio will take care of visibility during rendering. It is not critical that the two ends of the handle be physically joined.

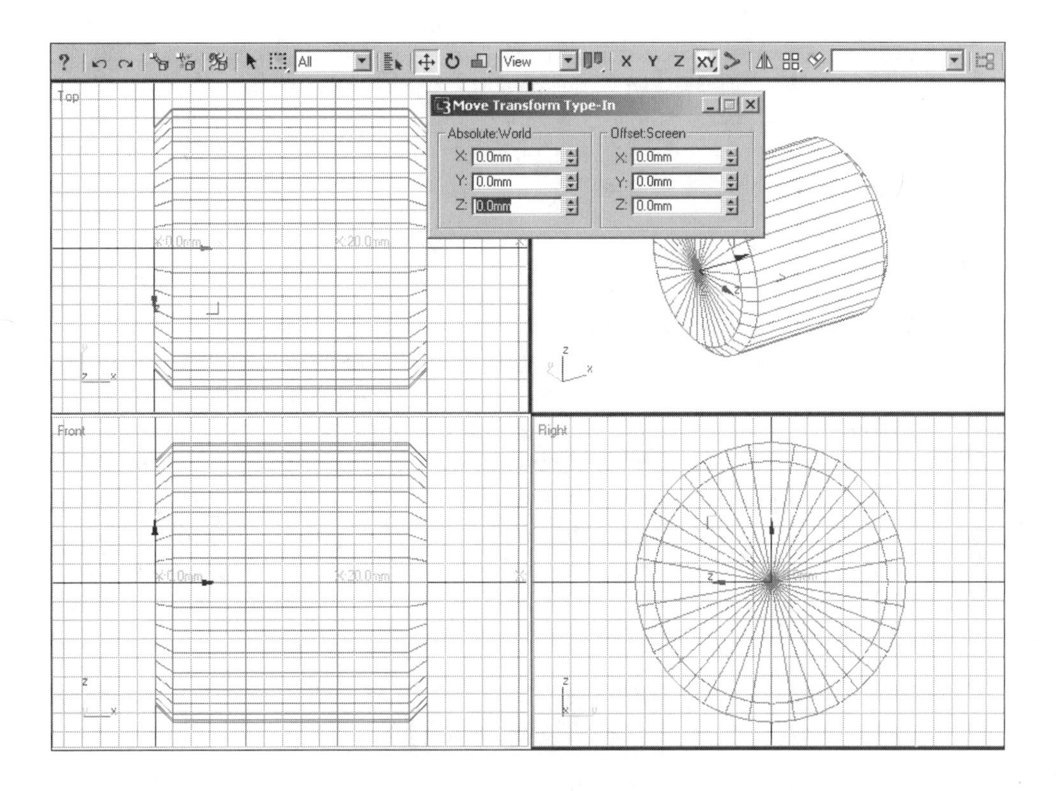

Figure 1.12 *Chamfer cylinder is moved to the world origin.*

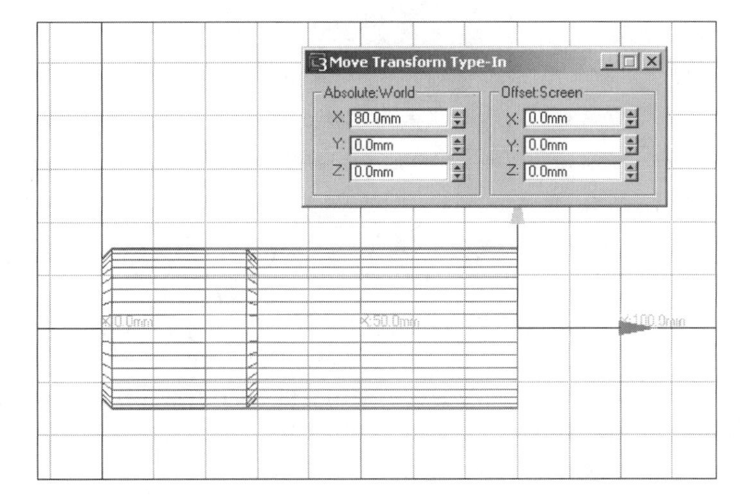

Figure 1.13 *Local origin of the cylindrical portion is at X=80 mm.*

Tip: Always name objects immediately after creating them. Don't accept default names. Even if you group the objects later (and assign a name to the group), assign your own names. Without meaningful names, you'll find it's almost impossible to efficiently select objects for modifying, hiding, locking, or making selection groups.

STEP 7

Create the Hex from a cylinder primitive of six sides. Do this in the Right Side view so that the object's local origin can be easily positioned at X=80 mm. Here's the tricky part: Hexagonal features in industry are usually specified by distance across flats. VIZ creates both cylinders and polygonal prisms (**Gengon** in the **Extended Primitives** menu) by corner-to-corner radius. It ends up that a corner-to-corner hexagon is 116.66 percent larger than a across-the-flats hexagon. Enter 8.75 mm as the correct **Radius** (7.5 x 1.166=8.75), 15 mm as the **Height**, and 6 for **Sides**. Again, turn off **Smooth** (Figure 1.14).

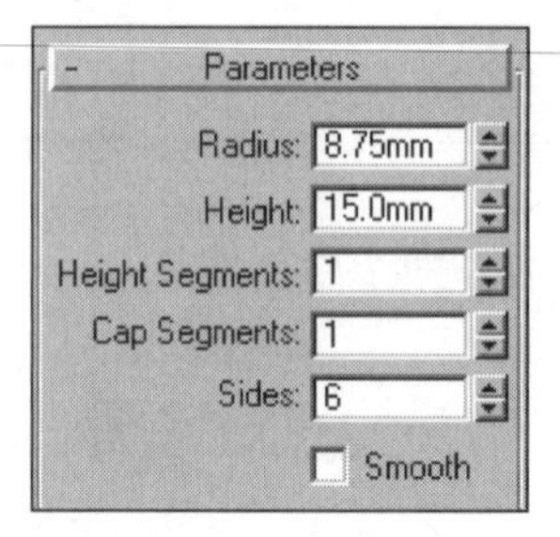

Figure 1.14 *Hex parameters. Hex is created in the right side view so the local origin is on the left of the object in the front view.*

STEP 8

Right–click on the **Select and Move** tool to send the hex to X=80 mm, Y=0, Z=0 (Figure 1.15) and Hex is in correct position. You may choose to rotate Hex 30 degrees so it appears in the same position as on the engineering drawing (flats on the side, corner at the top).

STEP 9

Create the threaded portion next (the neck will be filled in later). Note that the right side of the threaded portion is chamfered so we will take a similar approach as we took with the first two parts.

In the Left Side view, create the Thread Chamfer using a chamfer cylinder primitive. Enter 6 mm as the **Radius** as the **Height**, 2 mm **Fillet** and 36 for **Sides**. Again, turn off **Smooth** (Figure 1.16).

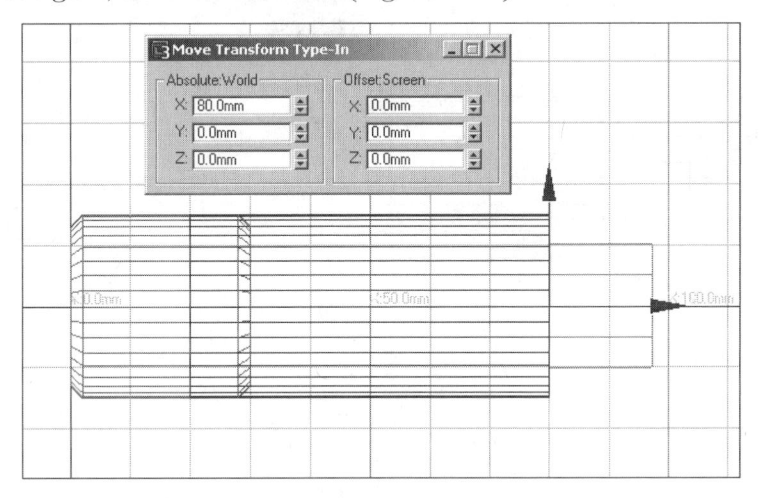

Figure 1.15 *Hex is moved to its position 80 mm to the right of the world origin and rotated 30 degrees.*

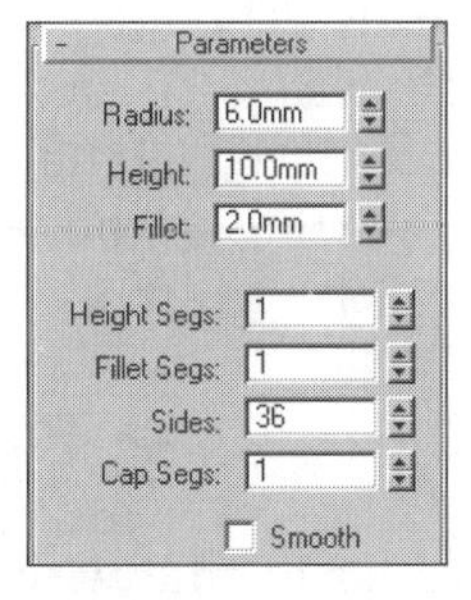

Figure 1.16 *Chamfer cylinder parameters form the thread chamfer portion of the handle.*

STEP 10

Right–click on the **Select and Move** tool and position the chamfer cylinder at X=139 (80+15+44=139). See Figure 1.17.

Tip: By selecting the best view, you can position the object origin so it is in the correct position. Remember that cylinders, cones, boxes, and pyramids are drawn *toward you*. This means that the origin will be on the far side as the object is created.

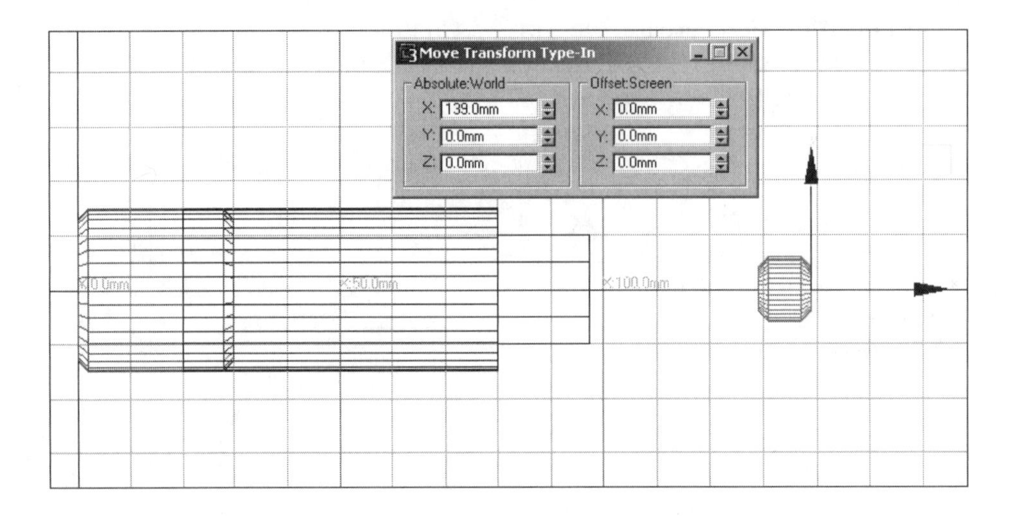

Figure 1.17 *Chamfer end located correctly along the world X–axis.*

STEP 11

Create the nonchamfered threads (Threads) in the Right Side view. Create a cylinder matching the specification of the chamfer cylinder (**Radius**=6, **Fillet**=2 mm, **Sides**=36) but with a **Height** of 35mm. This will make up the remainder of the threaded portion of the handle. When you position the cylinder at X=99 (80+15+4=99), your model should look like Figure 1.18.

STEP 12

Create the Ball at the end of the handle with a sphere primitive. Use **Radius**=6 and accept the default **Segments**. In this case, check the smooth option.

Move the Ball to its correct position at X=150, Y=0, Z=0 by right-clicking on the **Select and Move** tool with the Ball selected (Figure 1.19).

Tip: It is tempting to want to view your models in rendered form. If you can't live without a realistically rendered view, work in wireframe in Top-Front-Side views and render a pictorial User view.

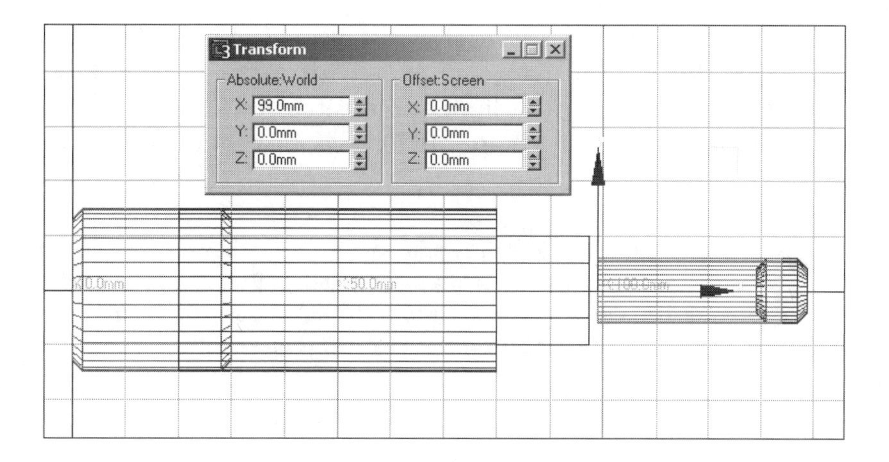

Figure 1.18 *Nonchamfered cylinder positioned so that its local origin is at X = 99 mm.*

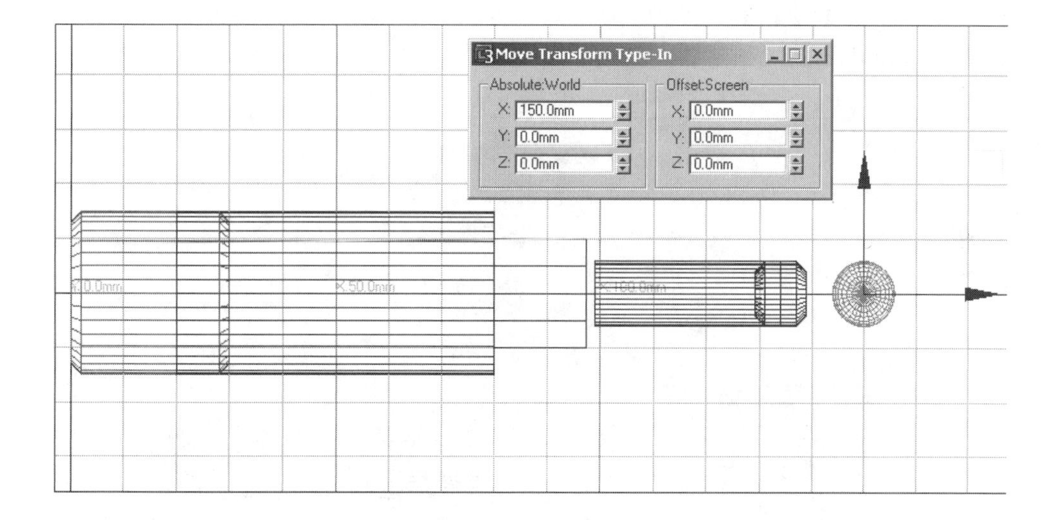

Figure 1.19 *Sphere positioned correctly along world X–axis.*

STEP 13

Create the Neck and Connector. From the engineering drawing you can see that the Neck is 2 mm in from the threaded portion. This makes the Neck 8 mm in diameter (12–2–2=8). The Connector is the same diameter.

Create a cylinder in either the Right *or* Left Side views **Radius**=4 and **Height**=10. Position at X=0,Y=0, and Z=0. With the **Select and Move** tool, drag the local X-axis until the cylinder connects the Hex and the Threads. Exact position isn't important here, as long as Hex and Threads are connected.

Tip: To get the highest quality print, save the file in TIF format at a high resolution (like 1024 x 768). Then in a program such as PhotoShop, resample the image at 150 dpi with interpolation off. Scale the resampled image to 50 percent of original size. The final print should be nearly photographic.

STEP 14

Hold the shift key down and drag a copy of the neck by its X-axis to connect the Thread Chamfer and the Ball. Name this new object Connector. By creating a copy of Neck you assure both Neck and Connector to be identical diameters. By moving Neck by its X-axis, you assure Connector to be at Y=0, Z=0. The completed handle in wireframe is shown in the front view in Figure 1.20.

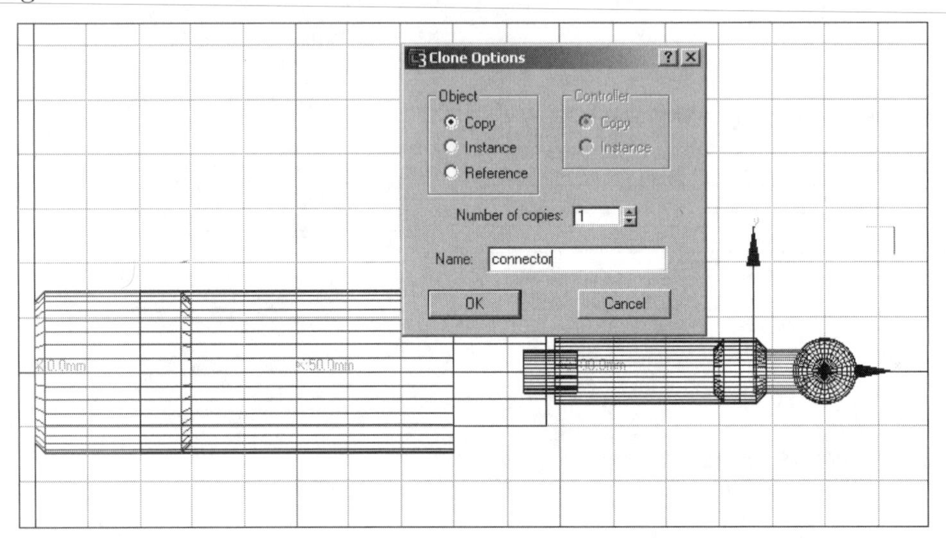

Figure 1.20 *Neck and Connector are positioned along the X-axis to complete the model.*

CHECK THE PARTS

If you have been diligent in naming parts as you created them you should be able to choose **Edit|Select By|Name** and be rewarded with a list of objects like that shown in Figure 1.21. If default names show up, select individually and rename.

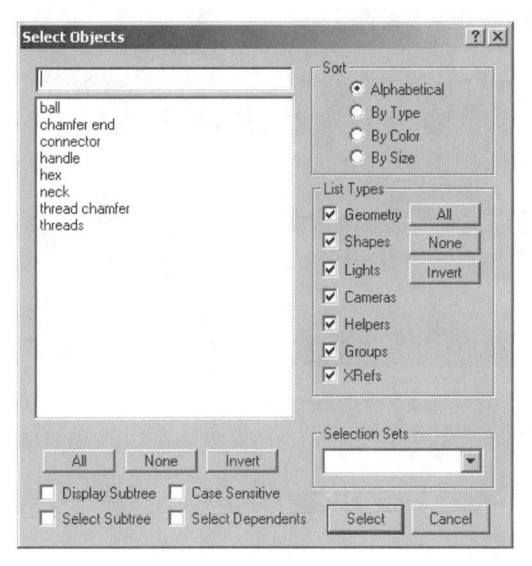

Figure 1.21 *Handle components have the names established on the geometric analysis sheet.*

GET A QUICK LOOK

We have been working in wireframe mode up to this point. But now it's time to get a more realistic look at the part. To see what your handle looks like, follow these steps:

STEP 1

Select all parts. Choose **Edit|Select All** or use the **Select** tool and drag a selection marquee around all parts.

STEP 2

Click on the color chip in the **Create** roll-out menu to bring up the color palette. Click on **AutoCAD ACI Palette** to bring up the extended color picker. Choose a medium gray from **GrayShades** at the bottom of the dialogue box. Click **OK** to assign the new color to the selected objects. This changes the color of your wireframe when not selected. These steps are outlined in Figure 1.22.

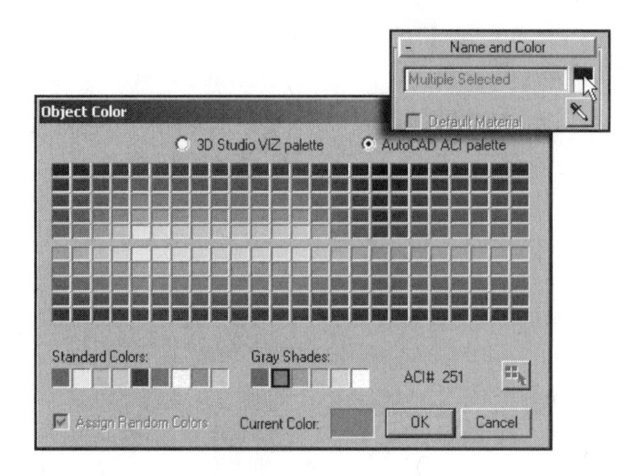

Figure 1.22 *Assign a color to selected objects.*

STEP 3

Make the front view active by right–clicking on it. Select the **Arc Rotate** (**Orbit View**) tool and pull on the orbit handles until the part is in a descriptive pictorial position (Figure 1.23).

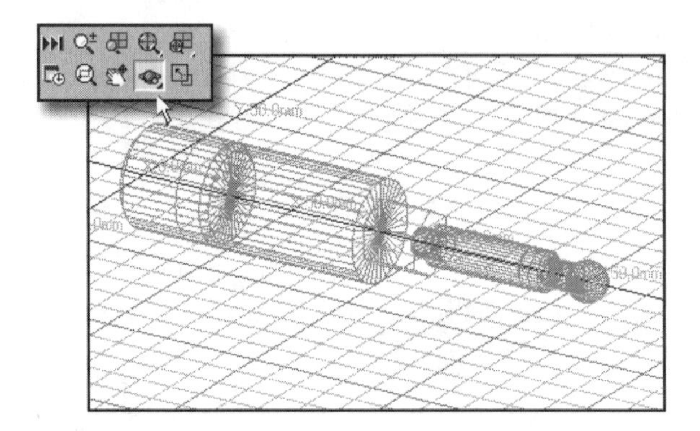

Figure 1.23 *The* **Orbit** *tool lets you change the viewpoint.*

STEP 4

The view label now reads User view because it isn't a Top-Front-Rear-Bottom-Side view. Right–click on the view label and deselect the **Show Grid** in the **View Options** menu. Do this again and select **Smooth + Highlights** (Figure 1.24).

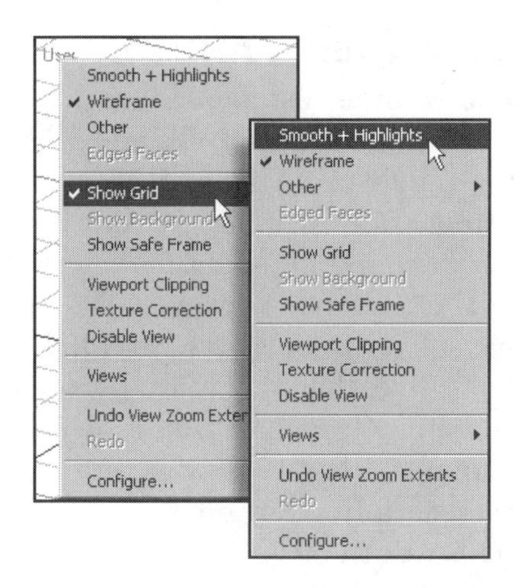

Figure 1.24 *Right-click on the view label to bring up the* ***View Options*** *menu.*

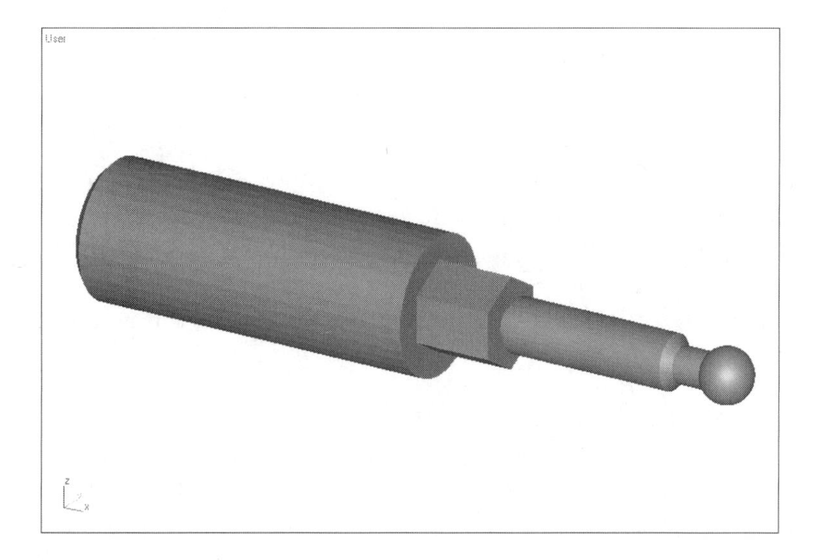

Figure 1.25 *Rendered handle in a pictorial viewport.*

STEP 5

Press the **W** key to display only the pictorial user view. Wow! That really looks like a handle! Choose the **Quick Render** tool and 3D Studio applies default lights and produces a realistically rendered user view (Figure 1.25).

STARTING LIGHTS, CAMERA, AND MATERIALS

You can view your model without establishing lights or a camera (as long as you don't have a black or white object). VIZ and MAX have ambient lighting that i9s sufficient for quick evaluation. Still, more realistic results are attainable by establishing your own camera and lights. Now is the time to turn to Chapter 14: Camera and Light Basics where this effective technique is covered. Complete the first two sections: The Basic Camera and Basic Lights. When you have set up your camera and lights, you will be rewarded with even more realism, like that in Figure 1.26. Note the difference between Figure 1.25 and Figure 1.26. The first is an *axonometric view*, formed when a principle orthogonal view is rotated into a User view. The second figure is a *perspective view*, formed by creating a camera object. Figure 1.26 has also turned on the **Smooth** option in the **Modify** menu for chamfer cylinders and cylinders. This removes the facets from these objects and produces smooth surfaces.

Note: Smoothing a hexagonal prism will remove the sharp edges.

MAKING YOUR FIRST PRINT

Even though a viewport with smooth and highlights turned on produces realistic results, even greater realism can be achieved by letting 3D Studio render the scene. The **Quick Render** command takes its settings from the **Render Design** command. So if you do a quick render, you have no control over the image size. To make a print the size you want, follow these steps:

STEP 1

Select **Rendering|Environment** and click on the **Background Color** chip. Change the color to white. Printing a full sheet of black is a waste of toner or ink. Don't do it.

STEP 2

Choose the **Render Design** command. Note that 3D Studio displays the **Render Dialogue Box**.

STEP 3

Select a standard 1.333 resolution or enter a custom size. In our case, choose 800 x 600. Click **Render**.

STEP 4

VIZ renders the scene with the basic gray material color you assigned to the parts. With the icons in the header bar of the render window (Figure 1.27) you can print or save the image.

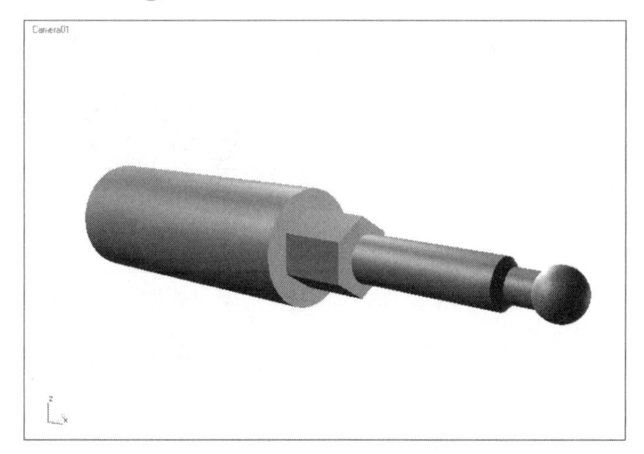

Figure 1.26 *The Camera view after establishing a camera and lights.*

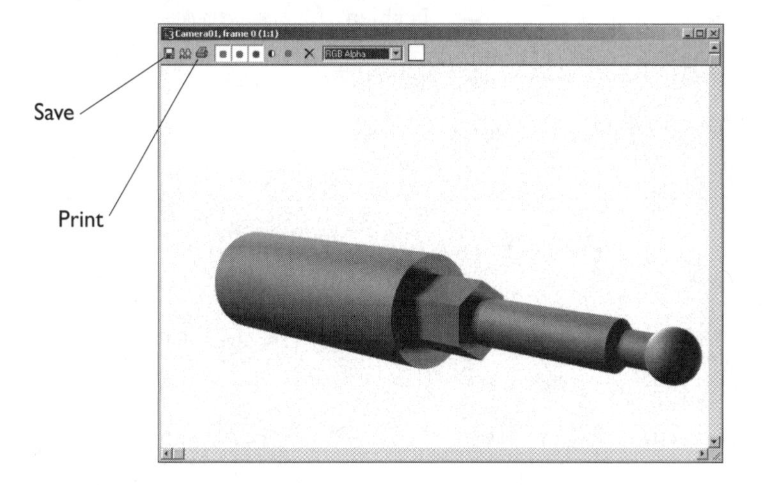

Figure 1.27 *After rendering, you have the option of saving or printing the scene.*

STEP 5

Click on the **Print** icon to print the file to your connected printer. Save rendered scenes only as necessary. These raster files can fill up valuable hard disk space. Remember, you can always re-render the scene.

PROBLEMS

Problems appropriate for solving with primitives are found on the following pages. Use the file *isogrid* found in the *tools* directory to plan your modeling strategies. Assign appropriate names to all objects both on your sketches and models.

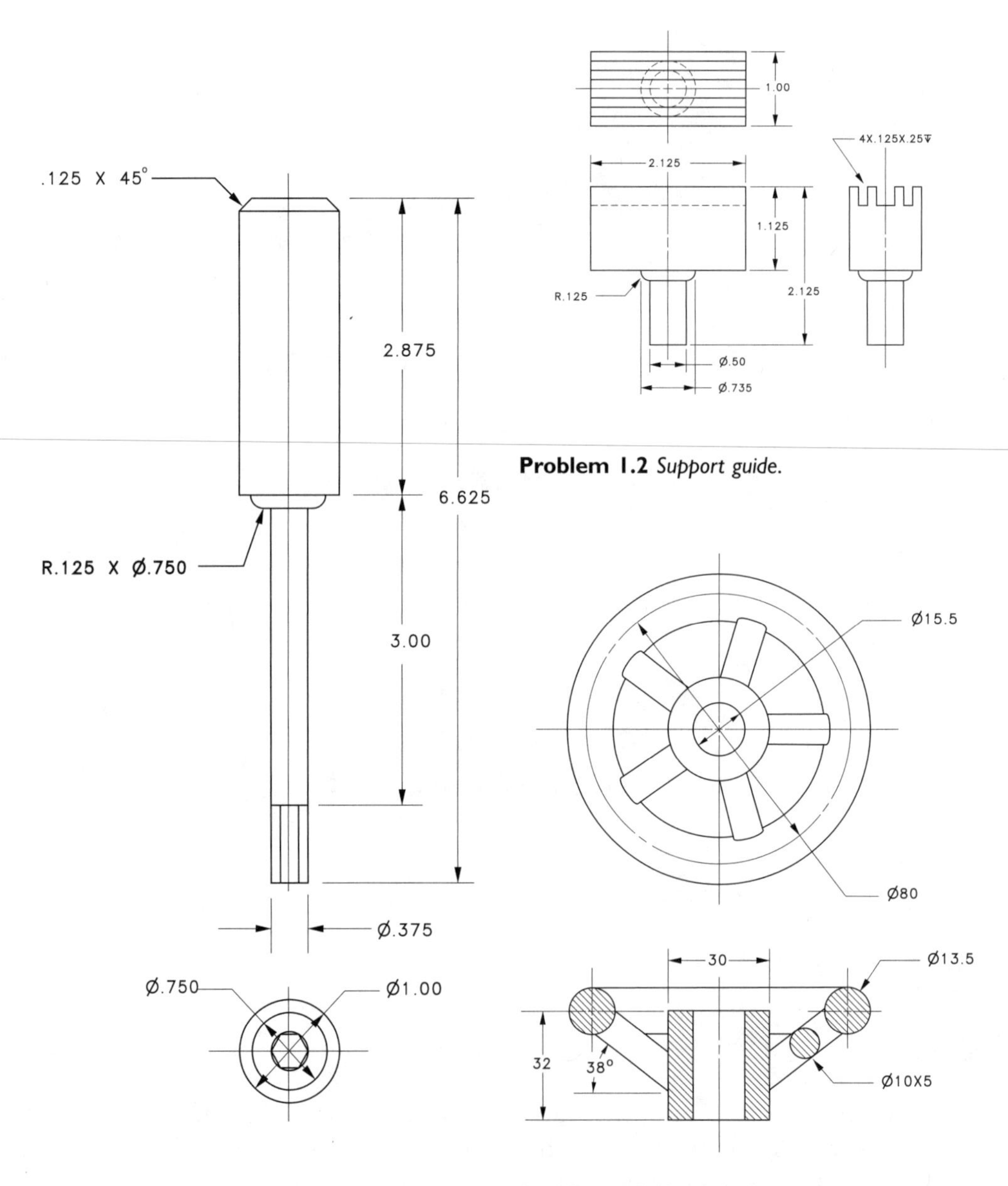

Problem 1.2 *Support guide.*

Problem 1.1 *Adjustment tool.*

Problem 1.3 *Handwheel.*

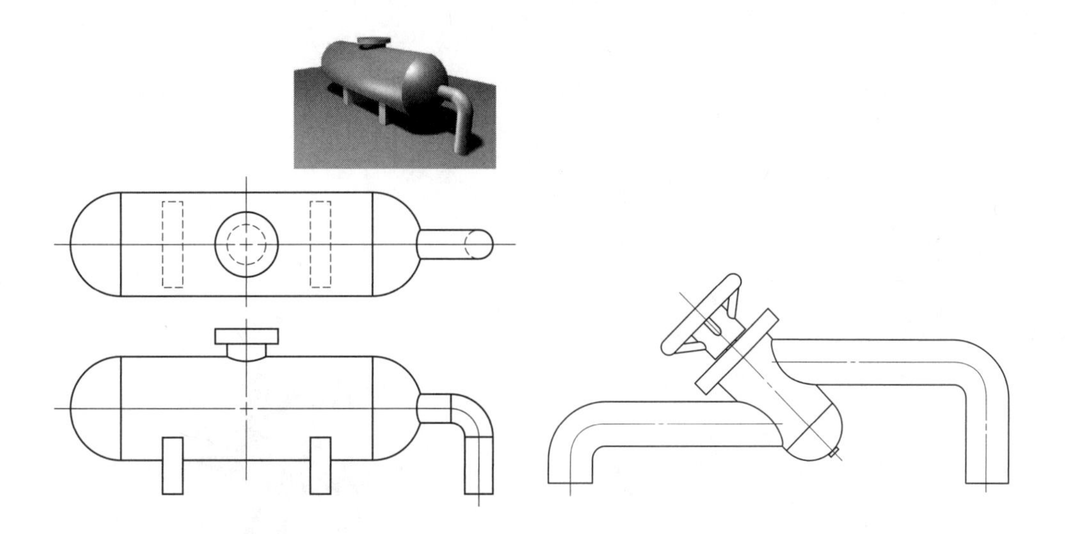

Problem 1.4 *Storage tank.*

Problem 1.5 *Water clean–out valve.*

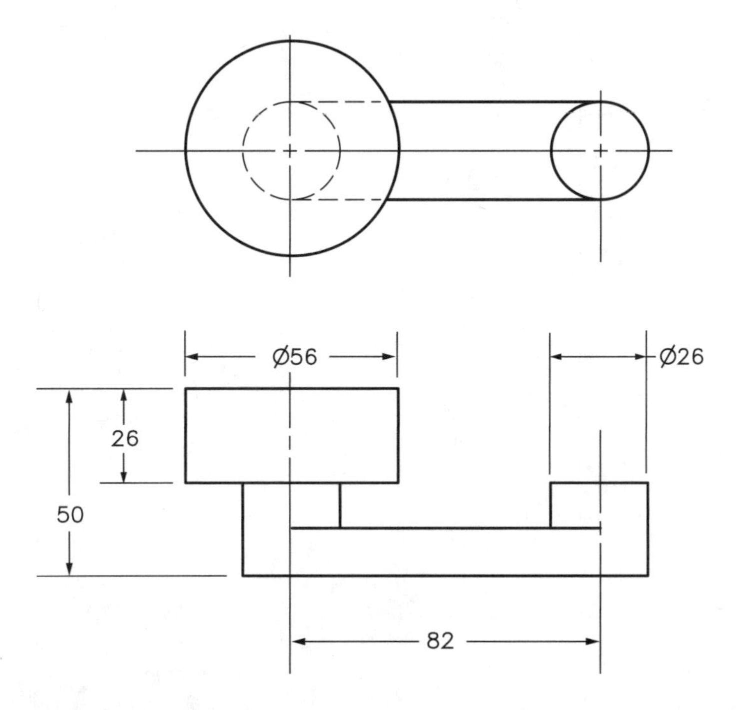

Problem 1.6 *Connection casting.*

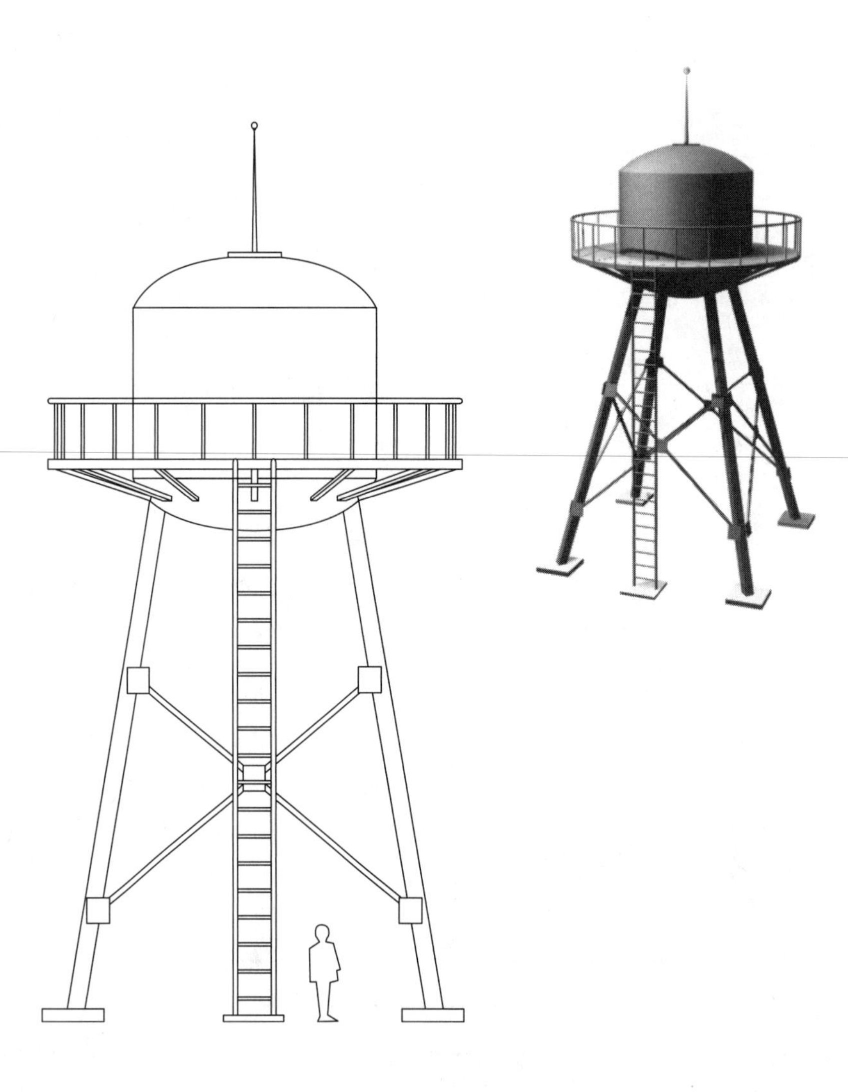

Problem 1.7 *Water tower. Figure is 6'-0" tall.*

Modeling with Extrusions

CHAPTER OVERVIEW

In Chapter 1 you solved a modeling problem by using geometric primitives available in the 3D Studio toolbox. By using readily available shapes you can shorten modeling time considerably. However, complex industrial parts usually require shapes that are not primitives.

In this chapter you will learn to recognize shapes that can be made using the **Extrude** modifier. Extrusion is a fundamental manufacturing process and many standard manufactured products contain extrusions. When material is drawn or forced through a die, a consistent cross section is created. This is characteristic of an extrusion: a consistent cross–sectional shape. Any object with a consistent cross section can be modeled effectively using 3D Studio's **Extrusion** modifier, not just shapes extruded through dies. In this chapter you will analyze shapes, learn to create accurate splines, and complete final extrusions.

KEY COMMANDS AND TERMS

- **Attach**—the modifier option that creates a single shape from two or more individual shapes. Nested shapes must first be attached to create extrusions. Overlapping shapes must first be attached to perform Boolean operations.
- **Boolean Subtraction**-—he operation that removes one shape from another. The first shape selected is the base shape. The second shape selected is removed from the first.
- **Boolean Union**—the operation that adds two shapes together. Order of selection is unimportant.
- **Cap Start and End**—the parameters that form end planes on closed extrusion shapes.

- **Distance**—the parameter that determines the length of the extrusion.
- **Extrusion**—the modifier option that creates a constant cross section 3D object from a 2D shape.
- **Flip Normals**—the parameter of the **Normal** modifier option that reverses the direction of polygon vector normals used for determining visibility.
- **Insert**—a modifier that inserts a vertex on the shape path.
- **Nested**—the condition where one shape is completely inside another. Nested extrusions will alternate positive and negative material.
- **Nurbs**—an efficient technique for defining extrusion shapes.
- **Segment**—the sub-object that allows portions of a shape between vertices to be selected.
- **Shape**—a geometric object used for defining the extrusion cross-section. A shape can be a spline or NURBS curve.
- **Spline**—a shape used for creating extrusion cross sections that passes through the defining points.
- **Spline (Sub–Object)**—the sub–object that allows an entire shape to be selected.
- **Steps**—the extrusion parameter that determines the number of times the extrusion shape is repeated along the extrusion distance.
- **Unify Normals**—the parameter of the **Normal** modifier option that points vector normals used for determining visibility in the same direction.
- **Vertex**—the sub–object that allows points on shapes to be selected.

IMPORTANCE OF MODELING WITH EXTRUSIONS

Extruded shapes are important components of modern manufactured goods. In fact, extrusion's reliance on polymers and plastics, materials particularly well–suited for extruding, means that extrusion continues to be a featured manufacturing process.

In 3D Studio, extrusion allows simple shapes to be drawn into 3D objects. This means that 2D engineering drawings where normal cross sections often appear in sectional views can form the basis of 3D models. You can **Insert** an engineering drawing and use a cross section as the shape for an extrusion.

Extrusion also can be more than an end in itself. An extrusion can be scaled along an axis, or individual vertices pulled, or twisted, or bent (see Chapter 7, Modeling with Modifiers) to produce a part that without the base extrusion would be difficult to model.

On a practical note, extrusion is important because objects with cavities, holes, and openings can be modeled without 3D Boolean operations. A multiple 3D Boolean, a part that is the result of successive Booleans, can be problematic in VIZ (see Chapter 5, Modeling with Booleans).

THE BASIC EXTRUSION PROCESS

An extrusion is made when a 2D shape moves along an axis or path. If all points of the shape are coplanar, the object is extruded perpendicular to the original shape. For example, consider the aluminum extrusion in Figure 2.1. This shape can be cut to any length, drilled, milled, and combined with other shapes to make any number of products. Note that the extrusion has two components: the shape and the path. Together they form the extrusion.

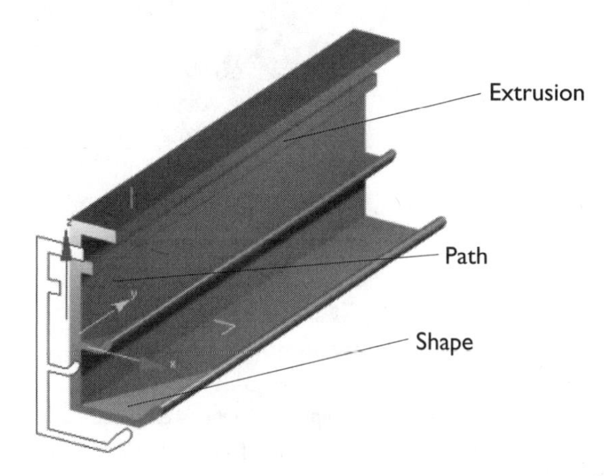

Figure 2.1 *An extrusion has two components: a shape and a path.*

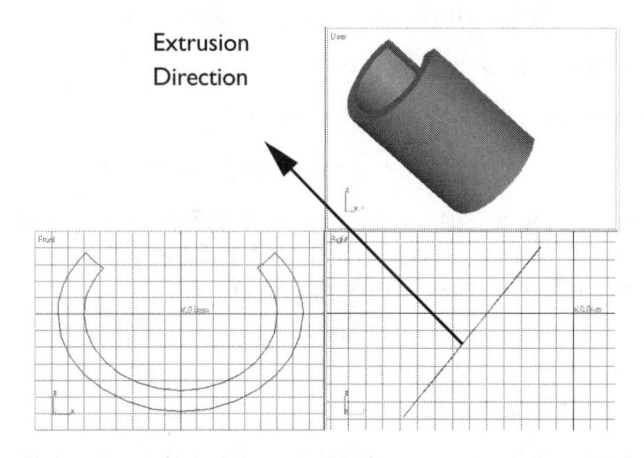

Figure 2.2 *An extrusion remains perpendicular to the shape when the shape is angular or oblique.*

When the shape is not parallel to one of the principle views (either angular or oblique) the extrusion is still created perpendicular to the shape (Figure 2.2). The shape, originally created in the front view and rotated backward in the side view is extruded perpendicular to the rotated cross section.

If all points on the shape do not lie on the same plane, the extrusion is performed perpendicular to the view in which the shape was created. In Figure 2.3 a shape has been created first in the front view, then edited so that all points on the spline do not lie in the same plane. You can see this non-coplanar condition in the side view. When extruded, the shape is extruded along the front view's Z–axis (Figure 2.4).

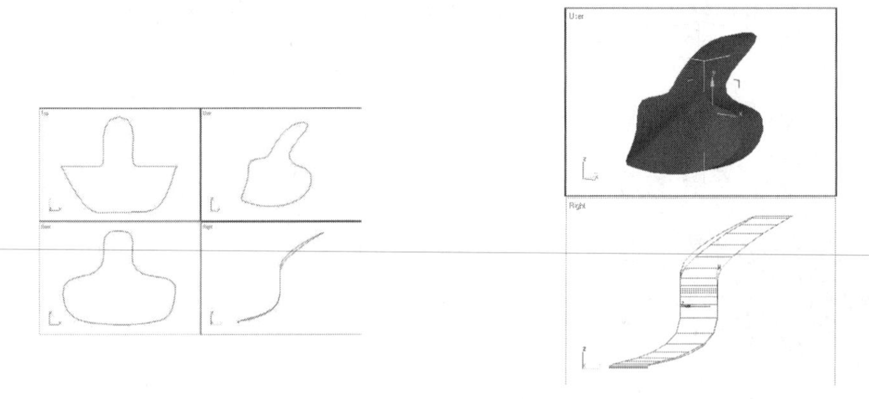

Figure 2.3 *A non-coplanar shape.*

Figure 2.4 *A non-coplanar shape is extruded perpendicular to the Z–axis of the view of creation.*

EXTRUSION GUIDELINES

There are a few guidelines that will help you plan your extrusions.

- Extrusion shapes can be spline or NURBS curves.
- Extrusion shapes can be closed or open.
- Closed extrusions can have closed (capped) ends while open shapes cannot.
- The depth of the extrusion is set in **Modify|Parameters|Amount** field and is expressed in current units set in **Tools|Drafting Settings|Units Setup** (VIZ) and in **Customize|Units Setup** in MAX.
- The extrusion is created from the shape positively in the direction of the shape's original local Z-axis. Reorienting the shape's pivot will not change the direction of extrusion.
- The default number of steps (number of times the shape is repeated from the beginning of the extrusion to the end) is 1 (one at start and one at end).

Increasing this number does aid visualizing in wireframe mode (and in later modifications), but increases the geometric complexity of your object.

- A PATCH output increases geometric complexity; a NURBS output decreases geometric complexity and allows your extrusion to be **Attached** to other NURBS curves.
- A spline extrusion cannot be attached to another spline extrusion.

Tip: The origin (pivot) of an extrusion will always be in the plane of the original cross–sectional shape. The extrusion amount (length) is raised toward you when looking at the shape. For example, if you create a number of cross sections in the top view and extrude them, they will all begin at world Z=0. This aids in positioning multiple extrusions.

- Visibility of extruded surfaces is determined by the direction of face normals (perpendiculars to surface polygons). You **Unify Normals** to point all normals the same direction. You **Flip Normals** to point them in the opposite direction.

TOOLBOX SHAPES FOR EXTRUSIONS

The basis of any extrusion is the cross–sectional shape. To minimize repositioning, it is important to analyze the view in which the cross section appears true shape. This keeps you from creating the extrusion in one view only to have to rotate and move it later.

Any of the spline or NURBS primitives in the **Create|Shape** toolbox can become the shape for an extrusion, although you will rely heavily on **Line**, **Circle**, **Arc**, **Rectangle**, and **Donut**. **Circle** will produce a solid cylinder, capped on both ends; **Arc** will produce a portion of a thin roll, open on both ends; **Rectangle** will produce a right four-sided prism, capped on the ends, with the option of rounded corners; **Donut** produces a tube with inner and outer diameters, capped on the ends. These standard shapes are shown in Figure 2.5.

The **Line** shape has been saved for last because it is more difficult while being infinitely flexible. A spline line passes through the points that define it; a **NURBS Line** can use this **Point Curve** method or the **Control Vertex** (CV) method. In this chapter we will discuss spline lines. In Chapter 6. Modeling with NURBS, **NURBS Curves** will be discussed in detail.

Note: In Figure 2.5, the arc extrusion demonstrates that nonclosed shapes do not have capped ends and the effect of viewing the inside of a surface when normals are pointing out. The far side of the arc extrusion really is there, you are just looking at the inside. Both inside and outside can be seen at render time by choosing **Force 2-Sided** in the **Render Scene** dialogue box.

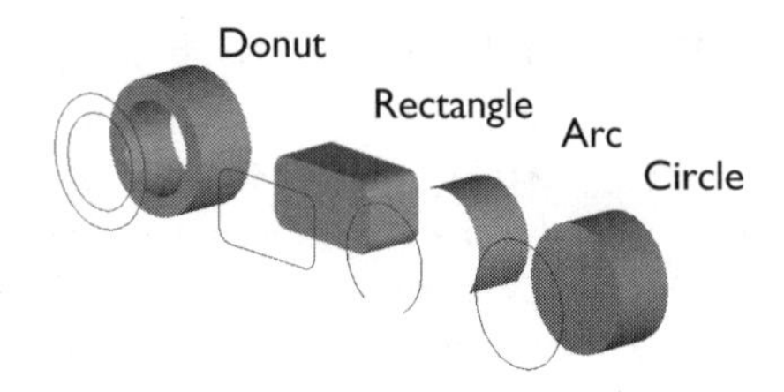

Figure 2.5 *Extrusions made from VIZ 2D toolbox shapes.*

A spline line has two types of vertices or points: corner points and curve points.

- A *corner vertex* is used to make sharp corners. To do this, simply click where you want the corner.
- A *curve vertex* is used to control the shape of a curve. To do this, click and drag.

You will almost never create a spline shape exactly the way you want it at the start. You will move points, reshape the curve, add points, and delete points. In fact, being overly concerned with the exact shape can actually slow down your modeling. It is often faster to approximate the shape and then quickly edit it, and 3D Studio will automatically close a freehand line shape if you put the last point on top of the first. Remember, click and drag for a smooth curve, just click for a sharp corner. Figure 2.6 shows a spline line shape being automatically closed. This shape can be edited to arrive at the smoother and more symmetrical shape in Figure 2.7, producing an effective extrusion.

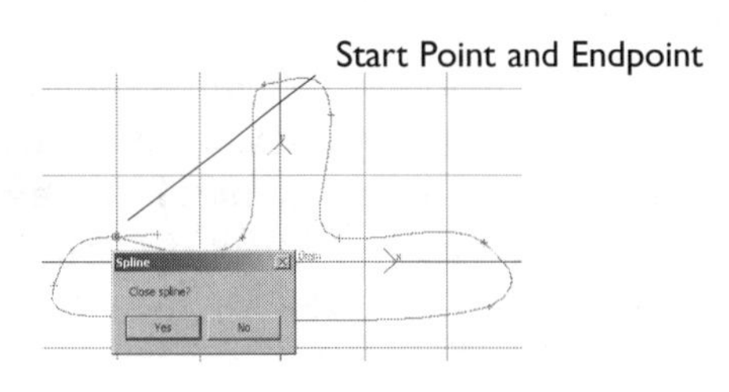

Figure 2.6 *A spline shape will automatically close.*

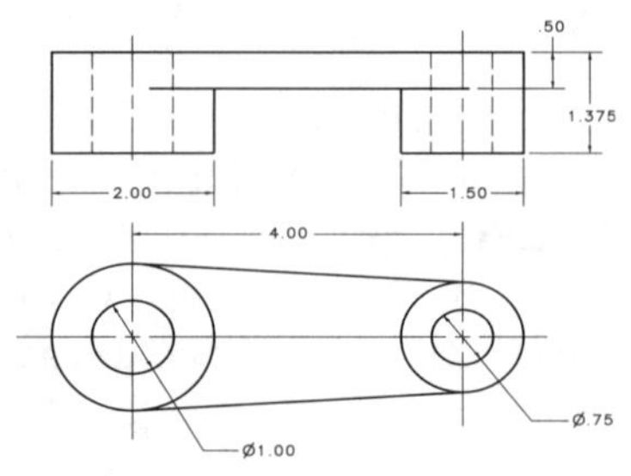

Figure 2.7 *After editing, the shape produces an effective extrusion.*

EDITING LINE SHAPES

Because it is relatively difficult to create a freehand line shape in finished form without at least some editing, you will need to practice making a number of extrusion profiles. Several practice objects can be found among the problems at the end of this chapter. There is no substitute for experience so after reading these guidelines, complete the problems until you are familiar with line editing techniques.

Move Points. To move a point, choose **Modify|Edit Spline|Sub Object|Vertex** and click on the vertex you want to move. Click on the **XY Lock** in the menu bar to freely move horizontally and vertically (or one of the other locks as desired).

Add Corner Point. To add a corner point, choose **Modify|Edit Spline|Sub Object|Vertex** and **Insert** in the **Geometry** rollout menu. Click on the line where you want the corner point. The cursor picks up the line and you can move this corner point to the desired location. Click to put the corner point down. Right–click to lose the tool and select another point.

Add Curve Point. Begin as you did with a corner point but click and drag where you want the curve point. Note: this will immediately reshape the curve (because you added another point) so be ready to do some serious editing. Right–click to set the point. Or, simply choose **Refine** and click where you want additional points; 3D Studio adds curve points that can be edited.

Delete Point. To delete a point, choose **Modify|Edit Spline|Sub Object|Vertex** and select the target point. Choose **Delete**; 3D Studio deletes the point.

Remove a Segment. To remove a segment, choose **Modify|Edit Spline|Sub Object|Segment** and select the target segment. Choose **Delete**; 3D Studio deletes the segment and the shape is no longer closed.

Close an Open Shape. To close an open shape, choose **Modify|Edit Spline|Sub Object|Vertex** and move one open end on top of the other. VIZ asks if you want to weld the vertices; answer **Yes**. Or, choose **Modify|Edit Spline|Sub Object|Segment|Insert** and click on one open end and then the other; 3D Studio again asks if you want to weld the vertices.

Join Two Shapes. To join two shapes, choose **Modify|Edit Spline|Sub Object|Spline** and move one shape by an end vertex to the open end of the other shape; 3D Studio asks if you want to weld the vertices. Answer **Yes**.

Certain of the spline modifiers stay on until you toggle them off (like **Insert**) while other modifiers (like **Delete**) are chosen after the vertex or segment and must be chosen again and again.

NESTING SHAPES

By placing one shape inside another (nesting), you can create holes or openings in the extrusion. Figure 2.8 shows the extrusion used in the last two figures with holes in its arms. The following steps demonstrate how this was done.

Figure 2.8 *Nested shapes create openings in the extrusion.*

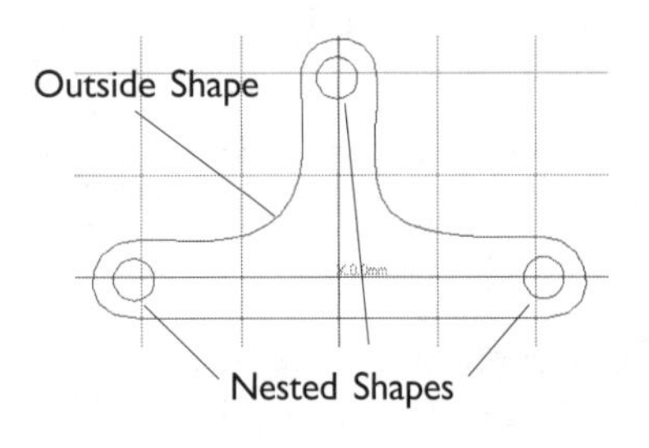

Figure 2.9 *Circles (holes) and outside shape.*

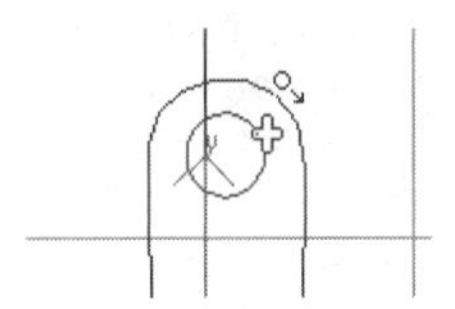

Figure 2.10 *Attach nested shapes so they become holes.*

STEP 1

Create the holes. Make sure the extrusion profile is at 0, 0, 0 so the holes will be in the same plane. Create three identical holes centered on the ends of the extrusion's arms (Figure 2.9).

STEP 2

Attach the holes. Select the outer extrusion profile. Choose **Modify|Edit Spline|Attach** and move the cursor to one of the holes. The cursor will change to a cross and a small attach identifier will appear (Figure 2.10). Click on the first hole and continue until all holes are attached. The profile and its holes are now one shape.

STEP 3

Extrude the attached shape. With the extrusion shape selected choose **Modify|Extrude** and enter an appropriate number in the **Amount** field. Make sure **Cap Start** and **Cap End** are checked. You should have results similar to Figure 2.8.

BOOLEAN OPERATIONS ON SHAPES

A Boolean operator combines, subtracts, or intersects one geometric form with another and in this way, almost any extrusion shape can be created. The two we will introduce first are Boolean subtraction and union.

Boolean subtraction removes the second shape identified from the first. You always want to select the shape you want left first and then the shape that is acting as the cutting tool. For example, Figure 2.11 has a number of circles arrayed around a ring. To create the external teeth formed by the arrayed circles, follow these steps.

STEP 1

Attach the ring and circle. Choose **Modify|Edit Spline|Attach.** It doesn't matter in which order the two are selected.

STEP 2

Subtract the circle from the ring. Choose **Sub Object|Spline** and **Boolean|Subtract** (the second icon) and select the ring. Rechoose **Boolean** and select the circle. The circle is subtracted from the ring (Figure 2.11).

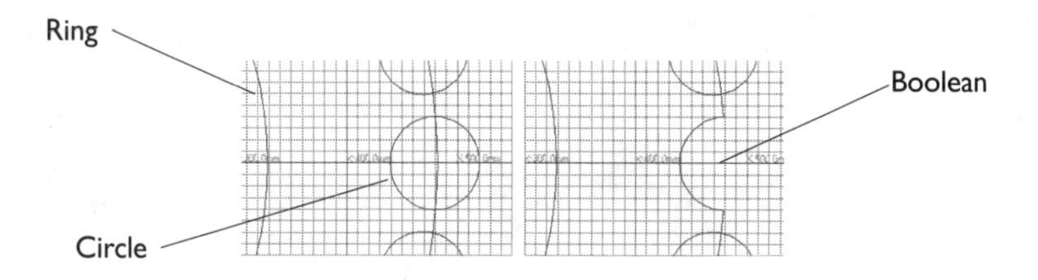

Figure 2.11 *Result of Boolean subtraction.*

Tip: When extruding circular splines it is important to assign a sufficient number of steps to the circle so that when combined later with other shapes all curves will be smooth. Select the circle and choose the **Modify** tool. In the **General** roll–out menu, enter **24** in the **Steps** data field. Do this with all circles as you create them.

A Boolean union combines the geometry of two shapes. In this case, it doesn't matter in which order the two are selected. Figure 2.12 shows an alignment tab in position. When the tab is unioned to the external tooth ring, the result is all the ring geometry and all the tab geometry.

STEP 1

Attach the tab. Select the ring and choose **Modify|Edit Spline|Attach**. Select the Tab.

Tip: To make the job of attaching all the circles easier, choose **Modify|Edit Spline|Attach Multiple** and attach all the circles to the ring at once. When you perform the Boolean subtraction, select the ring and then just march around its outside, selecting successive circles after the first subtraction.

STEP 2

Union the tab. With the attached ring and tab selected choose **Sub Object|Spline** and the **Boolean|Union** option. Select the Ring. Choose **Boolean|Union** again and select the tab. The finished union is shown in Figure 2.12 and the completed extrusion in Figure 2.13.

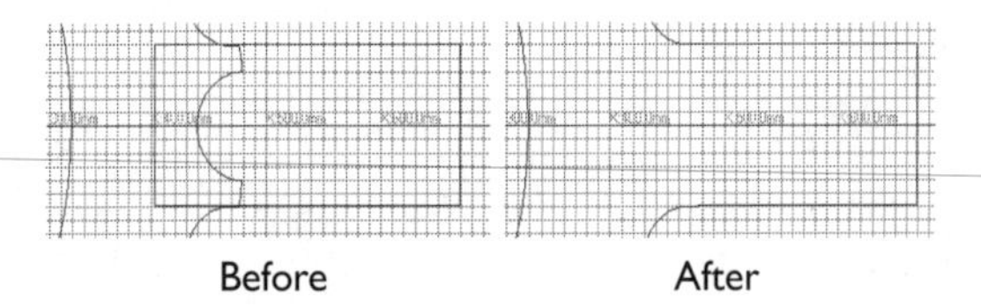

Figure 2.12 *Result of the Boolean union.*

Figure 2.13 *The finished extrusion.*

GEOMETRIC ANALYSIS

It may be that a part is appropriate for extrusion even when it is more complex than something formed simply by pushing material through an opening. For example, consider the views of a connecting bracket shown in Figure 2.14. On first inspection, extrusion might not seem the logical choice because the bracket isn't a uniform thickness. However, an analysis sketch like that in Figure 2.15 reveals that when considered as two closely related extrusions, the bracket fits nicely into extruded geometry. The front portion shares all the extrusion geometry with the rear portion with the addition of a nested opening between the holes.

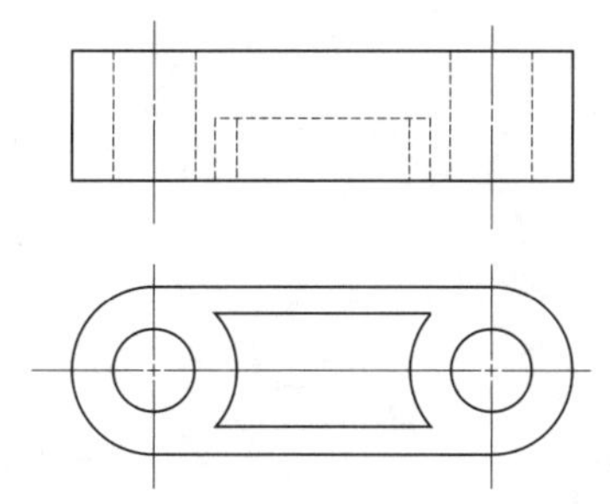

Figure 2.14 *Views of the connecting bracket.*

The approach is to create the entire extrusion profile (both front and rear geometry). Make a copy of the completed profile and edit both until they contain only the necessary shapes attaching and performing Booleans as necessary. On the left side of Figure 2.16 you can see the completed profile and the two separate extrusions that have been made from it. On the right side, the two extrusions have been aligned to form the final object.

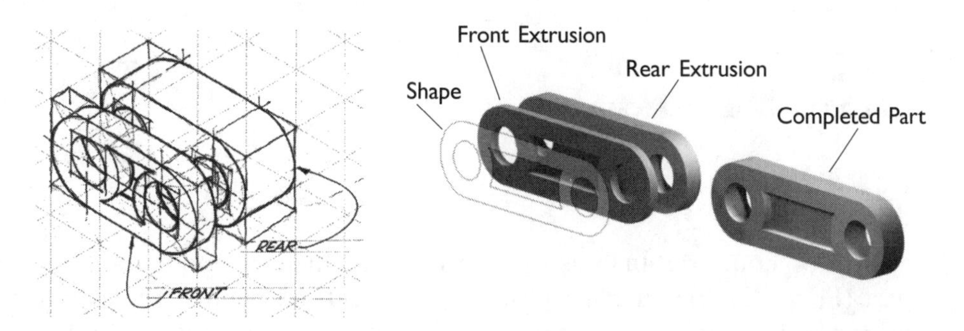

Figure 2.15 *A geometric planning sketch of the connecting bracket.*

Figure 2.16 *The connecting bracket is comprised of two extrusions made from common shapes.*

EXAMPLE: EXTRUDING THE FLANGE

The Flange in Figure 2.17 is a prime candidate for modeling by extrusion, and for combining multiple extrusions based on common shapes. It contains geometry that is identical as depth is changed, with holes that extend through the various components. Closer inspection of the part reveals that the Flange can be broken into three aligned components: the Base, the Body, and the Top.

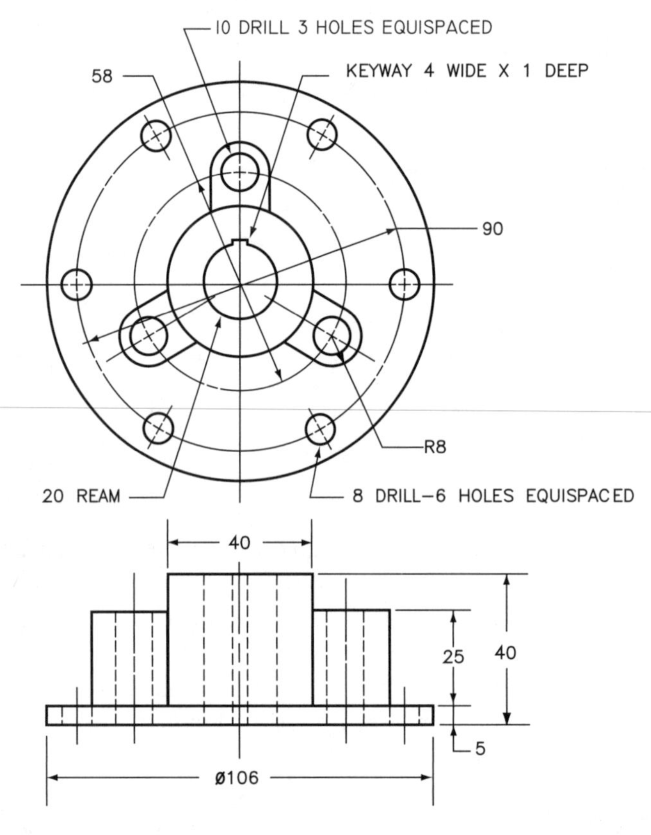

Figure 2.17 *Dimensioned views of the Flange.*

STEP 1

Perform a geometric analysis. As the three aligned components are pulled apart (Figure 2.18) you can see that certain features such as the center hole and its keyway and the 10 mm drilled holes pass completely through the Base and Body. The component we call the Top has only the center hole and the keyway. Make a mental note of the geometry that the three components share.

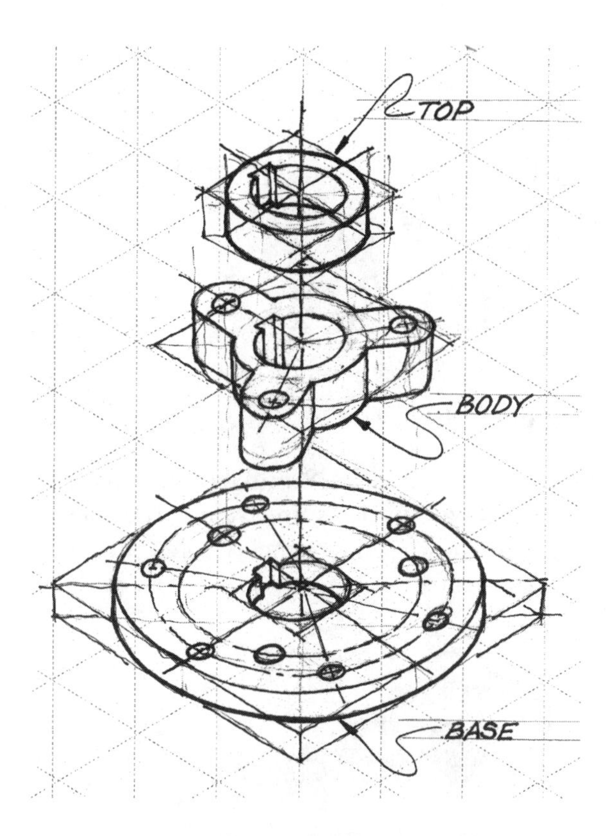

Figure 2.18 *Geometric analysis of the Flange reveals three aligned components.*

The benefits of sketching are numerous. Sketching (like that in Figure 2.18) helps you understand the geometry so you are more efficient working on the computer.

STEP 2

Create the Base plate. Begin by setting millimeters in **Tools|Drafting Settings|Units Setup** and displaying grid line every 10 mm in **Tools|Drafting Settings|Grid and Snap Settings**. In the Top View, create a spline circle of 53mm radius with its center at the world origin (Figure 2.19).

STEP 3

Create the first 8mm diameter hole. Move it to X56, Y0, Z0 (Figure 2.19) using the right–click option on the **Move and Select** tool.

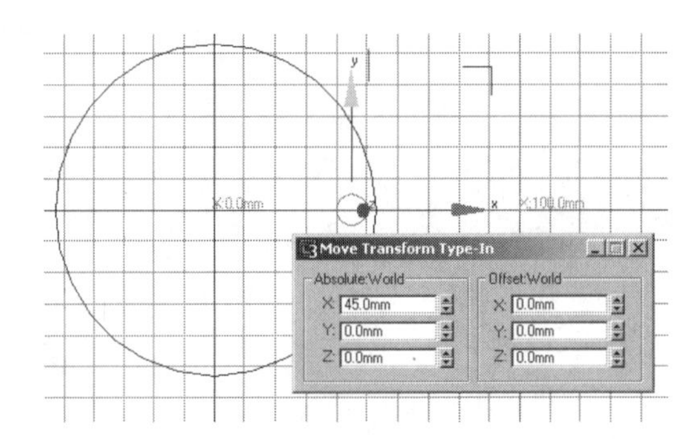

Figure 2.19 *First 8 mm hole on base is positioned as shown in Figure 2.17.*

Use **Hierarchy|Affect Pivot Only** to move the 8 mm hole's pivot to world 0,0,0 (Figure 2.20). Deselect the **Pivot** options and select the **Array** tool. Using the setting in Figure 2.21 array six evenly spaced holes (Figure 2.22).

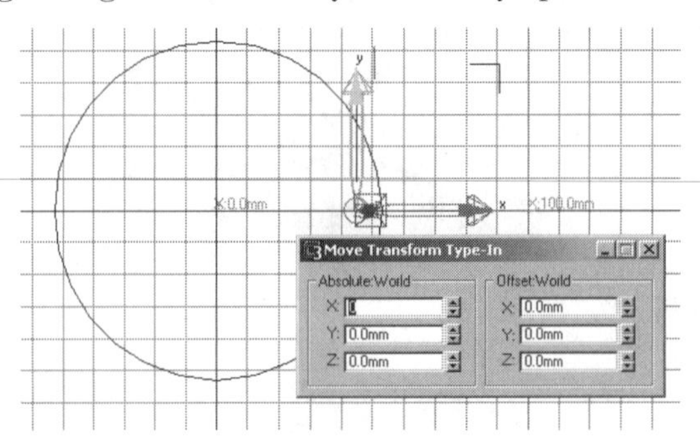

Figure 2.20 *Move pivot of hole to world origin.*

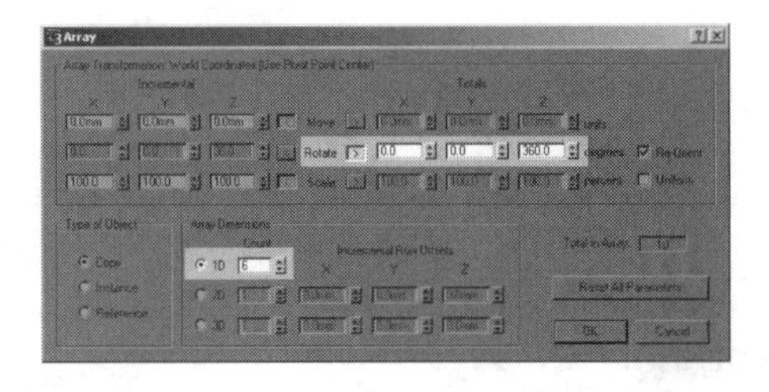

Figure 2.21 *Array a count of six holes about 360 degrees.*

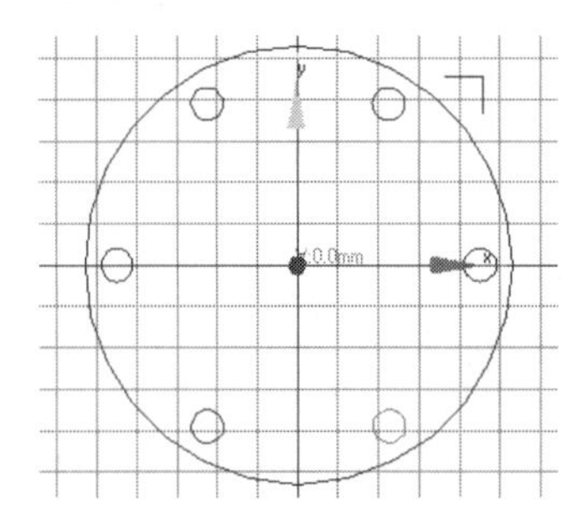

Figure 2.22 *Completed array on base.*

STEP 4

Create the 10 mm drilled holes. Position the first hole at X26, Y0, Z0. Move its pivot to 0, 0, 0 and rotate (right–click on **Rotate** tool) the circle –30 degrees about the world Z axis (Figure 2.23).

Array this hole as you did the 8 mm holes only this time change the count to 3.

STEP 5

Create the center hole and keyway. At 0, 0, 0 create a 20 mm diameter hole. At X0, Y10, Z0 create a 4 mm x 2 mm spline rectangle (Figure 2.24).

STEP 6

Attach the shapes. Select the center hole and pick the **Modify** tool. Select **Edit Spline** and the **Attach** option. Note that the cursor changes shape as attachable splines are found (Figure 2.25). Click on the keyway. You'll notice that both circle and rectangle are now selected. When attached, spline shapes can be acted on by Boolean operators.

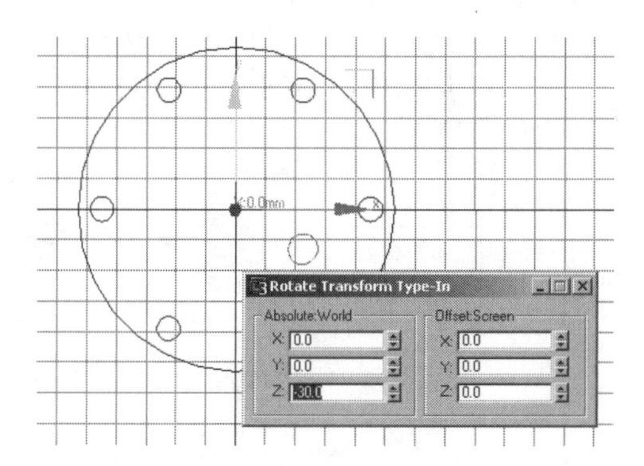

Figure 2.23 *The initial 10 mm drilled hole is positioned 30 degrees from the axis.*

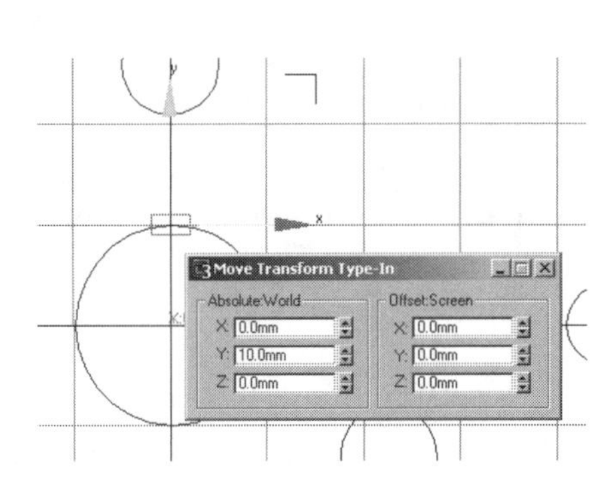

Figure 2.24 *The center hole and keyway in correct position.*

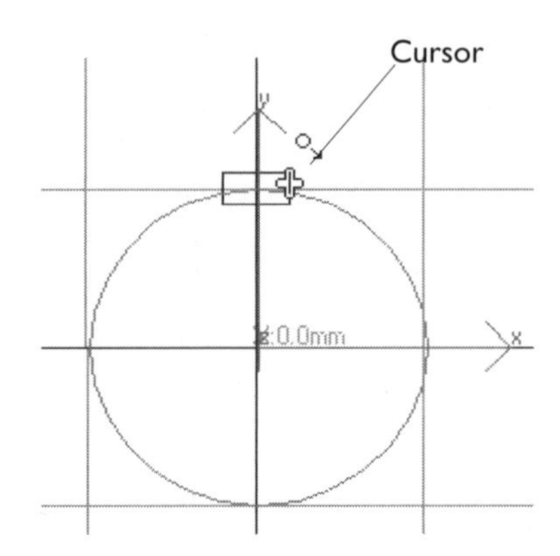

Figure 2.25 *Attaching spline shapes allows them to be acted on by Boolean operators.*

STEP 7

Union the center hole and keyway (Figure 2.26). In **Modify|Edit Spline** select the **Sub Object** modifier and choose **Spline**. Choose **Boolean** the **Geometry** roll–out menu and the **Union** option. Select the circle. Choose **Boolean** again and select the rectangle. The resulting shape is the union of the circle and the rectangle. Do not attach the keyed hole and base at this time. Save the file as *base.max* in a work directory.

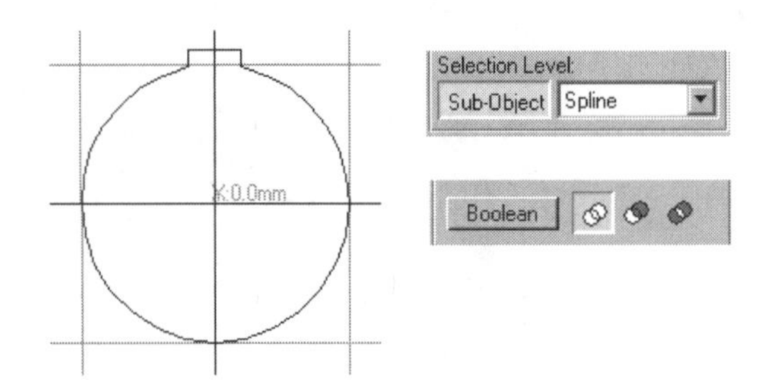

Figure 2.26 *The Boolean union operation joins the hole and keyway.*

STEP 8

Create the Body file. Because the three 10 mm holes and the keyed center hole are part of the Body, immediately save *base.max* as *body.max*. Delete the outer circle of the Base and the 8 mm holes because these are unnecessary for the Body. Save *body.max* again.

STEP 9

Create a first boss. Hide the keyed hole and 10 mm holes. We will be creating a first boss in a "three o'clock" position and rotate it into its final position 30 degrees below the centerline.

Create a spline circle radius 8 mm at X26, Y0, Z0. Create a spline rectangle 26 mm wide x 8 mm long and position it at X13, Y0, Z0 (Figure 2.27). Because the rectangle's pivot is at its center, by moving the center to X13 its right side will be coincidental with the circle's vertical diameter.

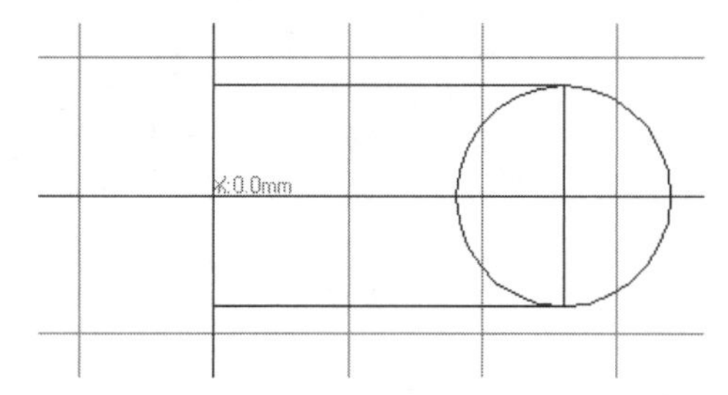

Figure 2.27 *The components of the first boss in correct position.*

STEP 10

Join the two boss shapes. Repeat the operations you completed in Steps 6 and 7. **Attach** the circle and rectangle. Choose the **Sub Object** modifier **Spline** and **Boolean Union** the circle and rectangle. The result is shown in Figure 2.28.

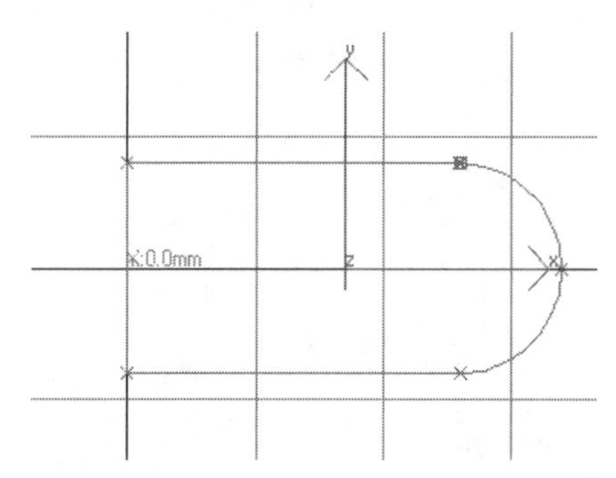

Figure 2.28 *The Boolean union operation joins the circle and rectangle.*

STEP 10

Array the boss. Relocate the boss's pivot to 0, 0, 0 and rotate it minus 30 degrees about the world Z axis. Complete an array of three bosses to place them in alignment with the 10 mm holes. Refer to Step 2. The completed array should look like Figure 2.29.

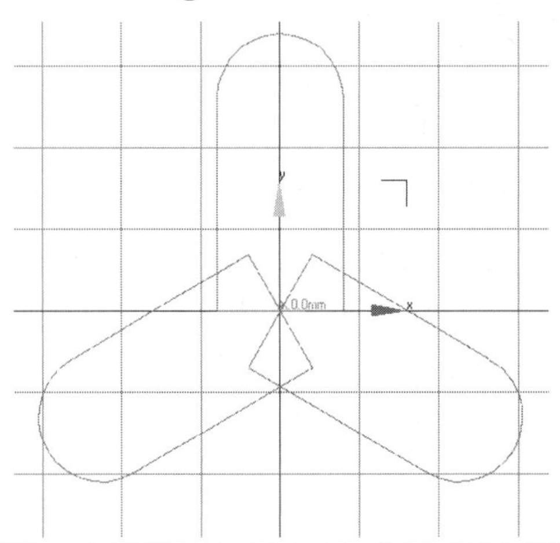

Figure 2.29 *Complete array of the bosses.*

STEP 12

Complete the Body. Create a 40 mm diameter circle at 0, 0, 0 representing the outer diameter of the Body's center. As done previously, attach each boss to the 40 mm diameter circle and perform three separate **Boolean|Union** operations to arrive at the shape in Figure 2.30.

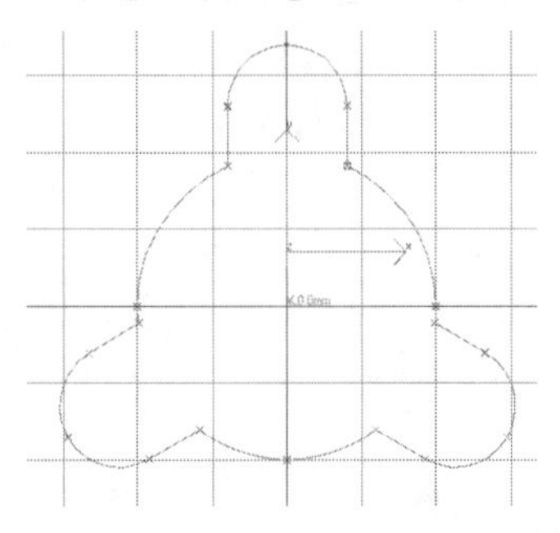

Figure 2.30 *The completed boss and cylinder.*

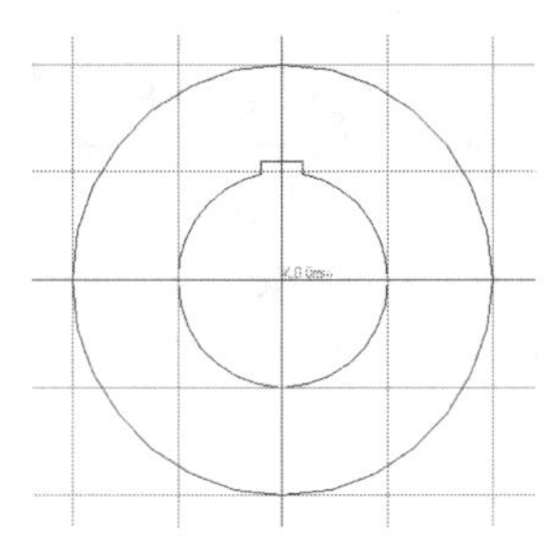

Figure 2.31 *Geometry necessary for the top component comes from the body.*

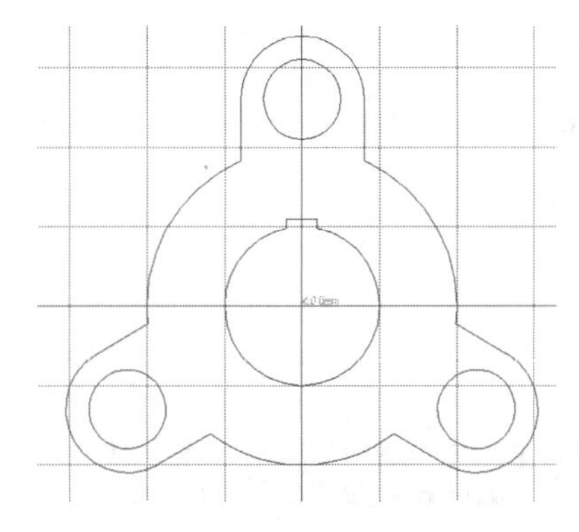

Figure 2.32 *The completed geometry for the body.*

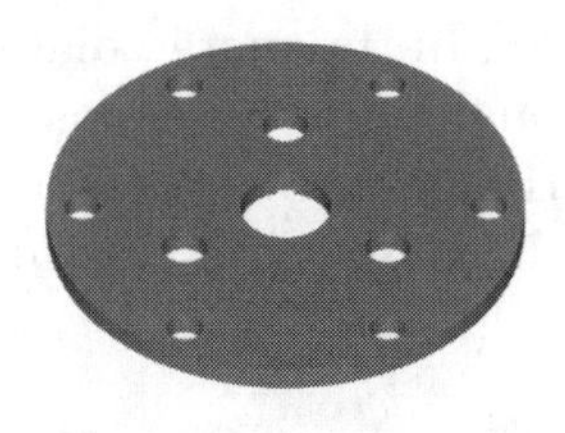

Figure 2.33 *The extruded base.*

STEP 13

Prepare the Top. Unhide the hole and keyway and hide all else. Create another 40 mm diameter spline circle at 0, 0, 0 (the first one was destroyed when it was unioned with the bosses). Save this file as *top.max*. Delete the boss geometry leaving the keyed hole and cylinder (Figure 2.31). Attach the outside circle and the keyed hole and name this shape "top". The Top is now ready to extrude and you can reload *body.max*.

STEP 14

Complete the Body. Display all geometry (Figure 2.32). **Attach** the boss profile, 10 mm holes, and keyed center hole. The body is now ready to be extruded. Name the attached shape "body" and save the file (*body.max*).

STEP 15

Extrude the Base (Figure 2.33). Open *base.max* and display T–F–S–User viewports. You left the base without attaching all the shapes (so you could make the body and top files) so now's the time to attach the outside of the base, the 8 mm holes, the 10 mm holes, and the keyed hole. Name this shape Base. The Base is ready to extrude.

Change the view option of the User viewport to **Smooth + Highlights** and rotate the viewpoint to get a pictorial view. Assign a gray color to the base shape. Click on the base shape and select the **Modify** tool and the **Extrude** option. Using Figure 2.17 as the source of depth information, enter 5 mm in the **Parameters|Amount** field.

Figure 2.34 *The extruded body.*

STEP 16

Extrude the Body. Load the file *body.max* and display four viewports as you did for the Base. From Figure 2.17 you can determine that the body is 25 mm from front to back. To assure that the shapes eventually can be joined, extrude the base 26 mm. The completed base is shown in Figure 2.34.

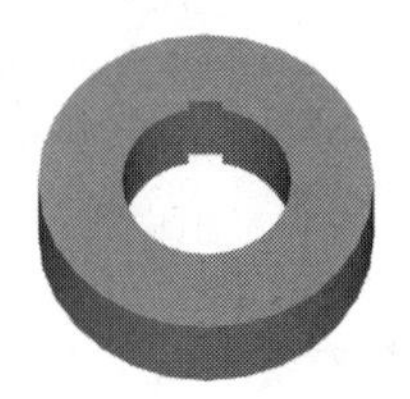

Figure 2.35 *The extruded top.*

STEP 17

Extrude the Top. Load the file *top.max* and display the four viewports. From Figure 2.17 you can determine the Top to be 10 mm front to back. To assure the Top intersects the Body, extrude the Top 11 mm. The completed Top is shown in Figure 2.35.

STEP 18

Merge the Body with the Base. The three files have been constructed consistently about 0, 0, 0. This allows the files to be easily combined. Load the file *base.max* and display the Front view with **Zoom Extents**. Note that because the Base shape was extruded in the Top view, the pivot (origin) of the Base is at world Z=0. The same will be true of the Body and Top, facilitating their final positioning. (An aside: you could have just as easily extruded the Body 30 mm and the Top 40 mm. This would have allowed the pivots of all three components to be coincidental at 0, 0, 0).

In *base.max* choose **File|Merge** and select *body.max*. In the merge dialogue box select the Body and click **OK**. The Body is brought into *base.max* with its pivot at 0, 0, 0. Move the Body to X0, Y0, Z4 (Figure 2.36). This allows the desired 25 mm of the Body to extend above the Base. Save this file as *flange.max*.

Tip: If you are using part of one file as the basis of another, save the file before attaching or combining shapes with Boolean operations. This way, if the operation is unsuccessful you can revert to the saved copy. MAX makes this easy with its Hold/Fetch option.

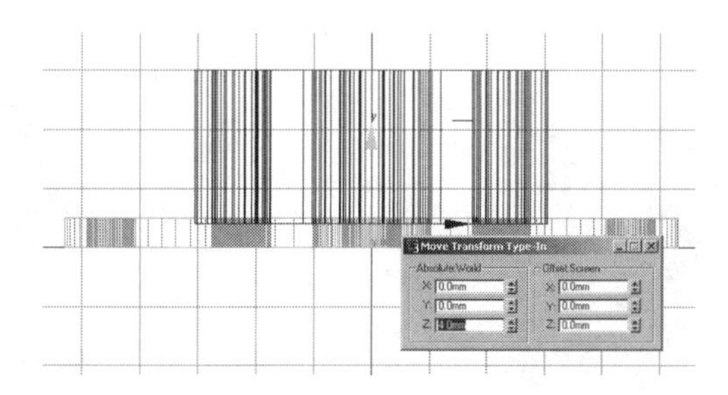

Figure 2.36 *The body merged into the base and correctly positioned.*

STEP 19

Merge the top with the Flange. Again, choose **File|Merge** and select the file *top.max*. The Top is also merged at 0, 0, 0. Reposition the Top at X0, Y0, Z29. This allows 10 mm of the top to extend above the base (Figure 2.37).

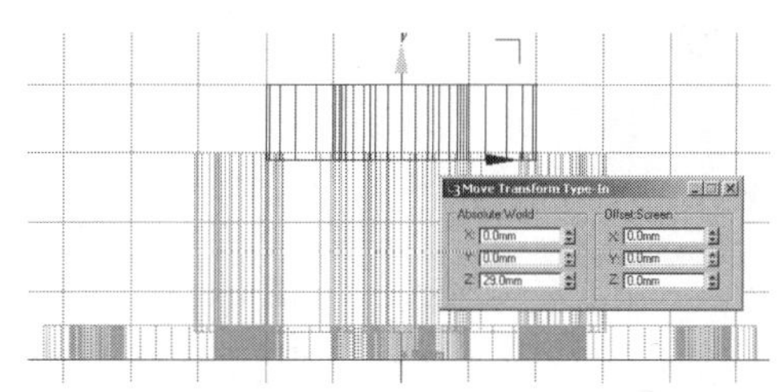

Figure 2.37 *The top is merged into the flange and correctly positioned.*

STEP 20

Display the complete Flange. Select the User view and press the **W** key. Right–click on the view label and turn off **Show Grid** and turn on **Smooth + Highlights**. Click on the **Zoom Extents** tool. Your completed flange should resemble Figure 2.38.

Tip: 3D Studio looks for files to merge based on the path established in **Tools|Configure Paths**. By default this will be *c:\3DSVIZ-MAX\Scenes*. Because you will be saving your designs in your own directories, you will always have to browse from this path to find the directories you want. If you consistently archive your designs to a certain directory, change this default path.

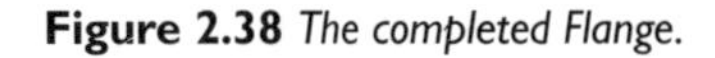

Figure 2.38 *The completed Flange.*

Figure 2.39 *Holes correctly extend through the Flange.*

If you remember, the keyed hole is shared by all three components. The three 10 mm holes are shared by the Body and the Base. By rotating the User view so that you are looking more downward you should be rewarded with a view like that of Figure 2.39. Note that the features correctly go completely through the Flange.

PROBLEMS

Problems appropriate for solving with extrusion are found on the following pages. Use the file *isogrid* found in the *tools*\ directory to plan your modeling strategies. Assign appropriate names to all objects both on your sketches and models. Assume an appropriate depth for each extrusion.

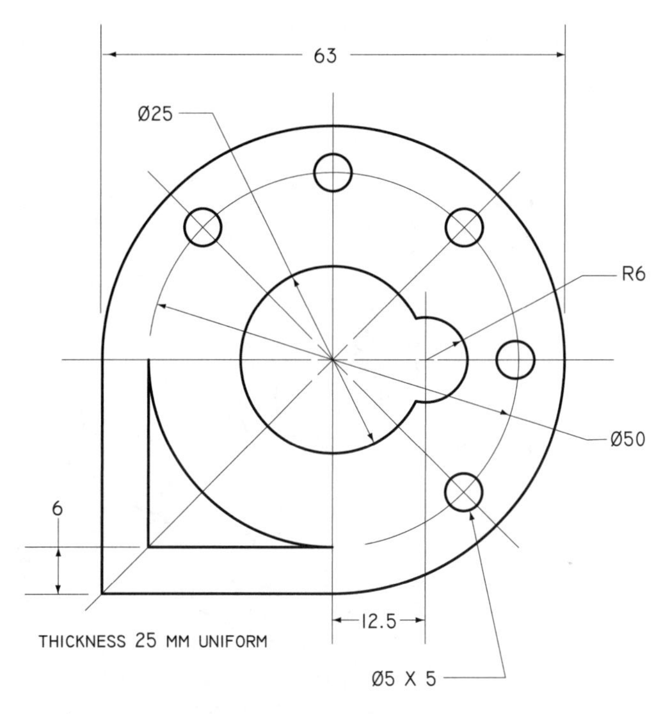

Problem 2.1 *Adapter plate.*

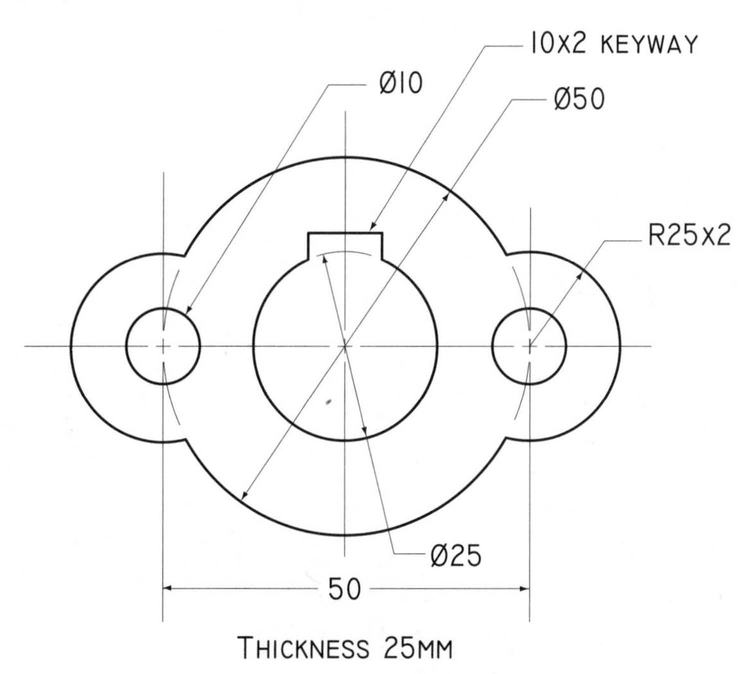

Problem 2.2 *Pump body.*

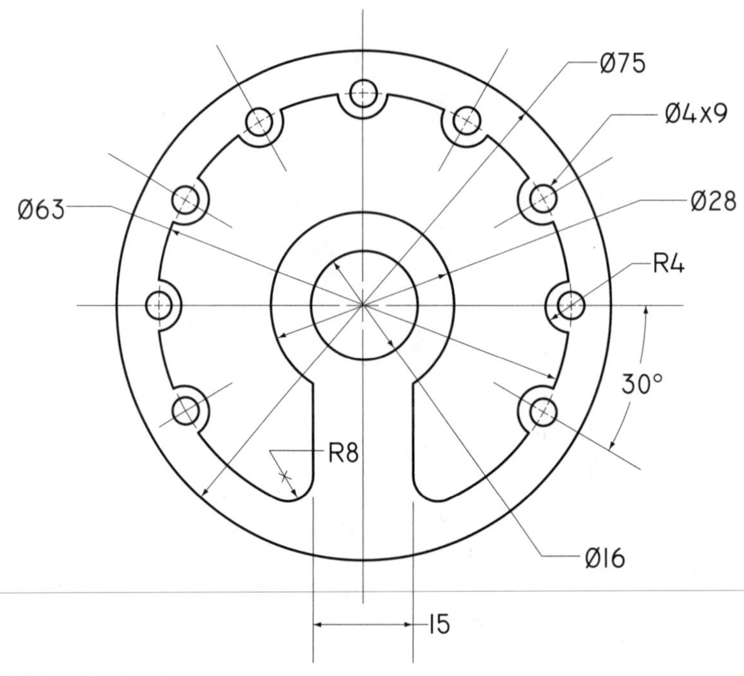

Problem 2.3 *Impeller housing.*

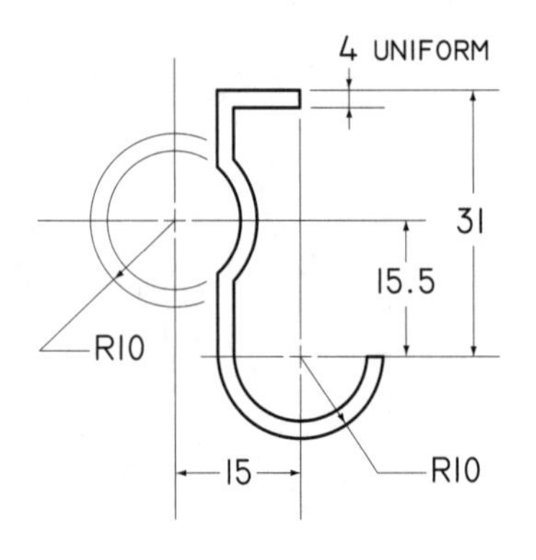

Problem 2.4 *Pipe hanger.*

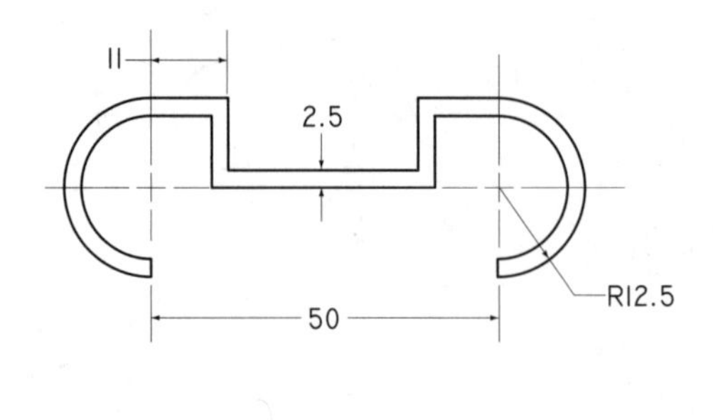

Problem 2.5 *Rail extrusion.*

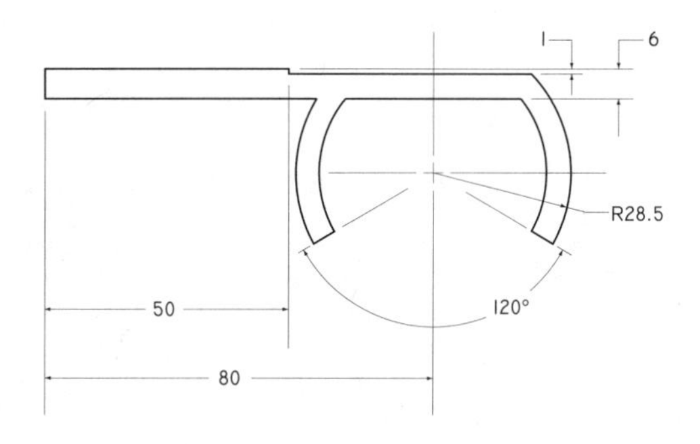

Problem 2.6 *Guide plate.*

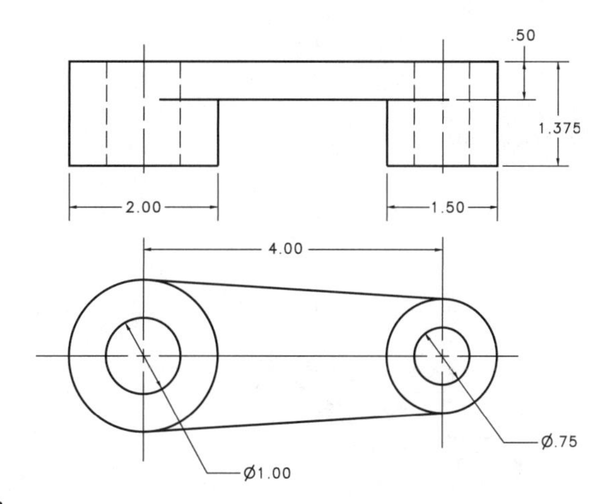

Problem 2.7 *Link arm.*

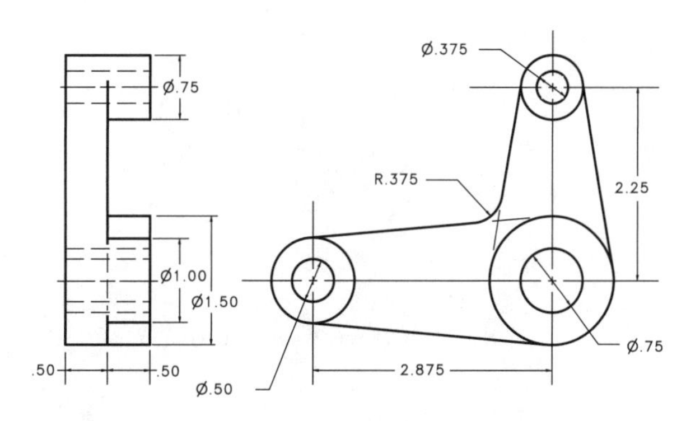

Problem 2.8 *Crank Arm.*

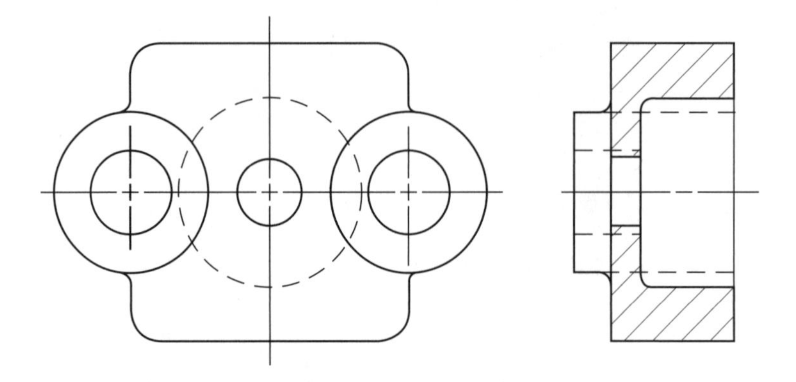

Problem 2.9 *Bearing Mount.*

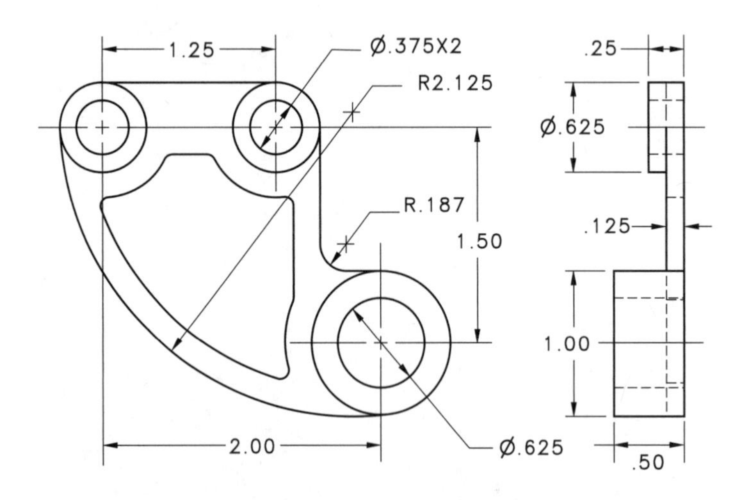

Problem 2.10 *Support Unit.*

Modeling with Lathes

CHAPTER OVERVIEW

After following the first two chapters you have developed skills in analyzing geometry, or identifying objects that can be modeled using primitives or extruded shapes. There is a class of geometry that can't be created easily from primitives or extrusions: swept or lathed shapes. The term *lathe* in 3D Studio describes any profile that is revolved or swept in a circular path about an axis. The term *lathe* also describes a machine tool that creates objects symmetrical about an axis so when performing the lathe operation, as was the case with extrusion, there is a direct parallel to a manufacturing process. A similar manufacturing process, spinning or shaping, also produces a circular symmetrical part by forcing a sheet of material over a mandrel.

In this chapter you will learn to recognize shapes that can be made using the **Lathe** modifier. Many manufactured goods are created on lathes. But beyond this turning process, any object that is symmetrical about an axis is suitable for lathe modeling. For example, storage tanks or vessels, glassware, jugs, bulbs, injection-molded containers all can be effectively modeled using the **Lathe** modifier.

KEY COMMANDS AND TERMS

- **Force 2-Sided**—the option in the **Rendering|Render** dialogue box that applies a material to both sides of a surface.
- **Insert**—the option that allows geometry from a select list of graphics programs to be merged with the 3D Studio design.
- **Lathe**—the shape modifier that sweeps a profile about an axis forming a surface of revolution.

- **Local Origin**—see **Pivot**.
- **Lathe (Sub–Object|Axis)**—the selection level of a lathed object that allows repositioning of the axis of symmetry.
- **Normal**—a vector perpendicular to a surface used for determining visibility.
- **Pivot**—the local origin of an object used as the basis for all translations, rotations, and scaling.
- **Profile—**the 2D shape object that defines a lathed object's cross–sectional shape.
- **Segment**—a **Lathe** parameter that sets the number of times the profile will be repeated around the object.
- **Smooth**—a **Lathe** parameter that interpolates between **Segments**, rendering the surface smooth.

IMPORTANCE OF MODELING WITH LATHES

Lathed shapes form the basis of many industrial models and like any other geometry created in 3D Studio, lathed shapes can be subsequently edited into even more sophisticated geometry. In comparison to extruded cylindrical shapes, lathed shapes can feature fillets and rounds where extrusions cannot. However, extrusions can feature holes and irregular cavities where lathed shapes cannot. Holes in lathed objects must be subtracted using compound Boolean operations (see Chapter 5, Modeling with Booleans).

The heart of any lathed shape is the cross–sectional profile and, as luck would have it, engineering designs are replete with these profiles. Sectional views are prime candidates for providing shapes that can be lathed. Detail views can be edited into usable shapes for modeling with lathes (Figure 3.1). In fact, there are but three conditions required for a successful lathe:

- The object is symmetrical about an axis.
- There is a contiguous cross–sectional shape.
- There is an axis about which that shape is swept in a circular path.

The symmetry is evident by the outer profile being identical on either side of a centerline. This may not be immediately evident because a design may have features that obscure its symmetry such as tabs, bosses, ribs, or flanges.

The shape needs to be contiguous, though it need not be closed. In fact, it is often more efficient to leave the interior of a part open so that a separate shape (with a separate material) can be added later to the inside of the model. This is appropriate in the case of a rough casting whose interior bore has been machined.

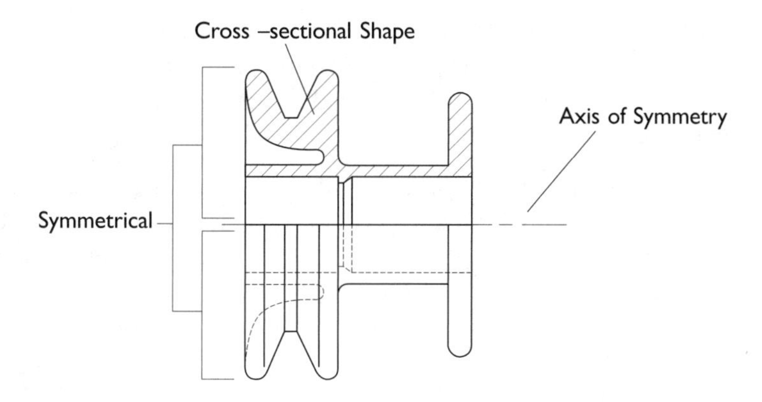

Figure 3.1 *Engineering drawings provide the shapes for lathe operations.*

The axis is important because the lathed object is symmetrical. By marking the axis with a line shape, the lathe's **Sub Object|Axis** can be relocated so that the finished lathe is the correct diameter.

BASIC LATHE PROCESS

A lathe object is formed when a shape object is swept in a circular arc perpendicular to an axis. Each position of the shape is called a **Segment**. Vertices on adjacent segments are connected with line segments, forming the surface. Because a lathe is always symmetrical, you need only half the complete shape. For example, in Figure 3.2 a shape and an axis are shown. In this case, one end of the shape touches the axis so when lathed, the resulting object is closed at that end. In the second example (Figure 3.3) the shape does not touch the axis, which leaves both ends open.

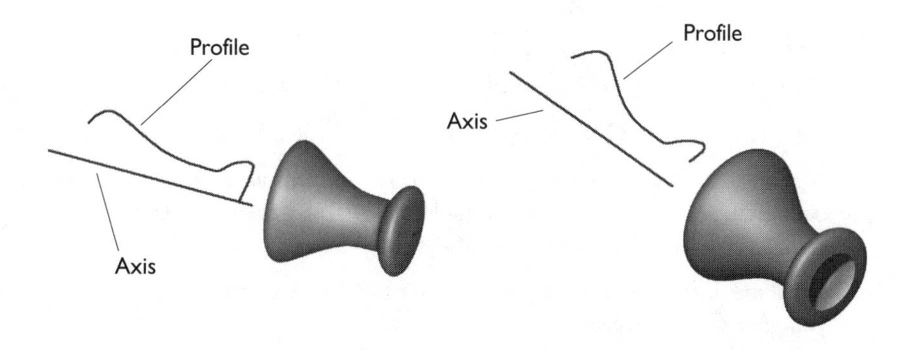

Figure 3.2 *Lathe shape and axis (one end closed).* **Figure 3.3** *Lathe shape and axis (both ends open).*

Other than for very small objects, the default number of segments (16, the number of times the shape is repeated around the circle by default) is hardly sufficient to define a smooth object. A value between 24 and 48 should produce realistic results.

Because of the way that 3D Studio determines surface visibility, a single lathed line will produce a single-sided surface of no thickness, much like thin sheet metal. The inside will not be rendered by default so as long as the surface is singularly convex, you'll always be looking at the outside. You can get around this as is described later in this chapter. Were the profile a closed or nearly closed shape (Figure 3.4), the lathed object will have wall thickness and 3D Studio can show interior detail.

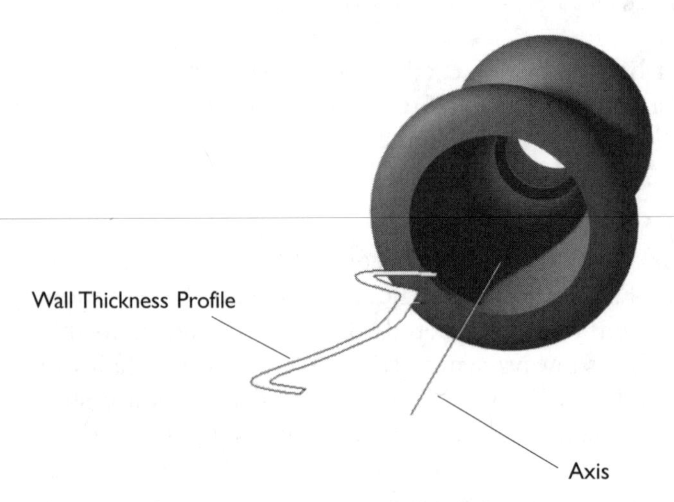

Figure 3.4 *Wall thickness reveals interior detail of the lathed shape.*

LATHE GUIDELINES

Consider these guidelines to help you plan your lathes.

- Any spline or NURBS curve, open or closed, can be lathed.
- Though space curves (curves where all points do not lie in the same plane) can be lathed, the results are unpredictable. In general, a **Lathe** is made from a cross section where all points lie in the same plane.
- The initial **Lathe** operation will be about the Y–axis of the viewport in which you created the profile.
- The initial **Lathe** operation will be about the shape's local origin (**Pivot**). You can move this origin to the axis of symmetry before the **Lathe** operation

(**Hierachy|Affect Pivot Only**). Or, you can move the axis of the **Lathe** (**Sub Object|Axis**) after the lathing operation.

- The initial setting of 16 segments/360 degrees must usually be set to 24 to 48. However, increasing segments increases geometric complexity when **Patch** or **Mesh** options are chosen.
- The **Lathe** can be created at any portion of a 360-degree circle. The sweep begins in the screen X-Y plane and progresses counterclockwise about the axis of symmetry. These partial lathes, however, do not form completely closed shapes.
- A patch or mesh **Lathe** can be directly converted to NURBS curves and surface.
- By checking the **Smooth** option, the lathed surface is interpolated between segments so that a smooth exterior is created. You will normally want this checked.
- Usually, vector normals will be pointed outward, revealing the external skin of the lathed object. You have the option of flipping the normals if necessary.
- The inside of a zero thickness, lathed surface can be seen at render time. Choose **Force 2-Sided** from the **Render Design** dialogue box.

GEOMETRIC ANALYSIS

Figure 3.5 shows a full sectional view of an object symmetrical about an axis. Note that the material on either side of the axis is identical—one the mirror image of the other and each equidistant from the central axis. An analysis sketch (Figure 3.6) shows that were a slice of the solid material swept in a circular path, maintaining a constant distance from the axis, a lathed surface of revolution would be formed.

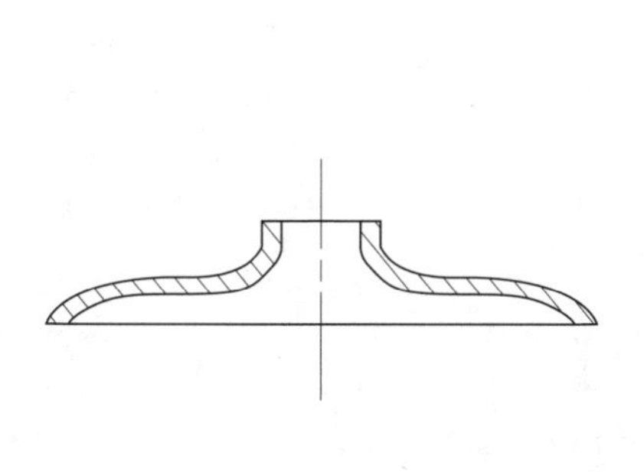

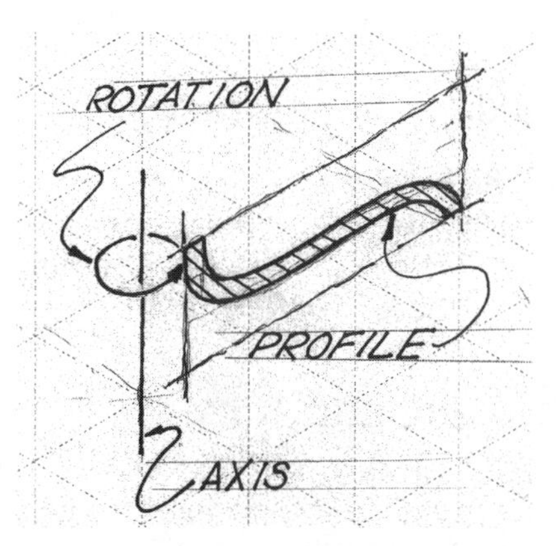

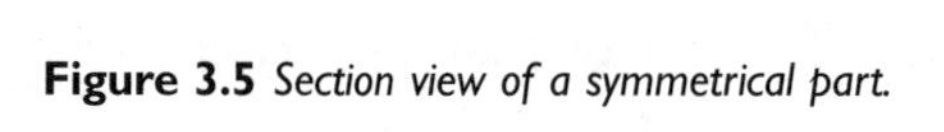

Figure 3.5 *Section view of a symmetrical part.*

Figure 3.6 *Analysis sketch showing how the part is modeled with the* ***Lathe*** *modifier.*

Tip: Positional geometry in VIZ functions as layout construction did in traditional engineering drawing. It is often more efficient to place several important features correctly in space and use this positional geometry to guide final line work than to attempt massive editing.

SHAPES FOR LATHES

The basis of any lathed object is the cross–sectional shape and almost any cross–sectional shape can be created right in 3D Studio. Review pages 29-32 where line shape operations are covered. There are three approaches to creating shapes for lathes.

If manufacturing accuracy is not paramount, you can create lathe profiles directly in 3D Studio.This is not because VIZ can't be that accurate, but rather because 3D Studio does not have many of the advanced 2D construction tools (trim to extension, automatic fillet, and so forth) of CAD programs such as AutoCAD.

- Create positional geometry (holes, fillets, rounds, surface limits) at correct world coordinate locations. Using **Shape|Line** and a **Grid** and **Snap** that makes sense for the design, create a continuous freehand curve using the positional geometry as a template. Click for a sharp corner; click and drag for a curved transition. Then, use **Modify|Edit Spline|Sub Object** and choose **Vertex** or **Segment** and fine–tune the curve to a final shape.
- Create the basic design of the shape using line, circle, and box shape primitives. Then, as you did in Chapter 2, **Attach** the shapes and Boolean them to arrive at a final profile. Edit vertices and segments as necessary.

A second approach is to use geometry created in another graphics program as the basis of your lathes. This has several advantages. First, you don't have to recreate the profile in 3D Studio, and you will save considerable modeling time. Second, sources of error that might be introduced by recreating the geometry are eliminated. And third, if part of an assembly, you know that all the parts will eventually fit together.

A third approach is to create profiles yourself in a drawing program and **Insert** them into 3D Studio. Do this if you already have experience in a program such as AutoCAD or Adobe Illustrator, or if you need the construction tools that these programs offer.

LATHE PROCEDURES

Recognizing geometry that can be lathed is fairly straightforward. First look for a centerline and then look for identical diameter geometry on either side (Figure 3.7). This part, a collar, is comprised of a 3.50–diameter cylinder and a

1.189–diameter cylinder having a .567–diameter hole and can be found in *problems/chapter3* on the accompanying CD-ROM. Only one-half of the symmetrical solid need be created. To model the part in Figure 3.7, follow these steps:

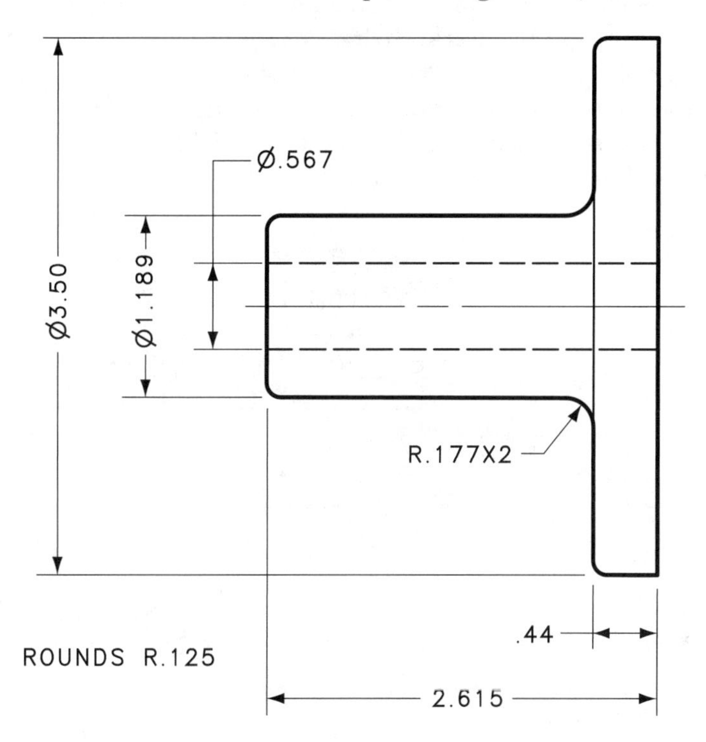

Figure 3.7 *A symmetrical part appropriate for* ***Lathe*** *modifier.*

STEP 1

Choose **Tools|Drafting Settings|Grid and Snap Settings** and **Feet W/Decimal Inches**; click **Default Units:Inches**. This establishes units that match Figure 3.7.

Note: Use **Customize|Units Setup** in MAX to establish the system of units appropriate for each problem.

STEP 2

Choose **Tools|Drafting Settings|Grid and Snap Settings|Home Grid**; set grid spacing to 0'0.1" and a major line every 10th unit. This establishes grid and snap that match Figure 3.7.

STEP 3

Create a 0.311 (length) x 2.615 (width) rectangle shape. Note: 3D Studio uses *length* to denote screen Y–axis size and *width* to denote screen X–axis size. Relocate its origin by **Hierarchy|Affect Pivot Only** and the **Move** transform. Type-in half the width (–X) and half the height (+Y). This puts the origin in the upper–left corner. Relocate this corner to X0, Y.5945 (Figure 3.8). This represents the top wall of the middle cylinder. The hole is between this wall and the Y=0 origin grid line.

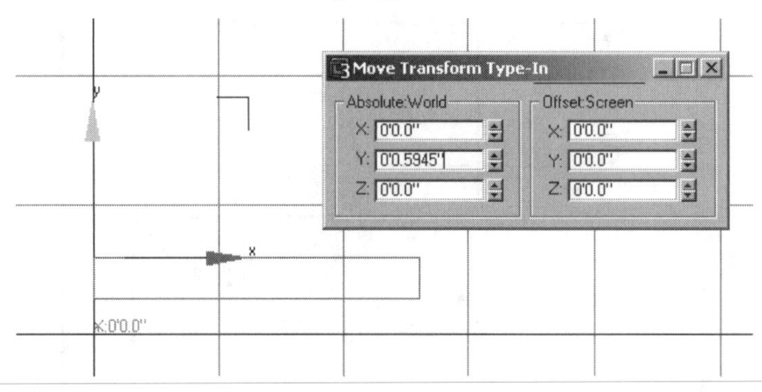

Figure 3.8 *Rectangle representing the top wall of the 1.189 diameter cylinder.*

STEP 4

Create a 1.75 (length) x .44 (width) box. Relocate its origin to the lower left corner using half the length (–Y) and half the width (–X). Position this box at X2.176, Y.2835 (Figure 3.9). This represents the upper portion of the Collar's flange.

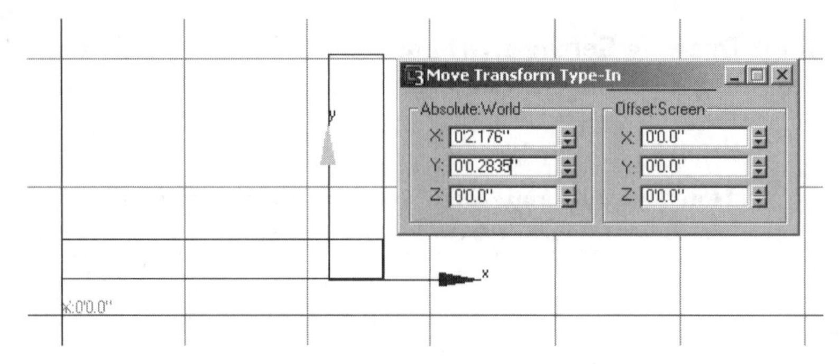

Figure 3.9 *Rectangle representing the Collar's flange.*

Note: One would think that the pivot could be moved to the vertex corner of the rectangle by using **Snap|Pivot** to grab the local origin and **Snap|Vertex** to snap the pivot to the corner. Unfortunately, the pivot snap does not snap to the pivot's 0,0 point so it is impossible to accurately relocate the pivot this way.

STEP 5

Create the .125 radius rounds as circles. One circle is at X.125, Y.4695 while the other is at X2.3, Y1.625. Then, do the same for the .1777 radius fillets (you determine the world location). Correctly positioned, they should appear as in Figure 3.10.

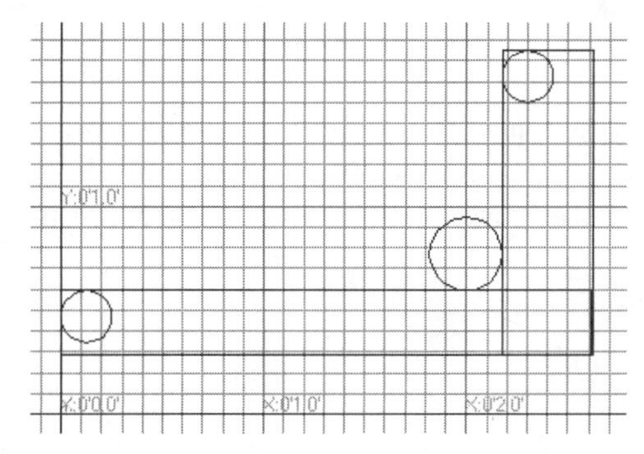

Figure 3.10 *Rounds in position and fillet in position.*

Tip: If you insert a shape created in another program such as AutoCAD or Adobe Illustrator, you will want to plan ahead. The X-Y coordinates of these two-dimensional shapes are mapped to 3D Studio X-Y world coordinates in the top view. So they are effectively inserted into 3D Studio top view.

STEP 6

Now you have to make a decision: whether or not you want to edit the existing geometry (ugh!) or use its geometrically correct position as a guide. I prefer to use **Snap|Vertex** and with the **Line** tool to quickly overdraw the construction, click at corners and click–drag at tangents. When you get back to the starting point, let 3D Studio close the spline (Figure 3.11). Note that a curve point is added at the apex of each fillet and round so that the curvature can be matched.

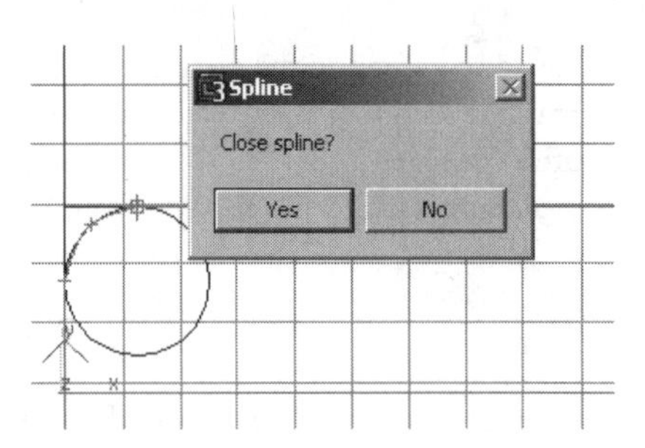

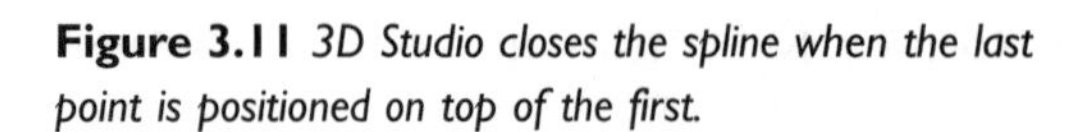

Figure 3.11 *3D Studio closes the spline when the last point is positioned on top of the first.*

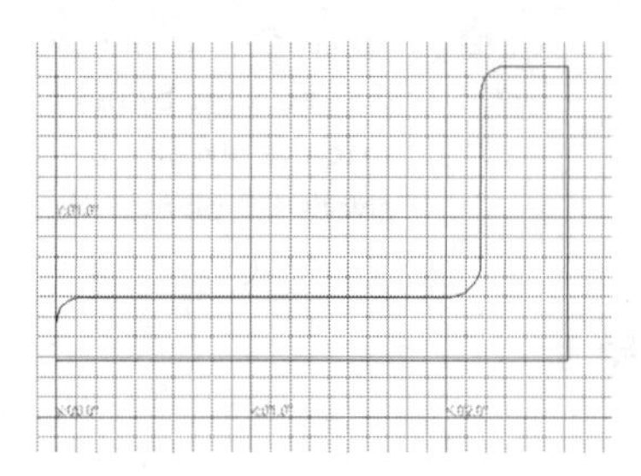

Figure 3.12 *The shape ready for the* ***Lathe*** *operation.*

STEP 7

Name your lathe shape *collar profile* and edit curve points to match the underlying circles. Hide the construction geometry. You should have the correct lathe profile (Figure 3.12) positioned the correct distance from the Y=0 axis.

STEP 8

Select the shape. If it is contiguous, the entire profile is highlighted. Choose **Modify|Lathe**; 3D Studio sweeps the profile about the screen Y–axis. Choose **Direction|X Axis** from the **Lathe** roll–out menu to reorient the axis of revolution to match the X–axis (Figure 3.13). Though revolved about the correct axis, the lathe has been performed about the shape's local axis (the geometric center) and is not the correct diameter.

STEP 9

The true axis of revolution should be at Y=0, so choose **Sub Object|Axis** and right–click the **Select and Move** tool. Enter 0.0 in the world Y-axis field. The collar is now correctly lathed (Figure 3.14).

STEP 10

Now is a good time to bump up the number of **Segments** to 36. Assure that the **Smooth** option is checked. After changing the color to a gray metal and rotating the view, you should be rewarded with a **Quick Render** of the collar similar to that in Figure 3.15. Refer to the beginning of Chapter 19, Still and Movie Output, for an introduction to rendering.

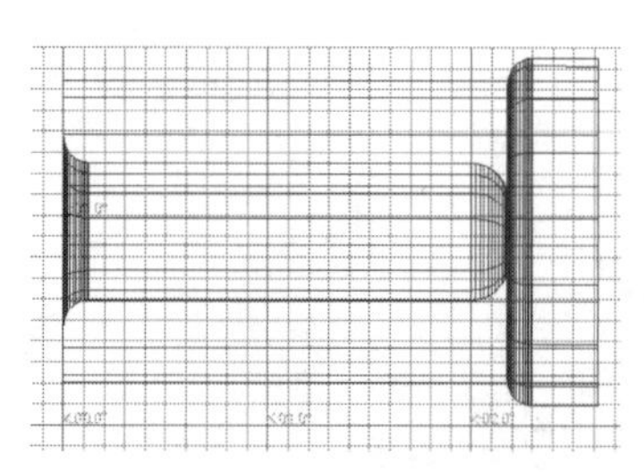

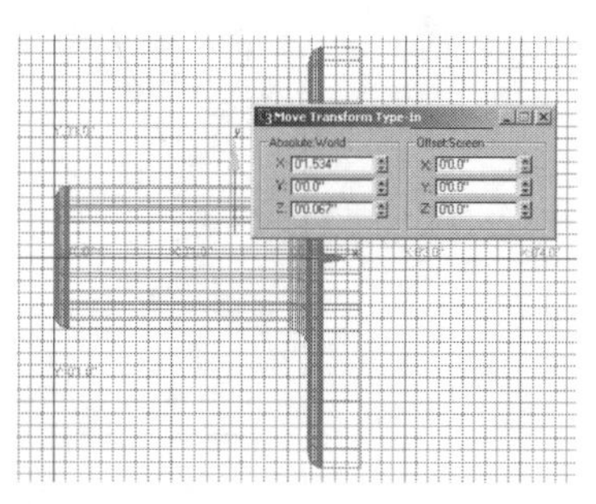

Figure 3.13 *The profile lathed about the X–axis.* **Figure 3.14** *Collar correctly lathed about Y=0.*

Figure 3.15 *The rendered Collar.*

EXAMPLE: LATHING THE PULLEY

The Pulley in Figure 3.16 is a prime candidate for modeling with a lathe. In this case the part is shown in half section and a scale is displayed. Up to this point we have used dimensioned views to guide our constructions. Yet in this case, the engineering drawing is at scale but none of the dimensions are notated.

This Pulley can be found on the CD-ROM in *problems\chapter3* if you would like to follow along. Make two prints, cutting the scale out of one to use on the other. Create dimensions on the print you didn't cut up by measuring with the scale (Figure 3.17). You have several options in modeling this part.

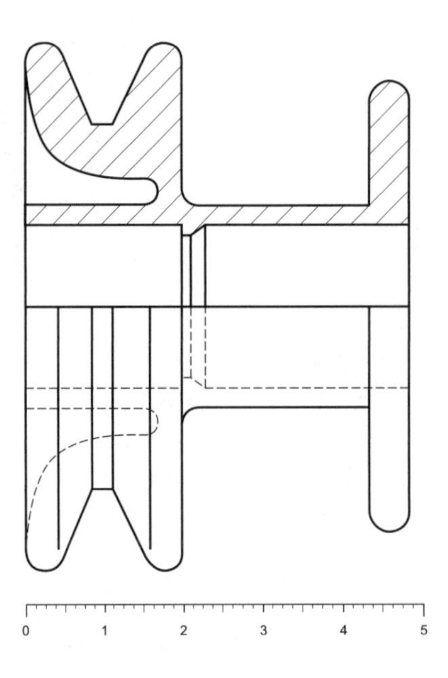

Figure 3.16 *The Pulley in half section.*

- Because the view is "Drawn to Scale," you could use the print as the basis of your construction. You could scan the drawing and somehow bring the scan into 3D Studio where it could be used as a template. But you don't know how to do that yet (this is covered in Chapter 10, Raster Material Maps).
- You could create the lathe profile in another program such as AutoCAD and **Insert** the profile and axis into 3D Studio.
- Or, you could use the dimensions determined by the scale to create accurate positional geometry, and as was done with the Collar previously, utilize **Snap|Vertex** to complete the profile. This is what we will do.

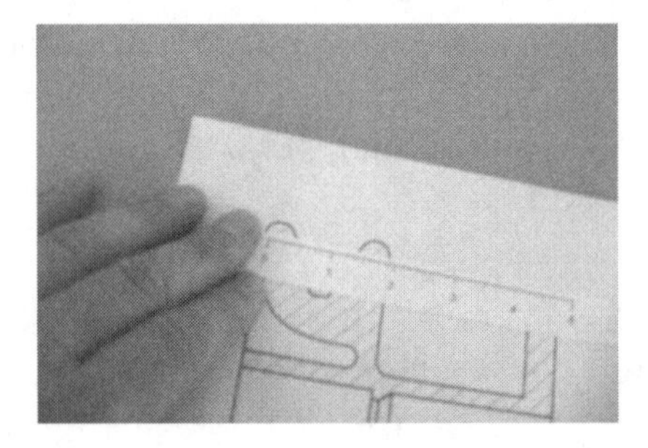

Figure 3.17 *The dimensions for the Pulley as determined by the accompanying scale.*

STEP 1

Using the dimensions you determined from the scale drawing of the Pulley, create accurate positional geometry in the Top view that describes the basic features (Figure 3.18). Position geometry limits (top, bottom, left, right) and correct diameter circles (fillets and rounds).

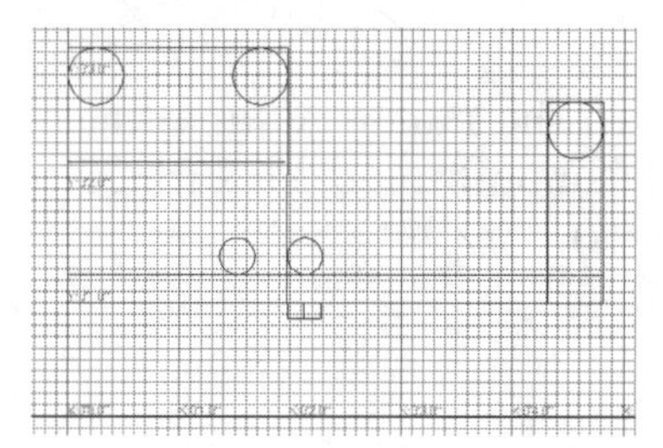

Figure 3.18 *Positional geometry for the Pulley.*

STEP 2

Using this positional geometry as a guide, and with **Snap|Vertex** active when appropriate, complete the upper profile in one continuous closed shape (Figure 3.19).

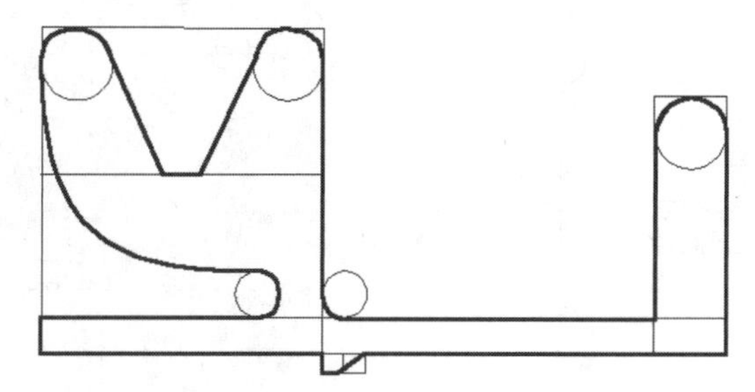

Figure 3.19 *Completed geometry for the Pulley.*

STEP 3

Select the profile and hide the positional geometry (**Display|Unselected|Off**). With the profile still selected, choose **Modify|Lathe** and click on the **X Axis** tab. Choose **Sub Object|Axis** and right-click on the **Select and Move** tool. Enter 0.0 in the **Absolute:World Y** axis field.

STEP 4

Flip the normals if necessary to view the Pulley from the outside. Enter 48 in the **Segments** field. Make sure **Smooth** is checked. You should be rewarded with a wireframe top view of the Pulley as displayed in Figure 3.20.

STEP 5

Assign a basic material to the Pulley. Assume that the Pulley has been finished with a moderately smooth metallic surface (see Chapter 9, Material Basics, for basic materials).

STEP 6

Establish a basic lighting scheme (see Chapter 14, Camera and Light Basics) containing an omni light and a target spot light, both to the front left of the object. Choose **Smooth + Highlights** by right–clicking on the Top view title. Rotate the view into a pictorial User view. You should see an accurate, realistic Pulley as shown in Figure 3.21.

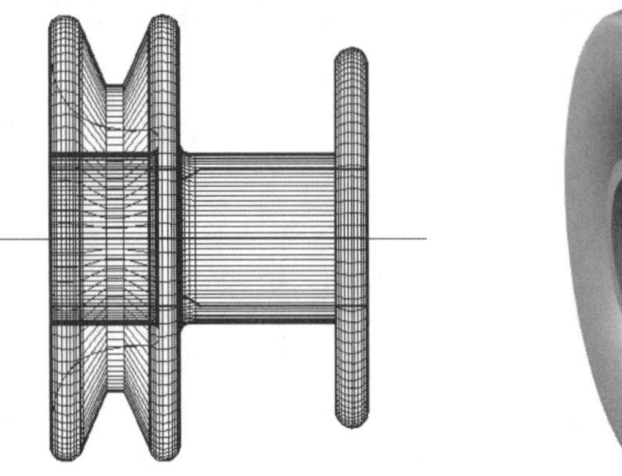

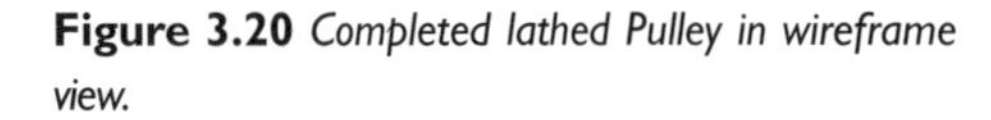

Figure 3.20 *Completed lathed Pulley in wireframe view.*

Figure 3.21 *Pulley rendered with lights and basic materials .*

PROBLEMS

Problems appropriate for solving with lathe objects are found on the following pages. Use the file *isogrid* found in the *tools* directory to plan your modeling strategies. Assign appropriate names to all objects both on your sketches and models.

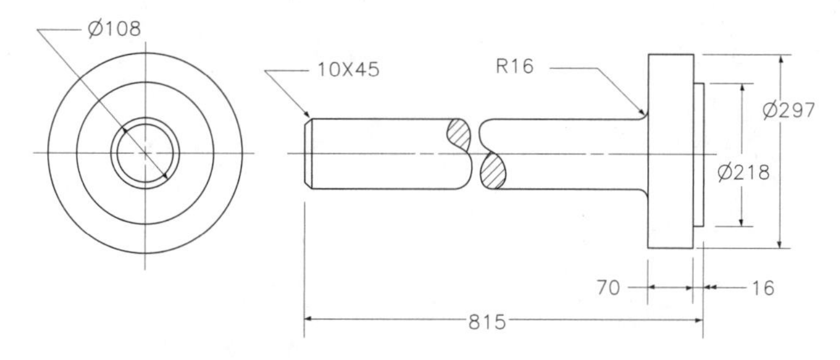

Problem 3.1 *Half Shaft..*

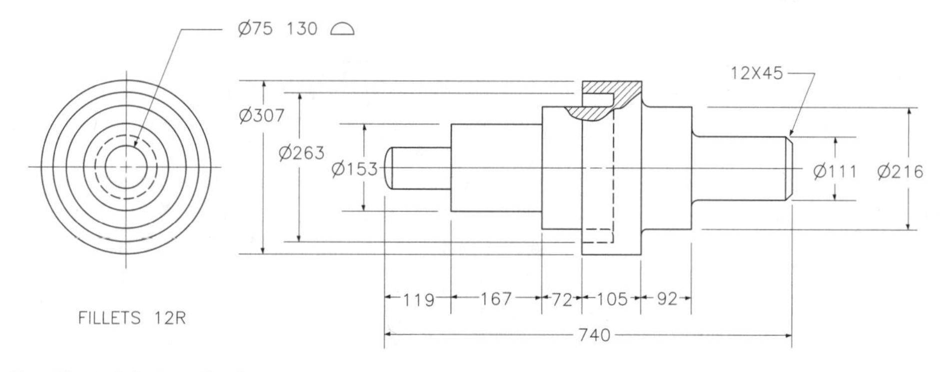

Problem 3.2 *Step Shaft.*

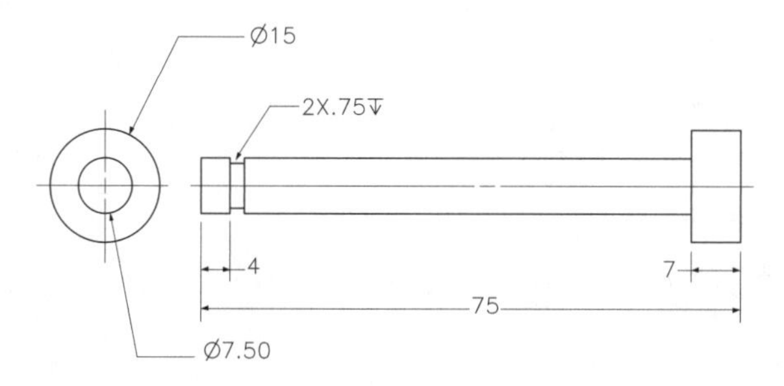

Problem 3.3 *Lead Screw.*

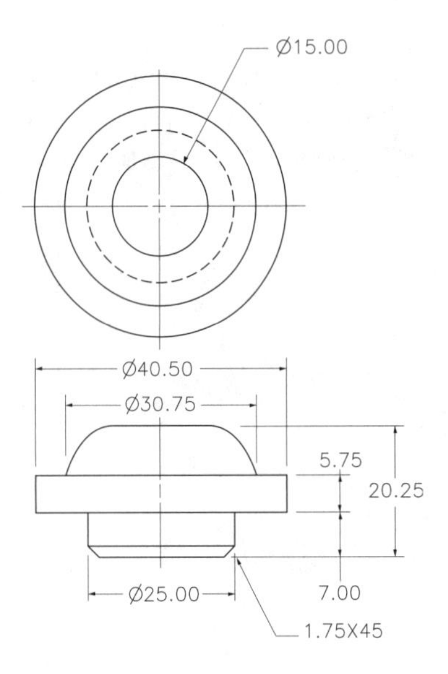

Problem 3.4 *Pressure Cap.*

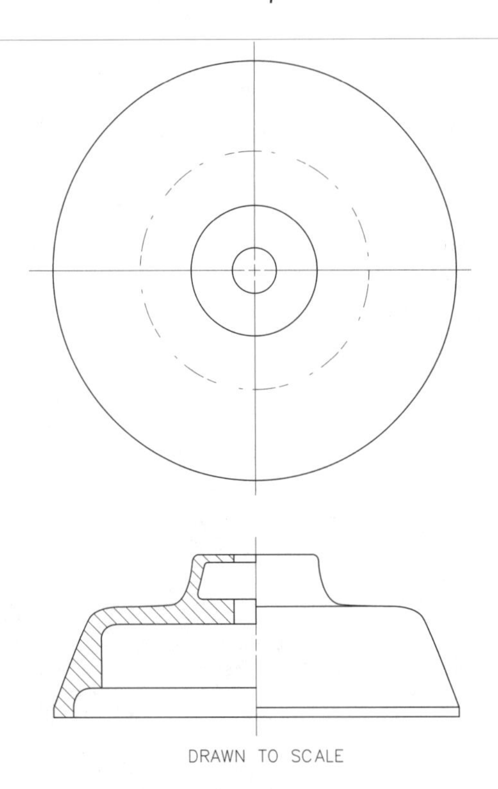

Problem 3.5 *Support Base*

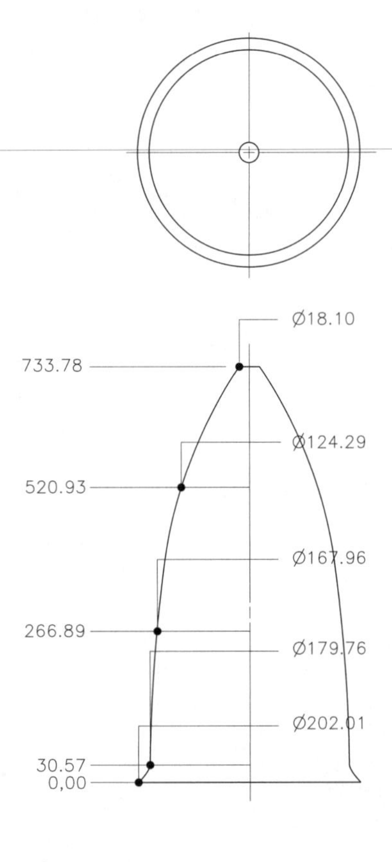

Problem 3.6 *Nose Cone.*

CHAPTER 4

Modeling with Lofts

CHAPTER OVERVIEW

In Chapter 2 you studied how extruding a shape or profile forms a three-dimensional object; the *extrusion* is formed by modifying a spline shape. This modeling technique is limited to objects exhibiting a consistent cross–sectional shape. Additionally, the shape is extruded along a straight axis, usually perpendicular to that shape, so the variety of objects you can create with the extrusion modifier is limited.

Chapter 3 presented a way of creating a three-dimensional object from a spline shape (as was done with extrusion) but with the limitation that the path be circular. We refer to this modification as a *lathe*. (In other programs this may be called a *sweep*.) Where an extrusion has a consistent cross section along a straight path, a lathe has a consistent cross section about a circular path. Modifying a spline shape forms the lathe.

A loft is a three-dimensional object formed as one or more cross–sectional shapes move along a path...any line or curve in any direction. Indeed, you could loft an extrusion or a lathe by limiting the path to straight or circular. But the strength in a loft is that the cross–sectional shape will follow any path. A loft, unlike an extrusion or lathe, is a *compound object* formed by a shape and a unique path.

KEY COMMANDS AND TERMS

- **Banking**—the option in **Skin Parameters** that keeps a shape aligned with a path when the path does not lie in a plane, assuring a consistent cross–section loft.
- **Compound Objects**—the **Create** option that forms objects from two or more 2D splines, 3D splines, or 3D objects.

- **Contour**—the option in **Skin Parameters** that keeps a shape aligned with the tangent of a path, assuring a consistent cross–sectional loft.
- **Deformations**—the **Modify** option that provides a mechanism for fine–tuning a loft.
- **Display**—the **Skin Parameter** option that controls the display of the loft.
- **Get Shape**—the **Creation Method** option that assigns a shape to an already selected path.
- **Get Path**—the **Creation Method** option that assigns a path to an already selected shape.
- **Path**—a 2D or 3D spline shape that defines the route a shape will take as it creates a loft.
- **Path Steps**—the number of times the shape will be repeated between path steps (vertices.)
- **Path Parameters**—the options that allow shapes to be placed along the path at a percentage, distance, steps, or to the current snap increment.
- **Shape**—a 2D or 3D spline or 3D object used to define the surface of the loft.
- **Shape Steps**—the number of elements along the surface of the loft between vertices of the shape.
- **Skin Parameters**—the options that control the smoothness and shape of the loft's surface.
- **Space Curve**—a spline curve that does not lie in a plane.
- **Steps**—values that determine the complexity of the loft mesh.

IMPORTANCE OF MODELING WITH LOFTS

Most industrial products adhere to linear and cylindrical manufacturing processes. Sheets and plates are rolled; tubes are spun, cast, or extruded; axles and shafts are turned; holes are drilled or reamed. But some industrial products cannot be made by these methods.

The loft compound object is appropriate for curved or bent tubes, hoses, or wires, or any circular cross section that follows an irregular path: ducts or connections that start with one shape and make a transition into another shape; and contoured edges, often formed by a shaper, or filleted or rounded profiles.

Take for example the pipe connecting the two storage tanks in Figure 4.1. It could be constructed from three separate pipes and joined by one of several methods. But because a loft will follow a path as it makes 45-degree bends, the pipe, with accurate intersections, can be formed in one operation. The path and shape are shown in Figure 4.2.

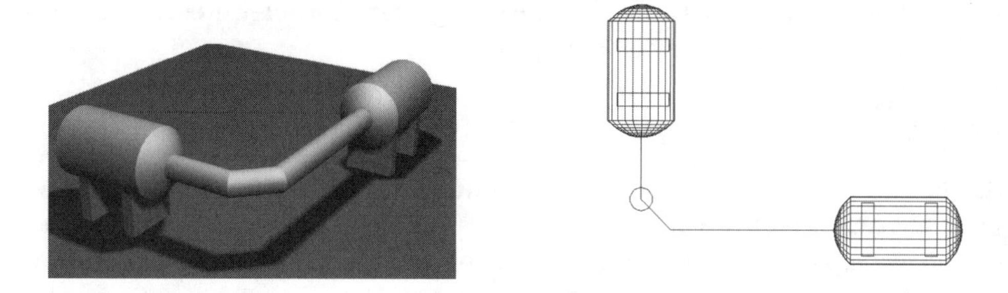

Figure 4.1 *A lofted pipe.*

Figure 4.2 *The path and shape.*

BASIC LOFT PROCESS

The loft compound object requires two things: a path and one or more shapes. Granted, some paths and shapes will loft more effectively than others. Consider the detail views of the connecting arm in Figure 4.3. The I-shaped rib connecting the two cylinders is a prime candidate for lofting. It has a closed cross–sectional shape and a path that defines the centerline. Follow these steps to model the connecting arm.

STEP 1

Create the path. In our case the path is the centerline normal in the Front view so that's the view we will use.

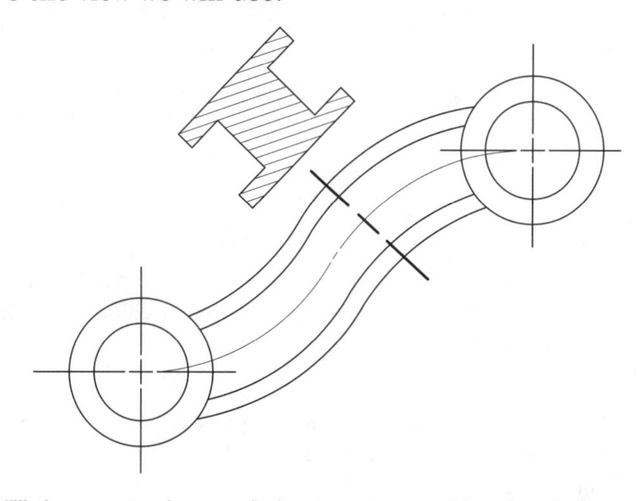

Figure 4.3 *The "I" rib connecting the two cylinders is a prime candidate for a **Loft**.*

STEP 2

Create the shape; 3D Studio will match the local origin (**Pivot**) of the shape perpendicular to the path at the starting vertex. You can create the path and shape in the same view (Figure 4.4).

STEP 3

Select the path. Choose **Create|Compound Object|Loft** and **Get Shape** from **Creation Method** roll–out. Select the **Copy** option so if you need to do the loft again you will have the loft shape and path to work with. As shown in Figure 4.5, 3D Studio creates the loft shape.

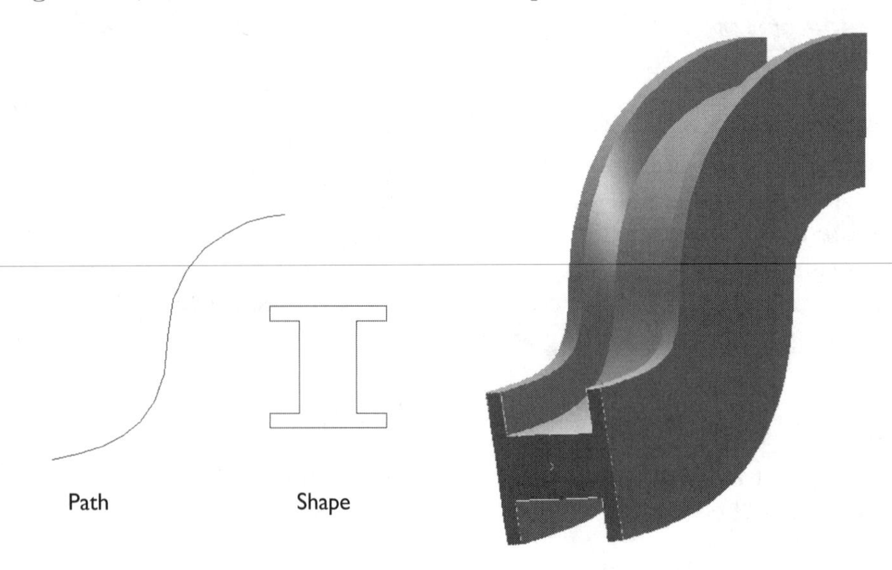

Figure 4.4 *The path and shape.*

Figure 4.5 *The lofted rib.*

STEP 4

Orient the shape. Move to the left side view. The loft has been created 90 degrees rotated from the desired position (Figure 4.6). With the loft object selected, choose **Modify|Sub Object|Shape** and select the shape by dragging a selection marquee around the end of the loft.

Note: 3D Studio defaults with the loft perpendicular to the view X-Y plane. You can change this to the view Y-Z or X-Y planes by toggling the buttons in the **Loft** roll–out menu.

Right–click on the **Rotate** tool and enter **90** into the **Offset:Local Z Axis** field. The loft is reoriented to the correct position (Figure 4.6).

STEP 5

Create the cylinders. Choose **Create|Standard Primitives|Tube** and create tubes using the dimensions from Figure 4.3. The completed connecting arm is shown in Figure 4.7.

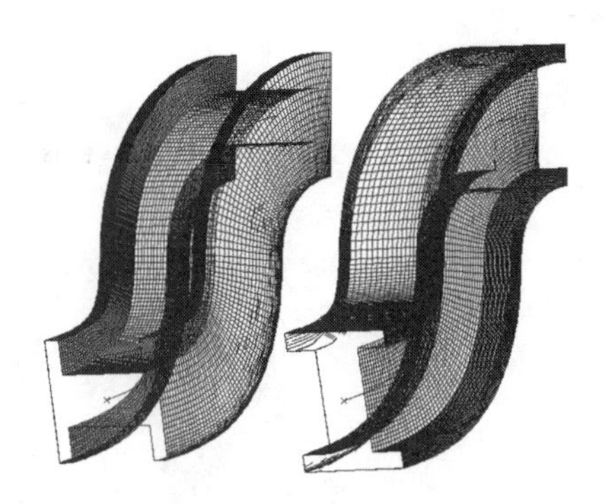

Figure 4.6 *The lofted rib before and after shape rotation.*

Figure 4.7 *The completed connecting arm.*

LOFT GUIDELINES

- Selecting the path first will move (copy) the shape to the path. Selecting the shape first will move (copy) the path to the shape.
- Be prepared to rotate or reflect the shape or change the axis of the loft after lofting—3D Studio will match path and shape pivots but this may not be what you desire.
- Once you create a loft object you can replace shapes or introduce new shapes at distance, percentage, vertex, or snap increment along the path.
- If a shape is invalid, the reason is displayed in the prompt line as you move the cursor over the shape.
- A mesh or patch loft can be converted to NURBS surfaces.
- Use **Modify|Deformations** to fine–tune a loft (see Chapter 7, Modeling with Modifiers).
- When **Move** is chosen, the path and shape become part of the loft. When **Copy** is chosen, the original path and shape are copied to their new locations.

When **Instance** is chosen, copies of the path and shape are moved to their new locations and any modification of one instance is reflected in all instances.

- A loft with varying cross sections must be planned carefully. Choose the method (percentage, snap, distance, steps) that matches the data you have for the part.
- An uncapped end will not show the inside of the loft unless **Force 2-Sided** is checked in the **Render Design** dialogue box.
- Use the minimum path and shape steps necessary to produce the part. This reduces geometric complexity and file size.
- Use **Contour** to keep shapes perpendicular to the path tangent.
- Use **Banking** to keep the cross–sectional shape aligned to a space curve as **Contour** maintains tangency.
- For a successful loft, make sure that the first vertex of each shape is in the same relative position (normally 0, 90, 180, or 270 degrees).

GEOMETRIC ANALYSIS

Study the air–conditioning duct shown in Figure 4.8. This is a common design where a duct of one shape must move smoothly into another shape. Transitions such as these are commonly found in air and material handling systems (food processing, oil and gas refineries, breweries and distilleries, heating, air conditioning, and ventilation). This is called a transition piece.

The transition piece begins at the top as a right circular cylinder and finishes at the bottom as a right rectangular box. The shape that smoothly blends between the two is not a regular shape, but one that a lofted object will create.

Observe that the duct has four shape steps: circle-circle-rectangle-rectangle. The critical design constraint is the exact position of each shape step. The first and last can be specified easily using **Percentage** path parameter (0 percent and 100 percent). For our purposes, we have a path (centerline) that has four vertices, one at each end and two correctly placed at the bottom of the cylinder and the top of the box (Figure 4.9).

STEP 1

Select the path. Choose **Create|Compound Object|Loft** and open the **Path Parameters** roll–out menu. Select **Path Steps**. This allows you to place a shape at each of the steps or vertices along the path.

STEP 2

Select the first shape. Under **Creation Method**, select **Get Shape** and click on the circle. VIZ assumes that you want this shape repeated at each of the steps or vertices (Figure 4.10).

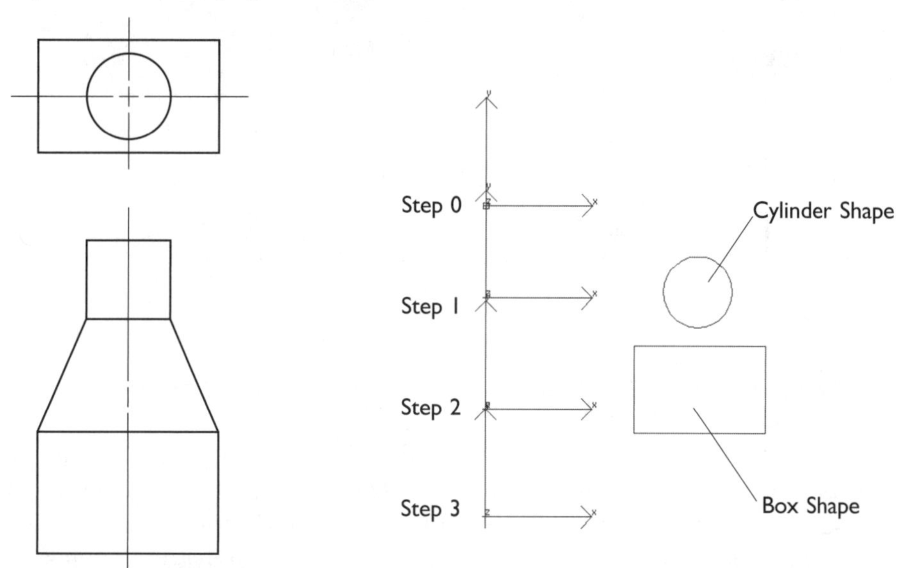

Figure 4.8 *An air–conditioning duct with transition piece.*

Figure 4.9 *Path with correctly placed vertices.*

STEP 3

Assign shapes to path steps. Note that the number of steps on the path (4) now appears next to the path spinner field. The first vertex is numbered zero (0) so the vertexes are numbered 0, 1, 2, 3.

Advance the spinner to step 1. The loft mesh disappears and a small yellow box appears at the second step. **Get Shape** is still selected so click on the circle again. Advance the spinner to 2 and select the box. Advance the spinner to 3 and select the box again. The transition piece is automatically lofted between the second and third steps (Figure 4.11).

Tip: VIZ automatically assigns succeeding shapes to match the current step. If the shape doesn't change, you can reduce the number of steps you explicitly assign shapes to by jumping over repeating shapes.

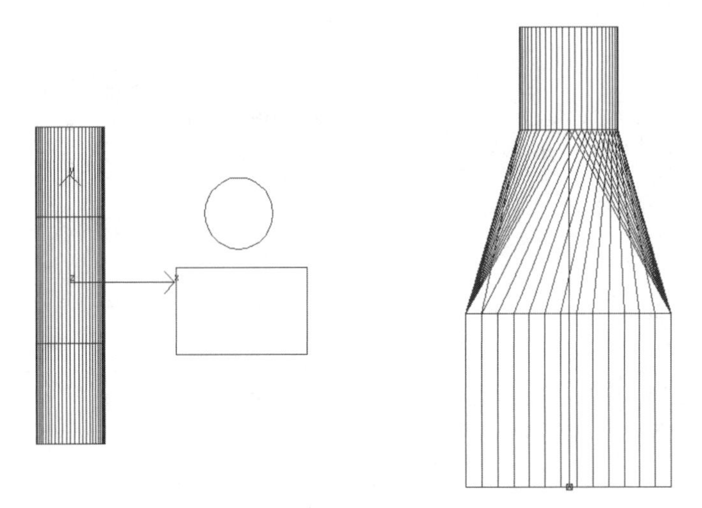

Figure 4.10 *First shape is assigned to all steps.* **Figure 4.11** *Lofted duct and transition piece.*

Note that 3D Studio may not have formed the transition the way you desired in its automatic matching of starting vertices in successive shapes. In many cases this is not important, as long as a smooth transition between the shapes has been formed. Figure 4.12 shows how the polygonal edges are twisted as the cylinder and box are matched. By rotating the individual shapes, you can align the polygons that form the transition (Figure 4.13).

STEP 4

Rotate the shapes. Rotate the view into a pictorial User View like that in Figure 4.12. This lets you see the polygonal elements better. Select the loft object and choose **Modify|Sub Object|Shape** and with the **Select** tool, click on the circle at the top of the transition (second step, bottom of cylinder).

The **Local** axis system is automatically chosen. Turn on the **Z Axis Lock** and click and drag the shape. The loft momentarily disappears and reappears when the mouse is released. Do this until the polygonal elements are aligned.

Click on the top shape (first step) and repeat the rotation process. When both circles are aligned, the transition appears as in Figure 4.13. The finished duct and its wireframe geometry are shown in Figure 4.14.

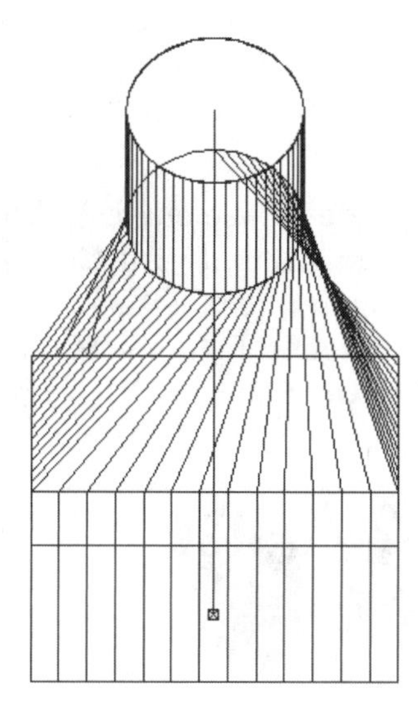

Figure 4.12 *Polygons of the transition piece are twisted.*

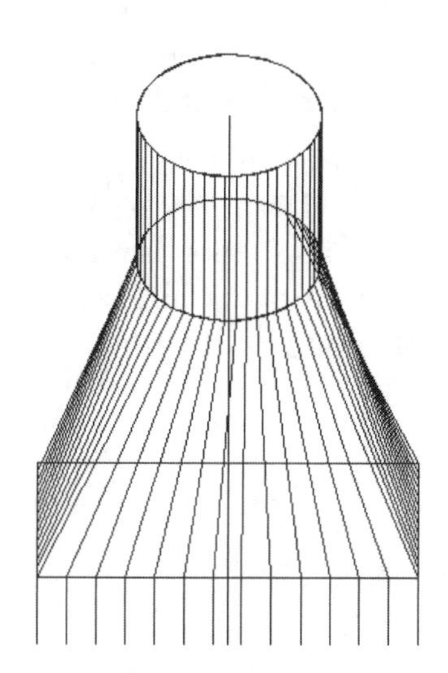

Figure 4.13 *Polygons of the transition piece are aligned.*

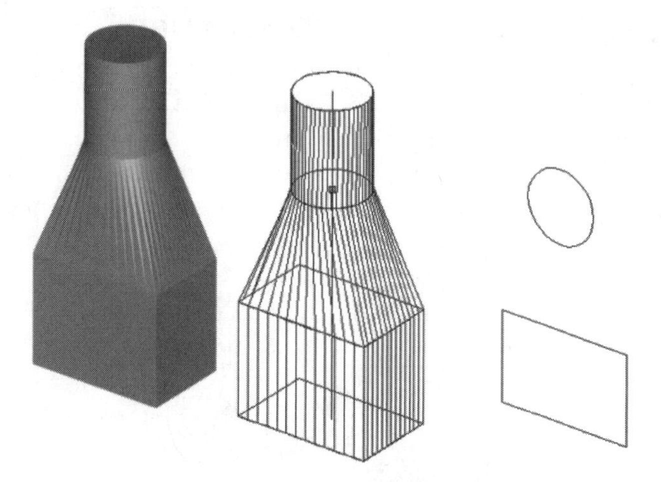

Figure 4.14 *Completed air–conditioning duct.*

LOFTING CONTROLS

Path Parameters have been introduced in the previous section. **Skin Parameters** determine the final appearance of the lofted object. Of particular interest are **Contour** and **Banking**.

When **Contour** is selected, the shape's pivot Z-axis is constantly aligned with the tangent of the path curve at each step. This results in a consistent cross section, something you usually want with a loft. Figure 4.15 shows the result of **Contour** being on and off.

Banking affects paths that do not lie in a single plane (space curves). When **Banking** is selected, shapes are rotated about the path and with **Contour** on, a consistent cross section is maintained. **Banking** rotates the shape about the local Z-axis as **Contour** aligns it to the path's tangent. Figure 4.16 shows the result of **Banking** being on and off (while **Contour** is on).

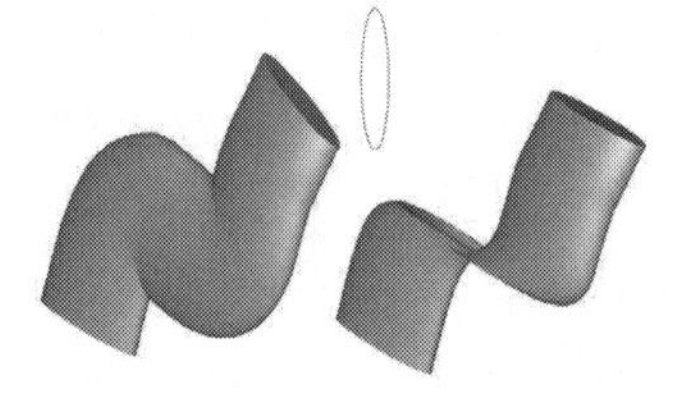

Figure 4.15 *Result of* ***Contour*** *being on and off.*

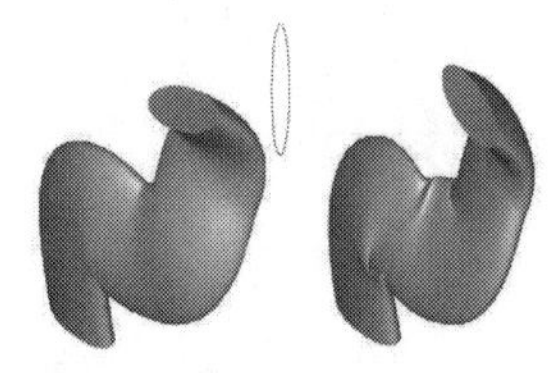

Figure 4.16 *Result of* ***Banking*** *being on and off.*

EXAMPLE: LOFTING A FILLETED FEATURE

The consistent cross–sectional shape of the Gear Cover in Figure 4.17 lends itself to lofting. The dimensioned views are found with the problems at the end of this chapter. The half section front view shows the true shape of the profile curve at both right and left limiting elements. The Top view reveals the path that will be used for the loft by a broken tangent line.

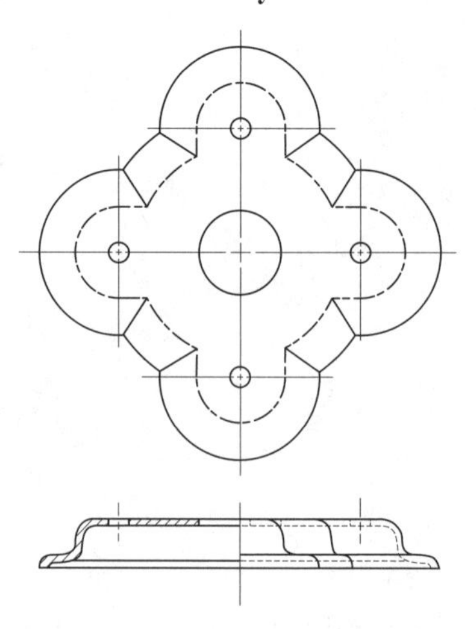

Figure 4.17 *Detail views of the gear cover.*

The strategy is to use the top plane of the cover twice. It would be nice to be able to loft the curved side and planar top at the same time but because the top surface varies in width and depth, this is impracticable. The top surface with its holes and the filleted and rounded perimeter are created as two objects. Because the same shape is used for the outside of the top plane and the path of the loft, the two objects will eventually match perfectly.

First, the shape and its holes are attached and extruded (see Chapter 2, Modeling with Extrusions) to the consistent thickness of the cover. Then, the outer shape of the top is used as the path for the loft. Figure 4.18 shows this in wireframe.

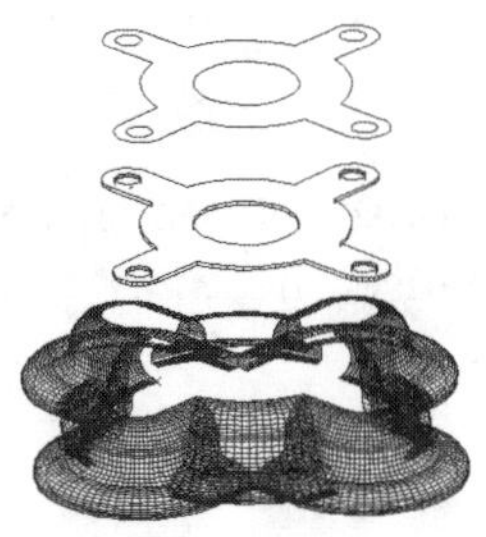

Figure 4.18 *The shapes, extrusion, and loft.*

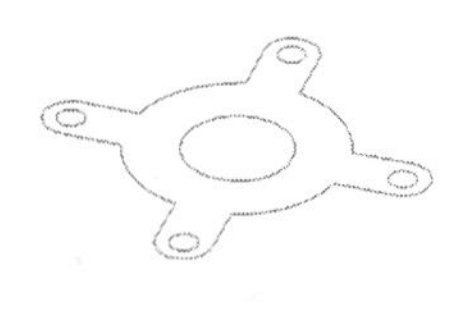

Figure 4.19 *The top shape and holes.*

STEP 1

Create the loft path. In the Top view, create the top plane outline and holes. Copy the top plane (Figure 4.19) and remove holes. This places the loft path at Z=0 world. Hide the top plane geometry.

STEP 2

Create the shape profile. Switch to the Front view. Begin at the upper–right end of the profile shape (**Snap|Vertex** will let you address the vertex on the outside of the curve). Note that this profile doesn't actually have a thickness (Figure 4.20), just a little lip on the bottom equal to its thickness.

STEP 3

Relocate the shape's pivot. Lofts are performed along the path relative to the shape's pivot. Because you want the upper right end of the shape to follow the path, choose **Hierarchy|Affect Pivot Only** and move the pivot to the end of the shape (Figure 4.21). Click this option off.

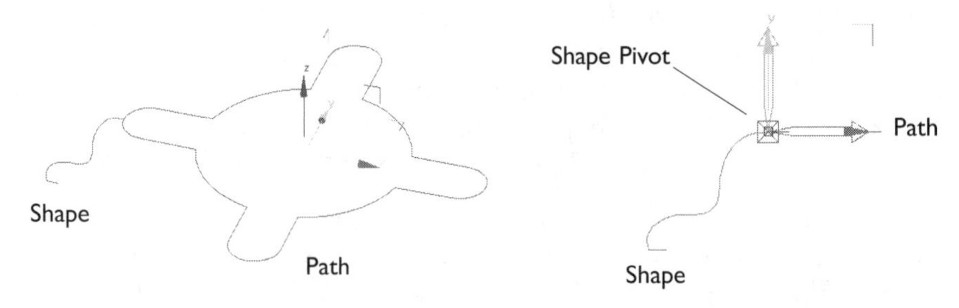

Figure 4.20 *Path and profile shapes in position*

Figure 4.21 *Shape pivot relocated.*

STEP 4

Loft the cover. Select the path and choose **Create|Compound Object|Loft|Creation Method|Get Shape**. Because the **Skin Parameters** is set to zero by default, you will have a loft like that in Figure 4.22. Turn on **Smooth Length** and **Smooth Width** and enter **24** in **Skin Parameters|Path Steps** then enter **12** in **Shape Steps**. You will be rewarded with a smooth loft (Figure 4.23).

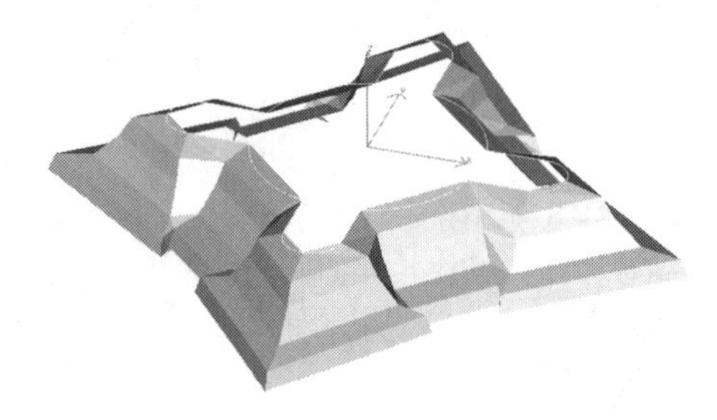

Figure 4.22 *Loft with steps set to default zero.*

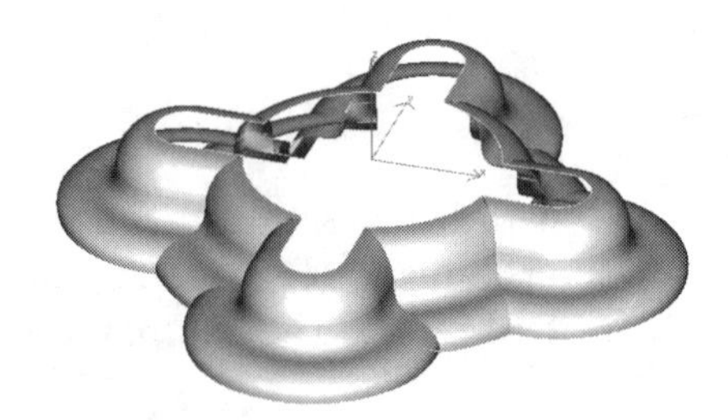

Figure 4.23 *Loft with appropriate steps.*

STEP 5

Extrude the top plane. Switch to the top view. Turn on the top plane and holes and hide the loft and loft path and shape. Select the outside shape and choose **Modify|Edit Spline|Attach** and select each of the five holes, one after the other.

Select **Extrude** and enter the thickness of the cover in **Parameters|Amount**. The top is extruded toward you from Z=0. The extrusion is shown hovering above the loft in Figure 4.24.

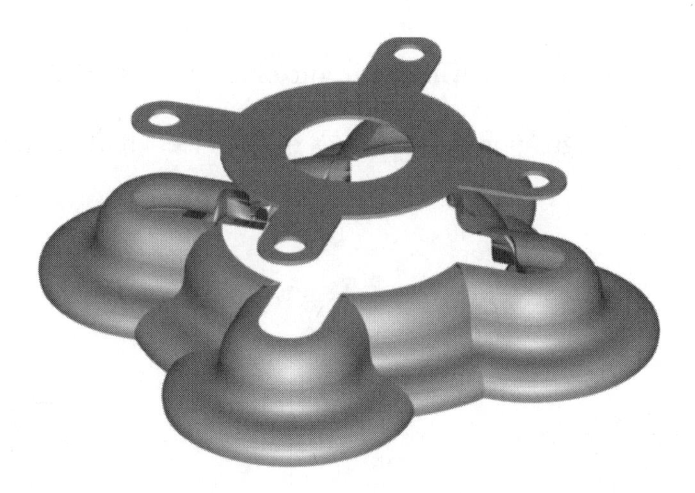

Figure 4.24 *Loft and extruded top plane.*

STEP 6

Align the top with the loft. Switch to the Front view. The top of the loft is at Z=0 world. The bottom of the extrusion is at Z=0. With the extrusion selected, right–click on the **Select and Move** tool and enter the negative of the cover's thickness in the **Absolute:World Z** axis field. This aligns the top surface of the extrusion with the top of the lofted perimeter. Figure 4.25 shows the final Gear Cover.

Figure 4.25 *Completed gear cover.*

PROBLEMS

Problems appropriate for solving with loft objects are found on the following pages. Use the file *isogrid* found in the *tools* directory to plan your modeling strategies. Assign appropriate names to all objects both on your sketches and models.

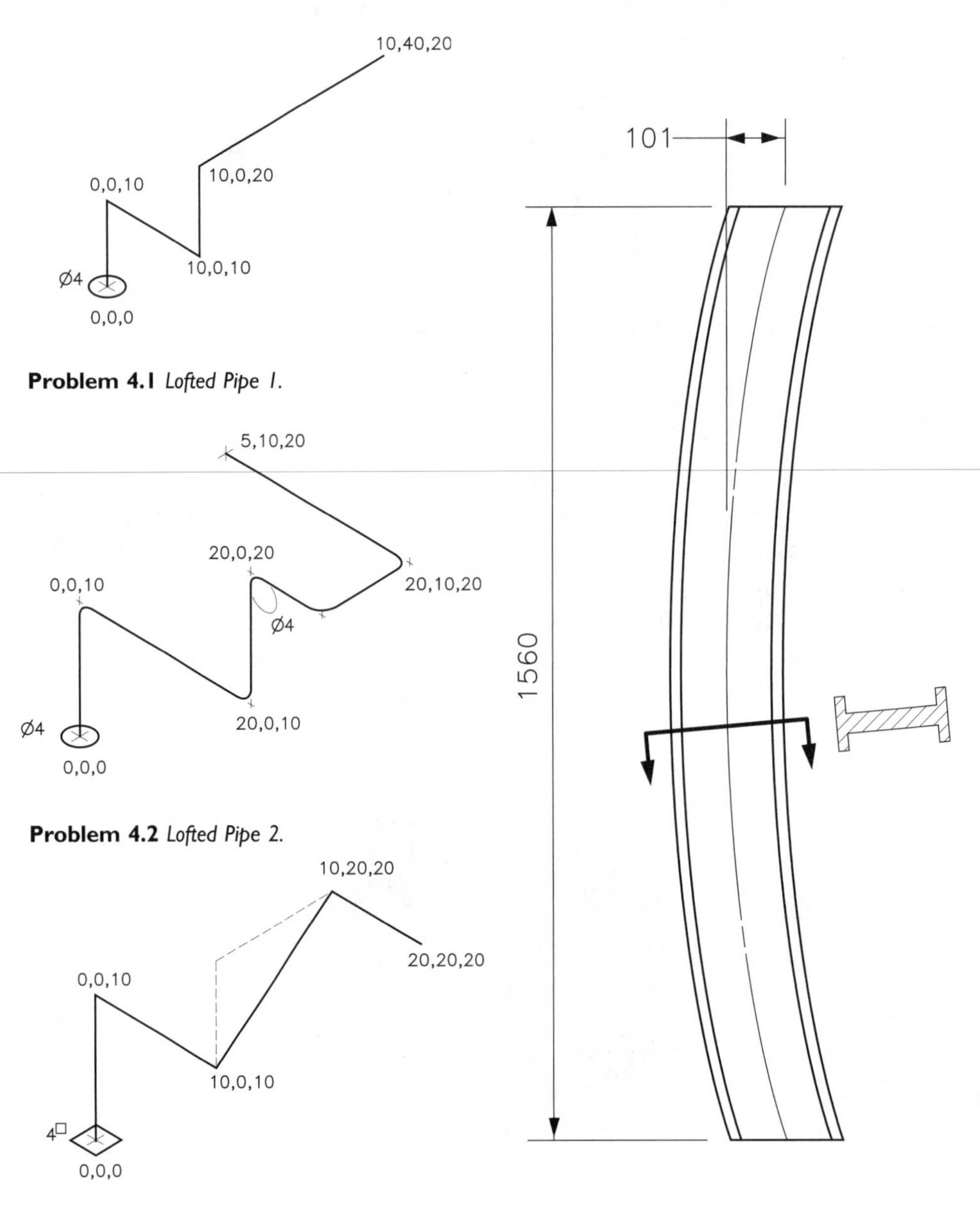

Problem 4.1 *Lofted Pipe 1.*

Problem 4.2 *Lofted Pipe 2.*

Problem 4.3 *Lofted Pipe 3.*

Problem 4.4 *Flex Beam.*

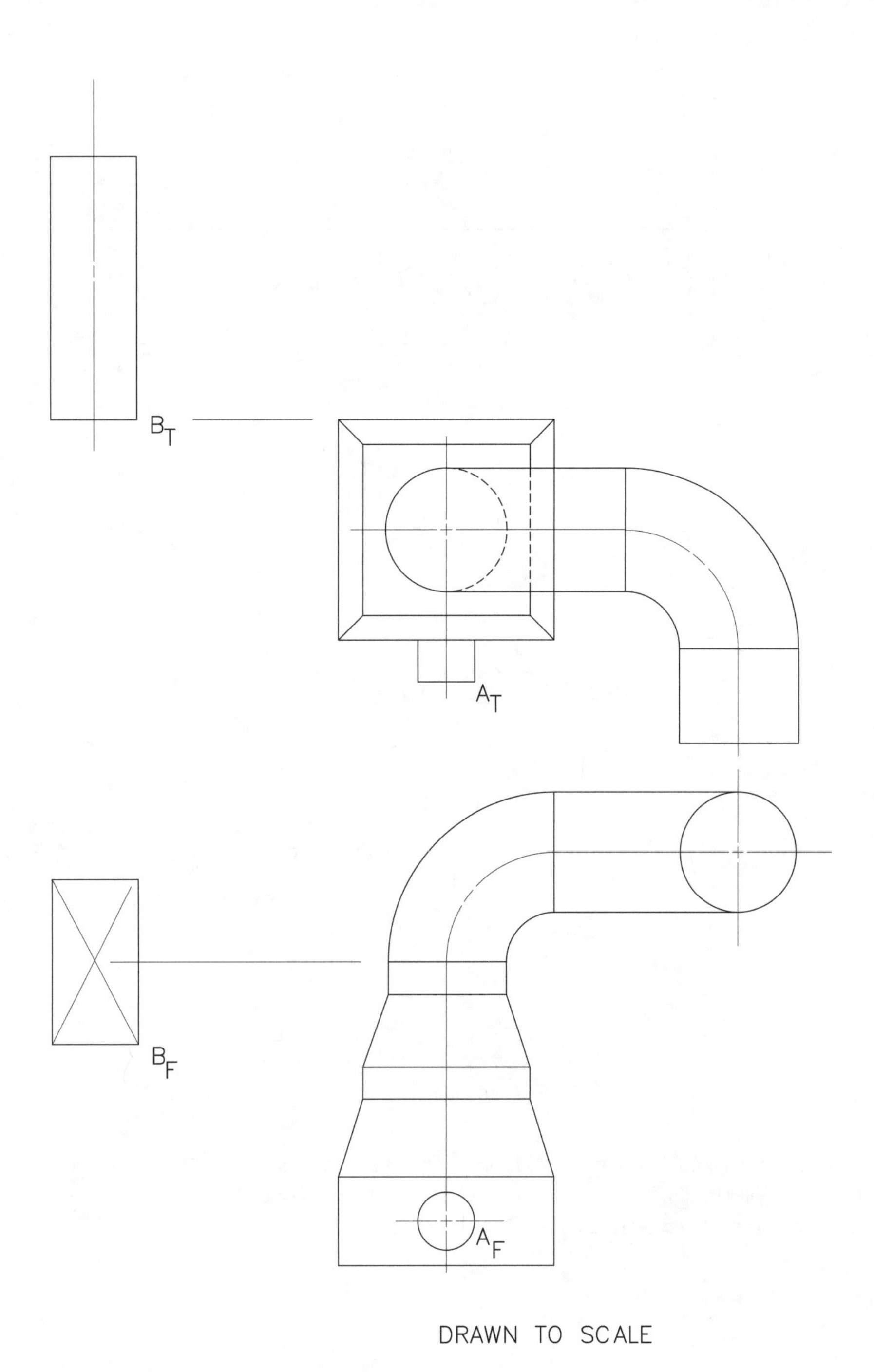

Problem 4.5 *Process piping. Complete the duct system as described. Connect cylindrical duct A and rectangular duct B with a short yet smooth transition.*

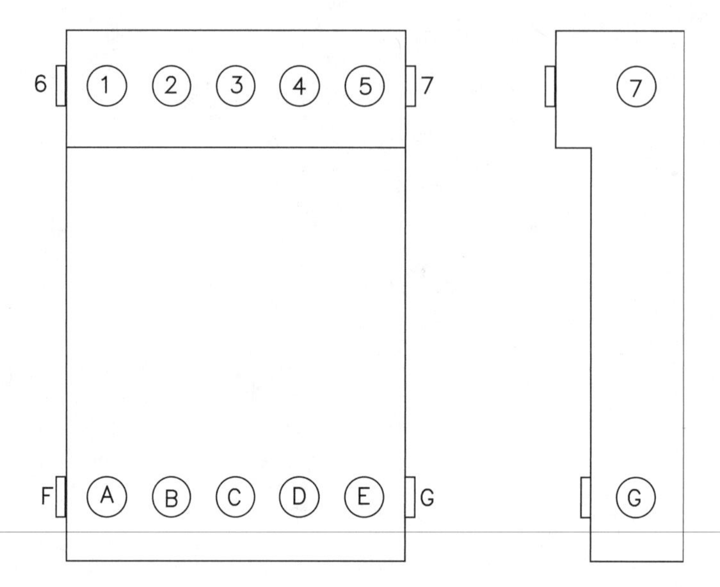

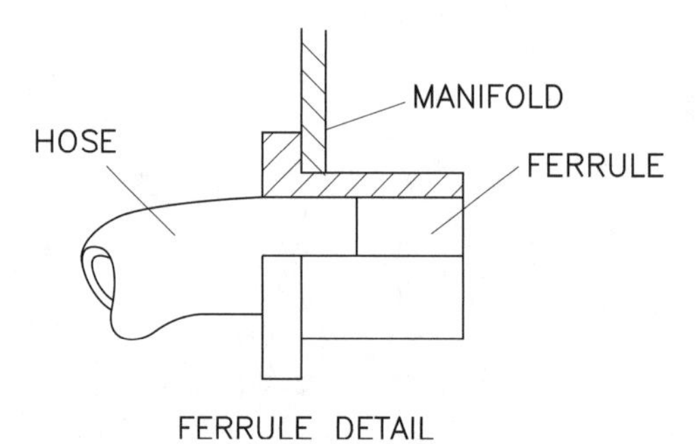

FERRULE DETAIL

Problem 4.6 *Hydraulic manifold. Model the manifold block by* **Extrusion**. *Model the hose ferrules using the* **Lathe** *modifier.* **Loft** *hoses to complete the connections as directed by your instructor. Assure no hose collisions or kinks. The shortest amount of hose is desirable.*

Modeling with Booleans

CHAPTER OVERVIEW

Boolean operations are important additions to primitives, extrusions, lathes, and lofts and 3D Studio provides methods for both Boolean 2D shapes and 3D objects. (Boolean 2D shapes were covered in Chapter 2, Modeling With Extrusions.) Three-dimensional Booleans are *compound objects* that add, subtract, or intersect 3D objects. With Booleans you can drill holes, mill away flat surfaces, counter sink or counter bore, or remove material much like the core of a casting. In other words, Boolean operations closely match welding and other industrial fastening methods when objects are combined; machine tool operations when material is subtracted; multiple extrusions when objects are intersected.

Tip: Booleans are not a panacea for poor planning. Each Boolean operation causes the surfaces affected by the operation to be divided and subdivided, increasing geometric complexity.

KEY COMMANDS AND TERMS

- **Boolean Intersection**—the operation that forms a surface from geometry shared by two objects. Order of operation is unimportant.
- **Bollean Subtraction**—the operation that forms a surface by subtracting geometry shared by two objects. The order of operation is important.
- **Boolean tree sketch**—a hierarchical sketch documenting each operand and operator necessary to create an object by Boolean operations.
- **Boolean Union**—the operation that forms a surface by joining geometry of two objects. The order of operation is unimportant.

- **Compound Objects**—the **Create** option that combines two objects to make a third.
- **Mirror**—the tool that reflects selected geometry at a distance about an axis with the option of copying.
- **Operator**—the Boolean function that evaluates two objects for union, subtraction, or intersection.
- **Operand**—an object operated upon by a Boolean operator. The first operand is referred to as "A" while the second is referred to as "B."
- **Resultant**—what is formed by applying a Boolean operator to two operands.

IMPORTANCE OF MODELING WITH BOOLEANS

A Boolean compound object is created by applying an *operator* (union, subtraction, intersection) to two *operands* (A or B). Only one operation can be applied at a time and can only involve two operands. By employing Boolean operations you can adopt modeling strategies that closely match manufacturing operations. For example, observe the welding drawing displayed in Figure 5.1. The part is comprised of four separate pieces that are welded together, making a final integral piece. One *could* devise a modeling scheme where the part might be made as a single piece of geometry. But since the design calls for four parts welded together, why not mimic the manufacturing process?

Figure 5.2 shows the four parts welded together in a Boolean union. By adding (welding) the individual parts together, you will be able to treat it as an individual object for materials mapping and animation. (This may be a reason not to Boolean!) If showing that the part is, in fact, the result of welding pieces together, you may even represent the welds with extruded geometry.

Note: Just to get the parts to *appear* that they have been welded doesn't require that Boolean operations actually be performed. Remember, 3D Studio will determine visibility at render time whether the parts are combined or are separate.

THE BASIC BOOLEAN UNION

A *Boolean union* combines all the geometry of one object with all the geometry of another. No geometry is discarded but shared geometry is combined. The objects do not have to overlap but when they do, the shared geometry is combined and a continuous surface with correct intersections is formed. The "T"–shaped handle in Figures 5.3 through 5.5 are examples of Boolean union.

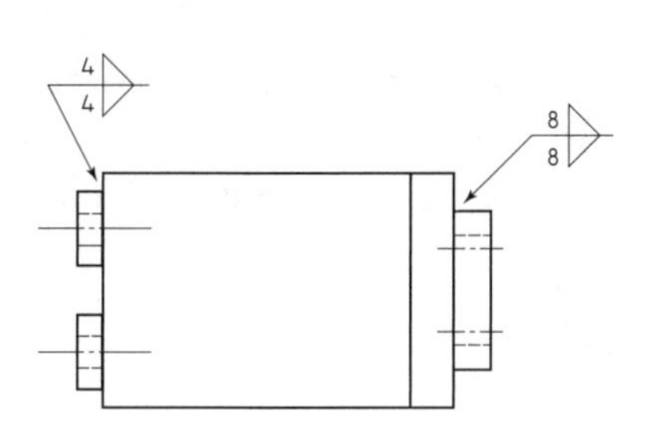

Figure 5.1 A weldment drawing.

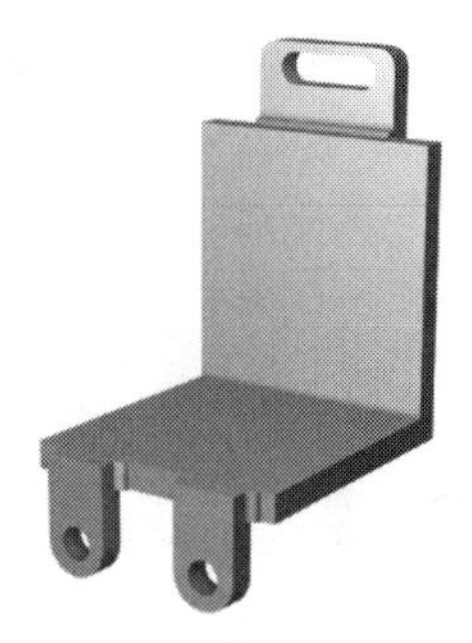

Figure 5.2 The weldment modeled and joined with Boolean union.

STEP 1

Create the parts. In this case, one part is a lathed handle and one part is a tube primitive (Figure 5.3).

STEP 2

Position the parts. Align the parts so that the handle intersects the tube at the desired location. (It may be convenient to perform this operation based on the world origin 0,0,0 and then move the result to its final location.)

Note: You must assure that the parts are in correct position in two adjacent views (like the Front and Right Side views in Figure 5.4).

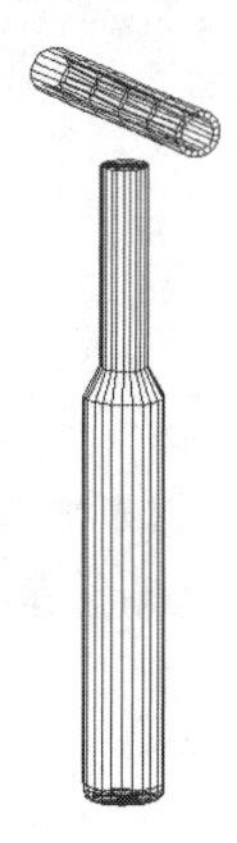

Figure 5.3 *Two parts to be unioned.*

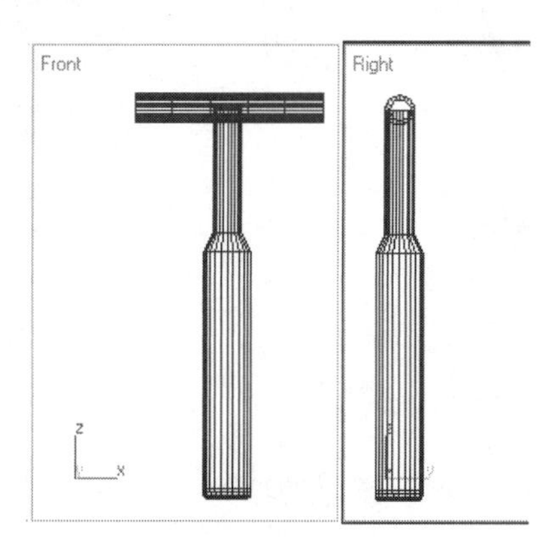

Figure 5.4 *Parts positioned for Boolean union.*

STEP 3

Select one part. With a **BooleanUnion**, it makes no difference what is the order of part selection. Choose **Create|Compound Objects|Boolean** and scroll down the **Boolean** roll-out to **Parameters|Operation** and choose **Union**.

STEP 4

Select the other part. Scroll back up the roll-out and choose **Pick Boolean|Pick Operand B**. The same options are available as have been previously discussed. Choose **Move** and no copies are left in the scene.

STEP 5

Shared geometry is combined and the intersection between the two surfaces is created (Figure 5.5). Vertices and elements are added as necessary to join the two.

Adhering to the rules of Boolean union, the part of the handle inside the tube is not removed (Figure 5.6). To cut the handle so that it perfectly matches the tube would add an unnecessary cost to manufacturing the part and weaken the eventual weldment.

Figure 5.5 *The completed "T" handle.* **Figure 5.6** *The handle is not removed inside the tube.*

THE BASIC BOOLEAN SUBTRACTION

A *Boolean subtraction* removes the geometry of one object from the geometry of another. The objects must overlap and the order (which is subtracted from which) is critical. The mounting plate in Figure 5.7 demonstrates common Boolean subtraction. Two holes and a slot are removed from an angled support.

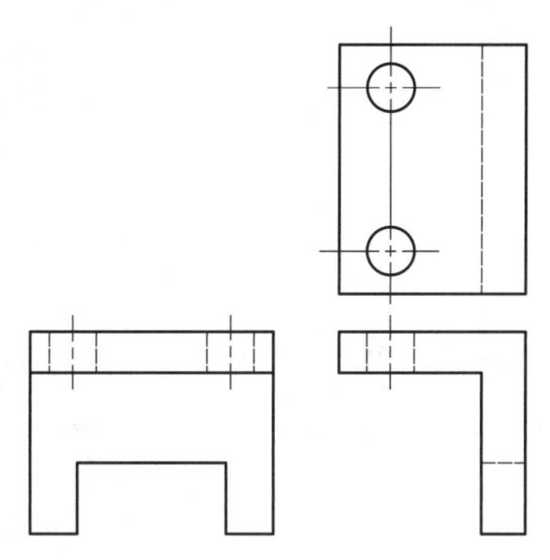

Figure 5.7 *A mounting plate.*

STEP 1

Create the base part. The base part is the object from which geometry will be subtracted. It is created in full height, width, and depth.

STEP 2

Create the tool shapes. Figure 5.8 shows the Cylinders (drill) and Box (mill) in position ready for Boolean subtraction. Note that each tool extends well past the limits of the base object.

STEP 3

Perform the Booleans. Select the base object. Choose **Create|Compound Objects|Boolean|** and **Parameters|Operation|Subtraction**. Choose **Pick Boolean** and **Pick Operand B**. Select the first Cylinder and the material common to the base object and the Cylinder is removed (Figure 5.9).

Figure 5.8 *Geometry in place.*

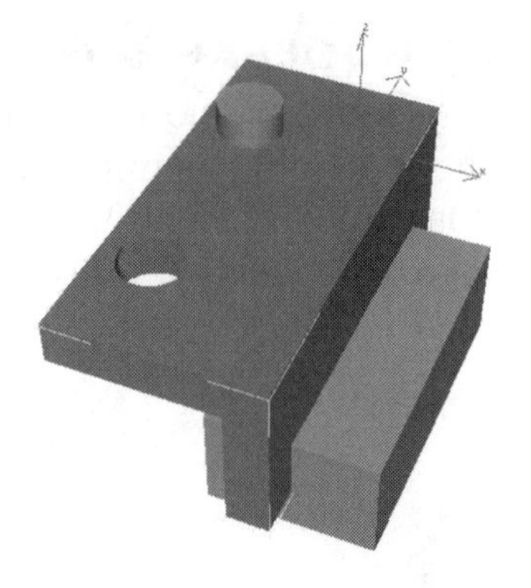

Figure 5.9 *First hole is removed.*

Note: Because the next two operations are also going to be A-B subtractions, you can select the next cylinder and box in order. However, greater success is achieved by not stringing these Booleans together. Deselect the base object after each Boolean, reselect, and choose the next tool.

STEP 4

Select the other subtraction tools. With the previous note in mind, perform the two remaining Boolean subtractions. The completed mounting plate is shown in Figure 5.10

Figure 5.10 *Completed mounting bracket.*

THE BASIC BOOLEAN INTERSECTION

A *Boolean intersection* takes geometry common to two objects and removes any not shared. The order of operand selection is unimportant. This is the most difficult Boolean to visualize and is reserved for cases when views are available that define the height, width, and depth profiles. A simple example demonstrates this intersection technique.

A cylindrical shaft having spherical ends is shown in Figure 5.11. Were the ends to be semispherical with radii equal to the radius of the cylinder, primitive construction could be entertained. But because a large sphere defines the profile of the cylinder's ends, a Boolean intersection is appropriate.

STEP 1

Create the parts. In the Right Side view create the cylinder with twenty–four sides, and with a length just longer than the cylinder's final length (use an integer value). Relocate the **Pivot** to the middle of the cylinder (integer length/2) and move the cylinder to world 0, 0, 0. In the Front view, create a forty–eight–sided sphere primitive. The **Pivot** of the sphere is at its center. Move the sphere to world 0, 0, 0 (Figure 5.12).

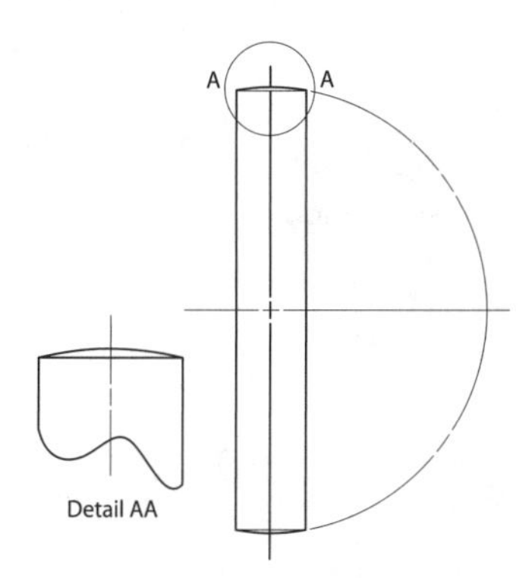

Figure 5.11 *Shaft with spherical ends.*

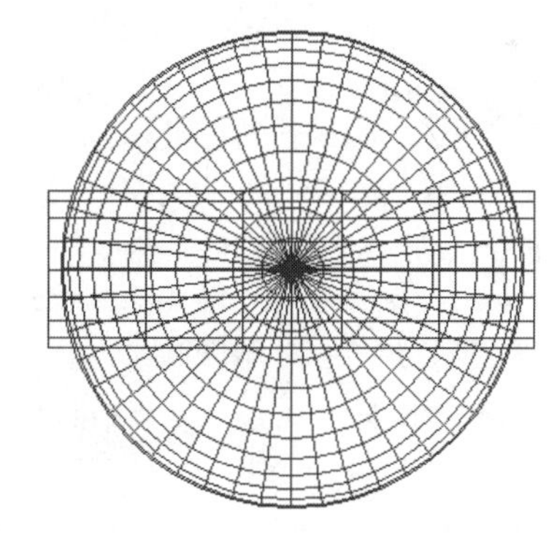

Figure 5.12 *Cylinder and sphere in position.*

STEP 2

Perform the Boolean intersection. Select the sphere or the cylinder. Choose **Create|Compound Objects|Boolean** and scroll down the **Boolean** roll-out to **Parameters|Operation|Intersection**. Choose **Pick Boolean** and **Pick Operand B**. Select the other object.

Note that 3D Studio discards any portion of the cylinder outside the sphere and any portion of the sphere outside the cylinder. The spherical ends are formed from this intersection (Figure 5.13) and because the cylinder was within the sphere, the cylinder is not discarded.

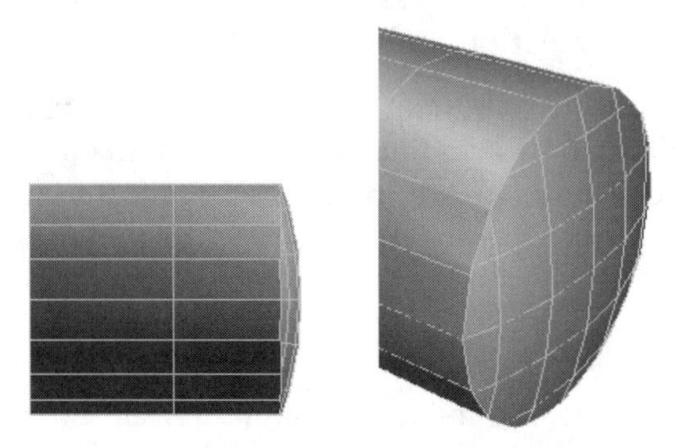

Figure 5.13 *The completed intersection with edged faces for emphasis.*

BOOLEAN GUIDELINES

- The key to successful Boolean modeling is planning. The tool for planning is a Boolean tree sketch.
- Always give 3D Studio an unambiguous set of geometry on which to perform Booleans. Avoid perfectly aligned edges where 3D Studio must make a decision whether or not to keep faces, vertices, or segments.
- Unify normals before performing Booleans and flip if necessary. 3D Studio determines what to keep and what to discard based on what's inside and what's outside an object.
- Always look to minimize the number of Booleans performed on an object by forming objects with holes and profiles based on extrusions, lathes, and lofts.
- Make sure you have the desired surface complexity before performing Booleans. The Boolean engine may add surface elements that alter the way the surface appears.
- Because VIZ doesn't have the **Hold/Fetch** feature of MAX, perform Booleans using COPY. This leaves the original operands in the scene to use if the Boolean doesn't turn out the way you wanted.

Tip: Although VIZ and MAX both have **Undo** options, undoing a Boolean is not advisable. You may end up with operands that behave unpredictably. Work on copies of the operands and return to the originals if necessary.

- When operands have materials, the resultant combines the materials. If operand A and B have materials, a **Multi/Sub Object Material** is created (See Chapter 9, Material Basics and Chapter10, Raster Material Maps). If A has a material and B does not, the new object has A's material. If B has a material and A does not, the new object has B's material.
- You can modify operands after the Boolean operations by identifying the operand in the modifier stack (See Chapter 7, Modeling With Modifiers).

GEOMETRIC ANALYSIS

The key to performing successful Boolean modeling is a critical analysis of the geometry. Some geometric features are naturally the result of Boolean operations while others may not appear at first to be likely candidates. Later (Figure 5.24) we will create a Boolean tree sketch, critical in planning an efficient modeling strategy. This sketch will describe each operand (object) and each operation. The Boolean operations are identified as union (+), subtraction (–), and intersection(^).

Figure 5.14 shows a sketch of a cab for a piece of industrial equipment. You will often work form sketches or concept drawings before engineering detail drawings are made. On first inspection, the Cab might not appear a Boolean candidate. But when rough views are created (Figure 5.15) a Boolean technique surfaces. Because the views (done in AutoCAD) show exterior profiles, modeling of the overall shape by multiple Boolean intersections is suggested. The window opening suggests Boolean subtraction. The fenders suggest Boolean union.

Figure 5.14 *A sketch of an industrial cab.*

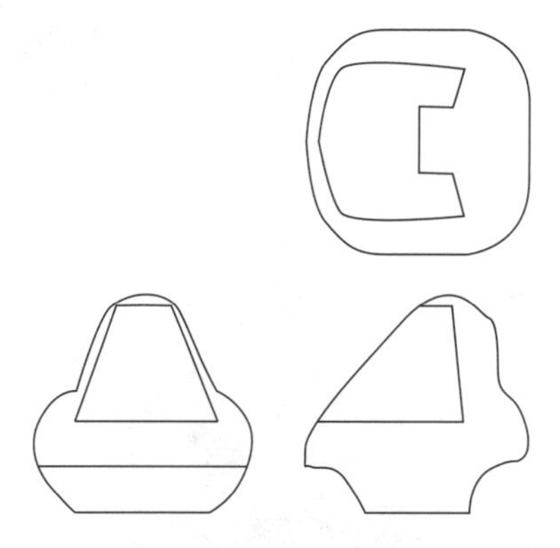

Figure 5.15 *Views of the industrial cab.*

Tip: Unless you have personally created the technical views, you can't assume that the shapes that come in from AutoCAD (or any other source) are continuous splines. Rather than spending hours editing shapes and welding vertices, it is usually more efficient to use the accurate views as an underlay. With appropriate snap settings, you can quickly draw over the imported views and be assured that you have valid geometry.

STEP 1

Bring the AutoCAD .dwg drawing into VIZ/MAX. You'll find this on the CD-ROM in *problems\chapter5*. In VIZ, **Insert|Other|AUTOCAD(.DWG)**. In MAX, **File|Import|AUTOCAD (.DWG)**. In either case insert the objects into the current scene as individual objects and accept the layer/color defaults. Geometry is inserted into the Top view (the two-dimensional X-Y coordinates are mapped to world X-Y coordinates). Delete the windshield geometry in all but the Front view. Redraw the curves as necessary using **Snap|Vertex**.

STEP 2

Position the views. Select the Front View of the Cab (profile and windshield). Right–click on the **Rotate** tool and enter 90 in the screen X-axis field. The Front View is aligned parallel to the front viewport. The extrusions, if created centered on 0, 0, 0, will be in correct position. If you are aligning the extrusions manually, you just need to make sure they overlap sufficiently for intersection.

Select the Left Side view of the Cab. Right–click on the **Rotate** tool and enter –90 in the screen Z-axis field. Enter –90 in the screen Y field. The Left Side view is aligned with the left viewport. Correctly extruded and positioned, the views appear in Figure 5.16.

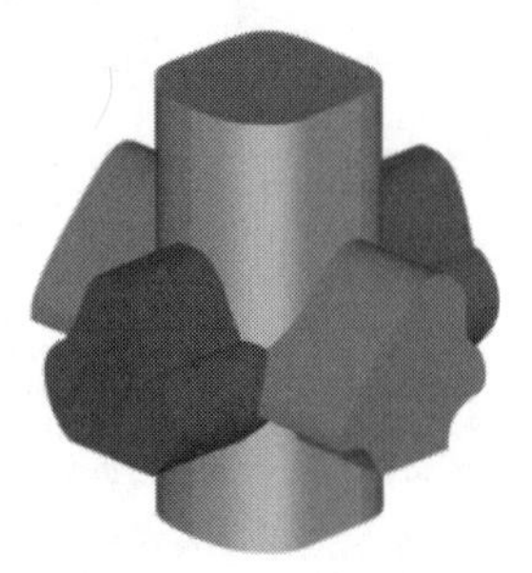

Figure 5.16 *Extruded views of the industrial cab*

STEP 3

Extrude the views. In order, select the views and extrude them to a distance greater than the height, width, or depth. The actual distance is unimportant as long as the extrusions hang out in all directions (Figure 5.16). Using X, Y, and Z positions, align each around 0,0,0 world.

STEP 4

Intersect the extrusions. Select the front extrusion. Choose **Create|Compound Objects|Boolean|Intersection** and then choose **Pick Boolean|Pick Operand B** and select the left extrusion. All material other than that shared by the two extrusions is discarded (Figure 5.17).

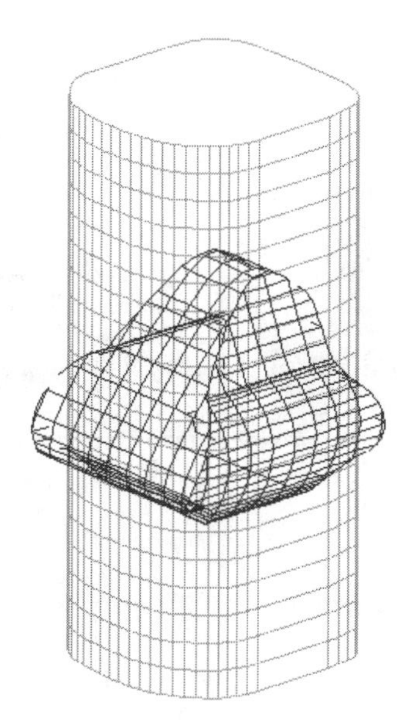

Figure 5.17 *Intersection of the front and left extrusions.*

Select the intersection. Repeat the previous intersection command from the previous example and choose and select the top extrusion. A second order intersection is created having only that material common to all three extrusions (Figure 5.18).

STEP 5

Cut the windshield. The windshield geometry is in the Front view but probably not in the correct position. Rotate this using the same numbers as the Front view. Move the windshield and enlarge its left edge so that it cuts completely through the Cab body. See Figure 5.19.

Extrude the windshield through the cab (Figure 5.20). Again, the actual distance is unimportant as long as it extends completely through the Cab body. Select the Cab (the result of two intersections) and this time choose **Create|Compound Objects|Boolean|Subtraction (A-B)**. Select the windshield extrusion as operand B. The windshield is cut from the body as shown in Figure 5.21.

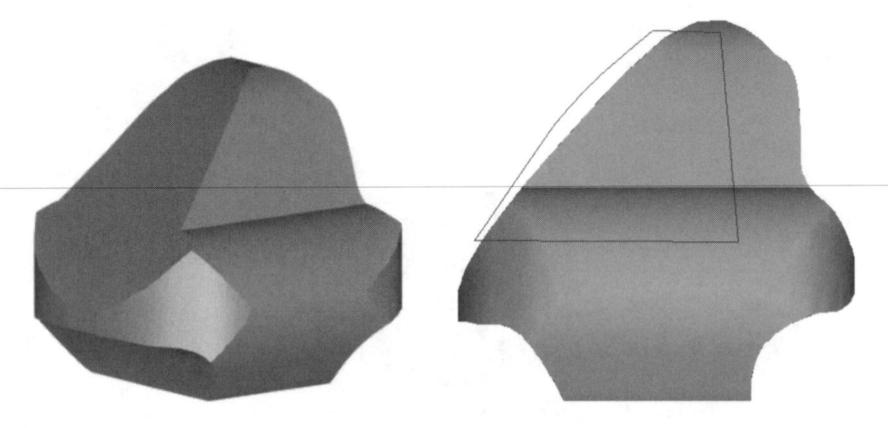

Figure 5.18 *Final intersection reveals geometry common to the three extrusions.*

Figure 5.19 *Windshield cuts through the body.*

Figure 5.20 *Extruded windshield.*

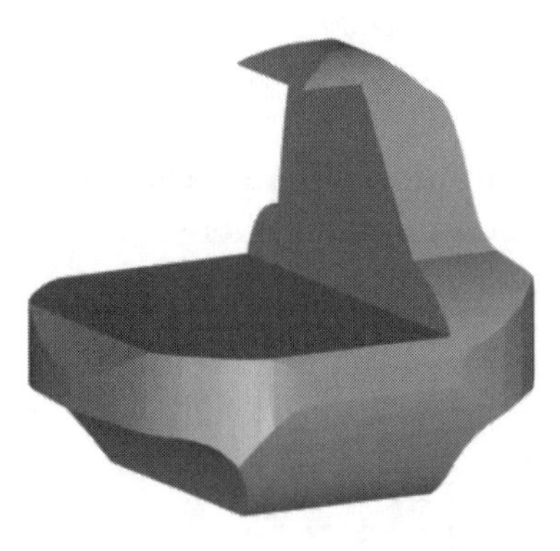

Figure 5.21 *Body with windshield removed.*

The Cab at this point may not look like much, but when unioned with the fenders and with detail and materials added, you can create a realistic concept model (Figure 5.22).

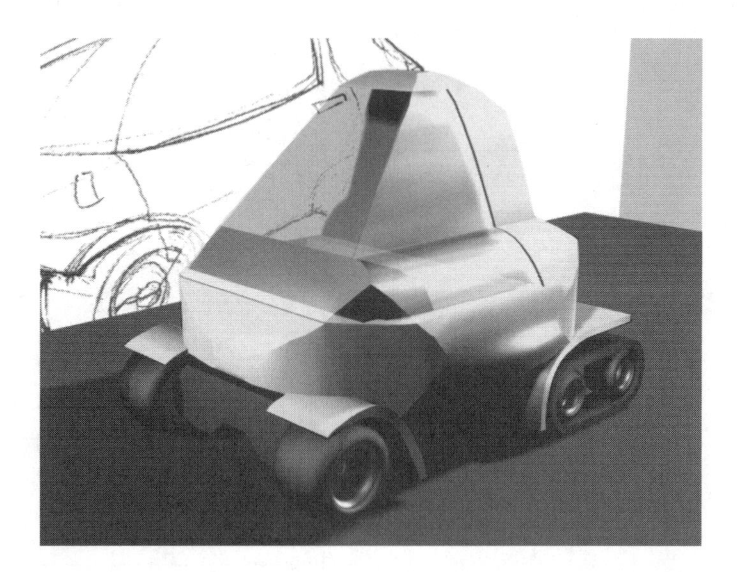

Figure 5.22 *Completed industrial Cab.*

EXAMPLE: THE SLIDE GUIDE BY BOOLEANS

Consider the Slide Guide in Figure 5.23. It contains three parts: a Flange with a hole below a Base that has curved ends and an opening on the left side and a Top with curved sides and cuts.

Now look at the pictorial sketch at the top of Figure 5.24. You can see the relationship of these three features. The Boolean tree sketch (Figure 5.24) describes each and every operand and operation required in its Boolean construction. You can find a blank title block sheet on the CD-ROM in *tools\title_block*. This example uses only primitives, though *second order primitives* (ones you create by extrusion, lathing, or lofting) are just as valid.

Tip: Complete a pictorial sketch of the object and then use a reduced photocopy to trace over for each of the steps. This way each branch of the tree sketch is at the same angle and scale.

STEP 1

Establish datums (refer to Figure 5.23). Consider the right side of the Slide Guide to be at Y=0 world. This brings the object toward you in the Front view. Consider the bottom of the Base to be at Z=0 world in the front view. This brings the object toward you in the Top view.

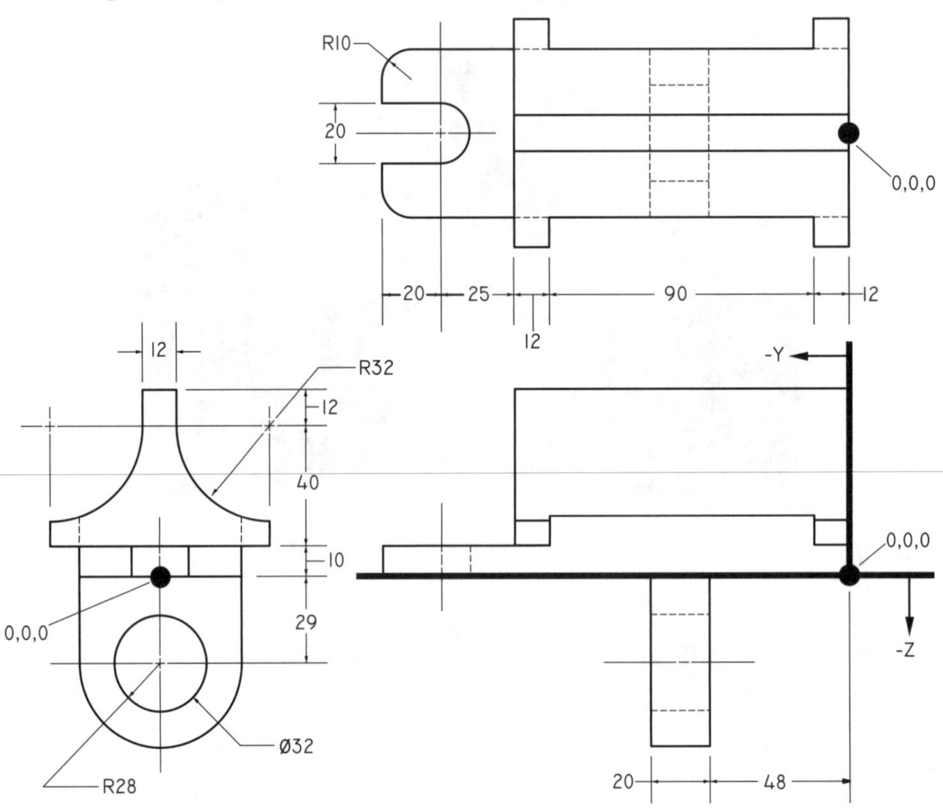

Figure 5.23 *The Slide Guide (Earl:* Graphics for Engineers*).*

The strategy for modeling the Slide Guide is to break it into the three component parts: Flange, Base, and Top. As each is modeled, refer to the Boolean tree sketch in Figure 5.15. We will be creating all the geometry and tools before performing the Booleans. This is not necessary, but will better show the development of this problem.

STEP 2

Create the Flange and hole tool. In the Front View create a 28 mm radius cylinder of 20 mm height at Y=48 world and Z=–29 world. This places the flange in front of and below the datum planes.

In the Front view, create a block representing the top of the Flange 29 mm long (Y) by 56 mm wide (X) by 20 mm height (Z). Move this block to 0, –48, 0 world. This aligns the back of the block with the back of the cylinder. Now some figuring: You want the bottom of the block at Z=–29 world (the middle of the cylinder). Since the block's **Pivot** is in the middle of the block, move the block to –14.5 (twenty nine divided by two) world Z.

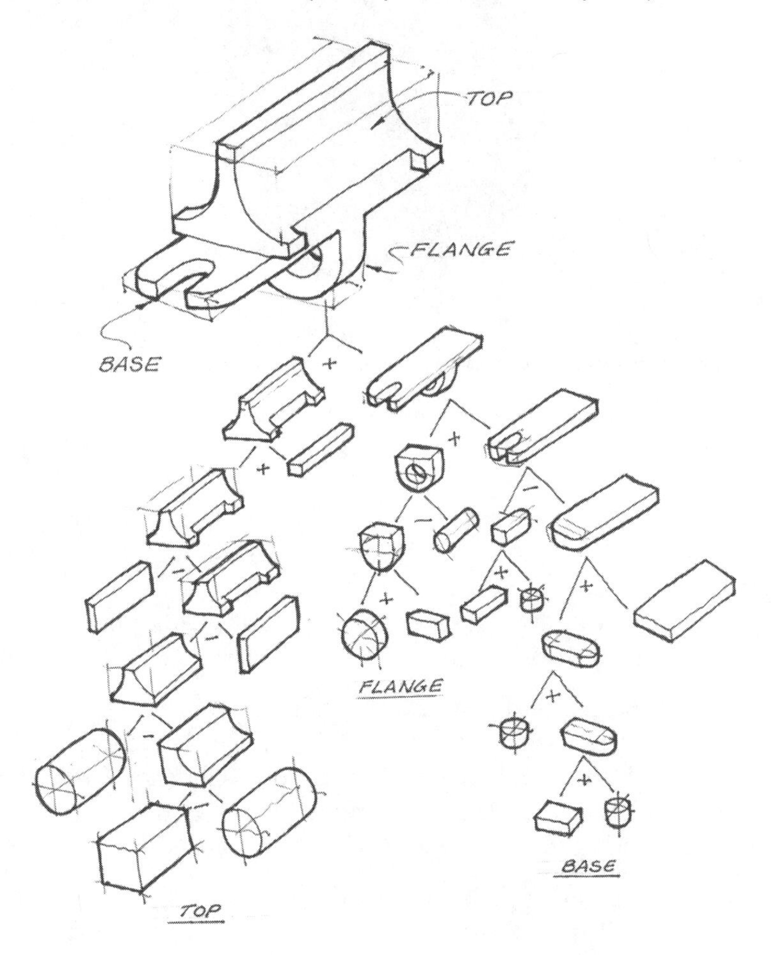

Figure 5.24 *A Boolean tree sketch of the Slide Guide.*

In the Front view, create the hole tool. Create a cylinder radius 16 mm and length 30 mm at 0, 0, –29 world. In the Right Side view, drag the cylinder by its horizontal axis until it extends through the Flange. Its exact position is unimportant, only that it extends through the Flange on both sides. Give the objects names such as *flange block*, *flange end*, and *flange drill*. The completed Flange geometry and tool are shown in Figure 5.25.

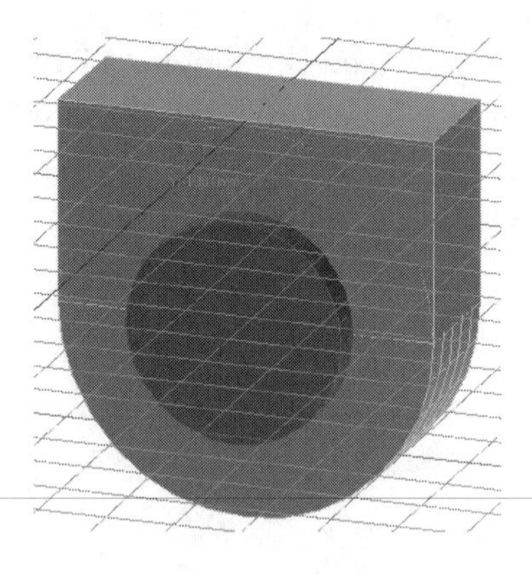

Figure 5.25 *Flange geometry and hole tool.*

STEP 3

Create the Base. In the Top view, create a box 149 mm x 56 mm x 10 mm. (This is the base minus the curved end that will be added later.) Move the box first to 0, 0, 0 world. Note that half the block extends behind Y=0 world. Move the block to 149/2=74.5 negative Y world. This places the back of the block at the Y=0 world datum. Call this part the *base*.

Create the rounded corners from two cylinders. Create a 10 mm radius cylinder 10mm high at X=–18, Y=–149, Z=0. Call this right radius. Select the cylinder. Choose **Mirror Selected Objects** tool. Choose **X Axis|Copy** and enter **36**mm (56–10–10=36) in the **Offset Field**. A copy of the first cylinder is placed on the other side of X=0 axis. Call this *left radius*.

Create the straight end from a box 36 mm x 10 mm x 10 mm. Move this box to X=0, Y=149 mm, Z=0. This completes the solid part of the base. Call this *end*.

Create the *slot tool* from a cylinder and a box. Create a 10 mm radius cylinder 15 mm high at X=0, Y=–139, Z=0. Create a 30 mm x 20 mm x 15mm box at X=0, Y=–154 mm, Z=0. Select both parts of the slot tool. Name one *slot box* and the other *slot curve*. Switch to the Front view. Grab the Y-axis and move the slot tool straight down until roughly half is above the Base and half is below. The completed Base and *slot tool* is shown in Figure 5.26.

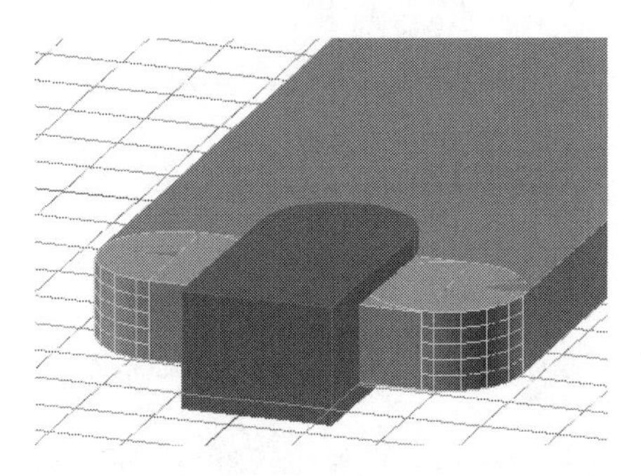

Figure 5.26 *Base geometry and slot tool.*

STEP 4

Create the Top. Switch to the Front view. Begin with a box 114 mm x 40 mm x 76 mm. Move this block to X=0, Y=0, Z=30 mm.

Note: The 12 mm x 12 mm x 114 mm extension above the curved cuts is not included in the Top. Included here, its sides would be curved, not vertical, when formed by the subtraction of the two large cylinders. It will be added later.

Create the big cylinder cutting tools that will cut the arcs from the sides. In the Front View create a 32–mm radius cylinder of 120 mm in height. Move this cylinder to X=38 mm, Y=0, Z=50 mm. Switch to the Right Side view and move the cylinder horizontally so that it extends equally on either side of the Top. In the Front View with the cylinder selected, choose **Mirror Selected Objects|X Axis|Copy|Offset=76mm**. Name one tool *right cutter* and the other *left cutter*.

Create the boxes that will cut the final slots from the Top. In the Top view create a box 90 mm x 12 mm x 20 mm. Move this box to X=38 mm, Y=57 mm, Z=0. In the Front View, move the cutting box until it is balanced on the area left by the cylinder (Figure 5.27).

Tip: After several Boolean constructions you may choose to model, Boolean, and then move the resultant to its final location, and do this for each of the subparts in turn. By creating all the operants in place before perfoming any of the Boolean operations you may avoid costly mistakes.

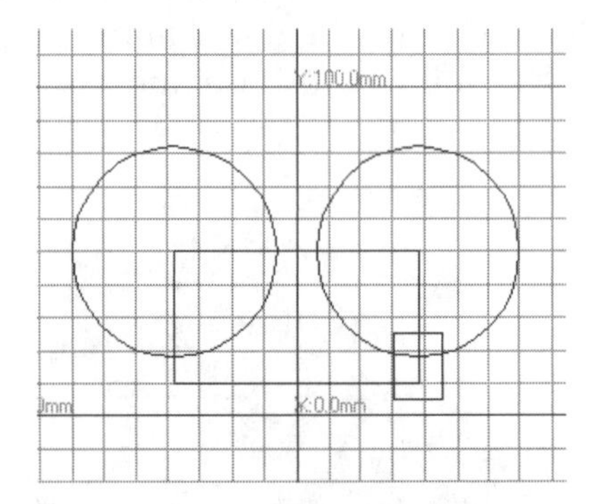

Figure 5.27 *Cutting cylinders and first cutting block properly positioned.*

Choose **Mirror Selected Objects|X Axis|Copy|Offset=–76mm**. The cutting block is mirrored along the X-axis the width of the Top. Name one tool *right block* and the other *left block*.

STEP 5

Create the Top extension. If you have been paying attention, you will remember that a 12 mm x 12 mm x 114 mm extension remains to be modeled on the Top. Create this box in the Front view and move it to X=0, Y=0, Z=56 mm. Once all the geometry and cutting tools are in place (Figure 5.28) the Boolean operations can commence.

STEP 6

Follow the Boolean tree. Refer back to Figure 5.24. Begin with the curved end of the base and perform the operations recorded in the Boolean tree. You may want to selectively hide and display the parts so you are working on the correct part of the tree. Figure 5.29 records these steps in order until a final, single part, is created.

THE BASE GROUP (FIGURE 5.29 STEPS A THROUGH E).

a. Union the right radius and the end.

b. Union the left radius and the resultant from step a.

c. Union the curved end to the Base. This is the Base Group.

Figure 5.28 *All geometry and tools in place.*

THE FLANGE GROUP (FIGURE 5.29 STEPS J THROUGH K).

j. Union the flange block with the flange end.

k. Subtract the flange drill from the resultant of step j.

THE TOP GROUP (FIGURE 5.29 STEPS F THROUGH I).

f. Subtract the right cutter from the top.

g. Subtract the left cutter from the resultant of step f.

h. Subtract the right block from the resultant of step g.

i. Subtract the left block from the resultant of step h. Union the extension to the resultant of step i.

JOIN THE THREE GROUPS (FIGURE 5.29 STEPS L AND M).

l. Union the Flange Group to the Base Group.

m. Union the Top Group to the resultant of step l.

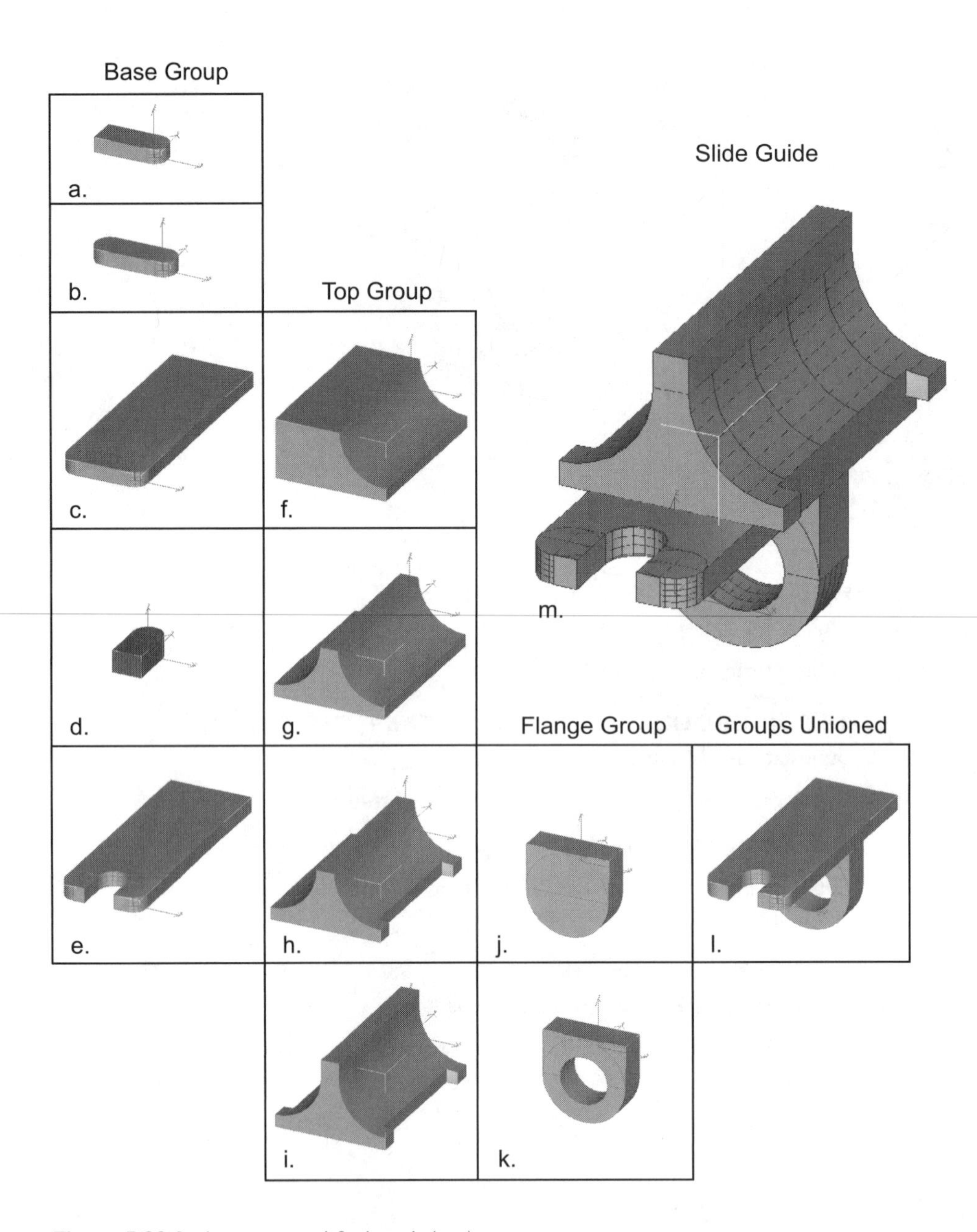

Figure 5.29 *Boolean steps and final result (a-m).*

PROBLEMS

Problems appropriate for solving with the Boolean compound objects are found on the following pages. Undimensioned views are to scale. Use the file *isogrid* found in the *tools* directory to plan your Boolean tree strategies. Assign appropriate names to all objects both on your sketches and models.

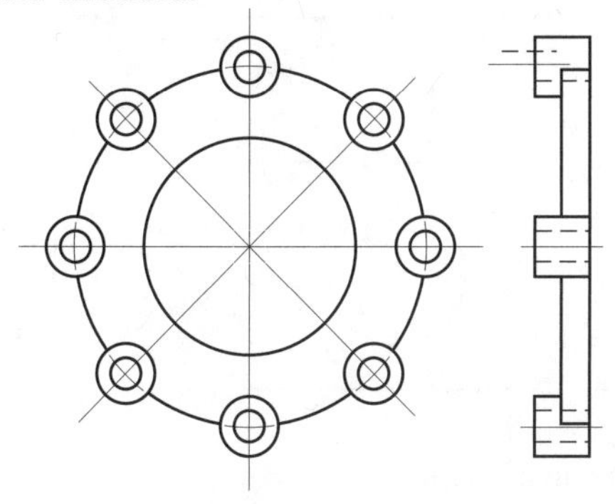

Problem 5.1 *Alignment Plate.*

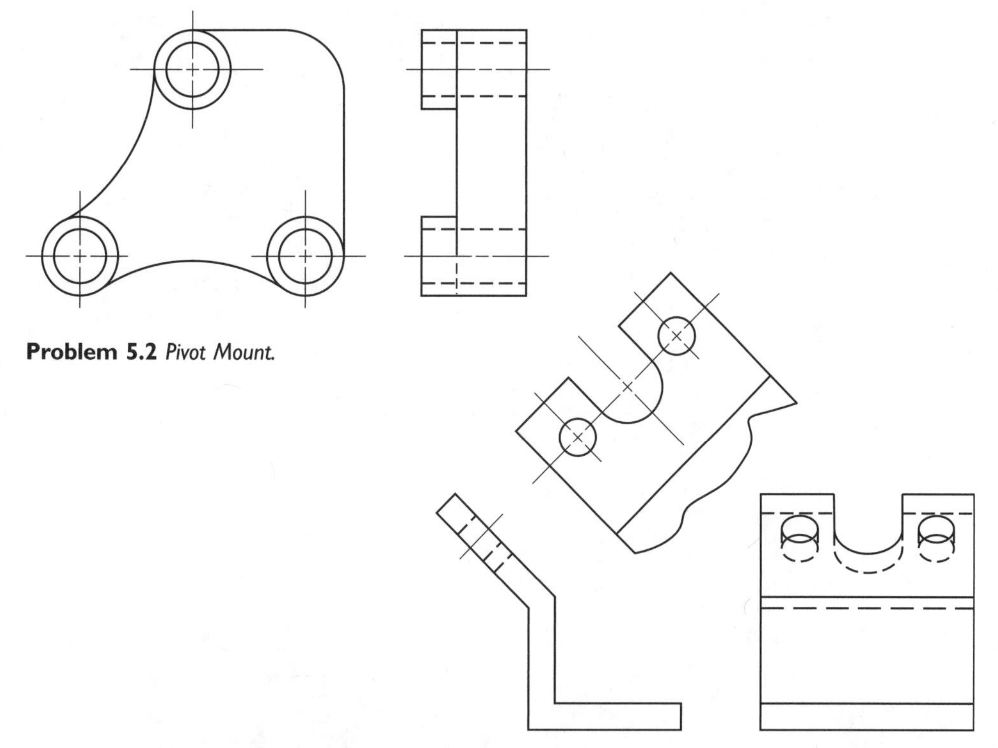

Problem 5.2 *Pivot Mount.*

Problem 5.3 *Angle Brace.*

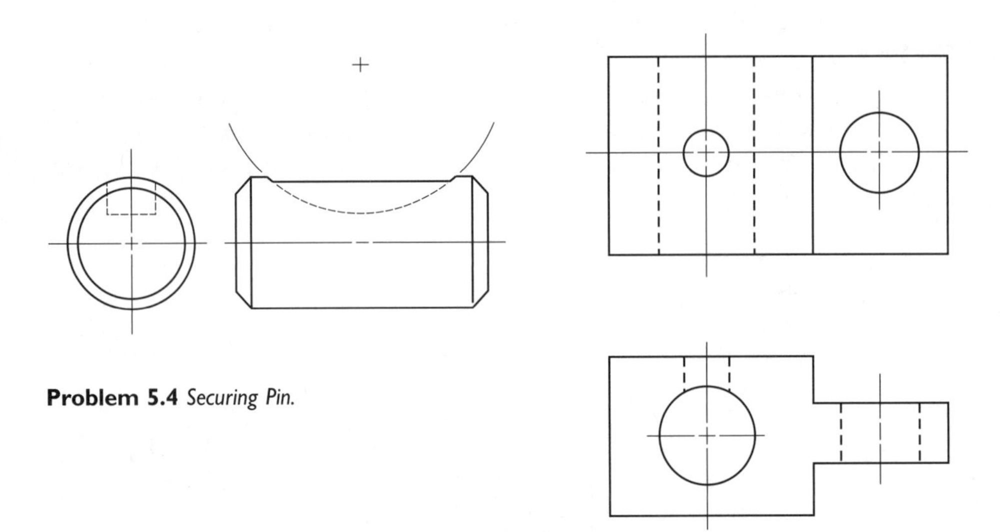

Problem 5.4 *Securing Pin.*

Problem 5.5 *Top Block.*

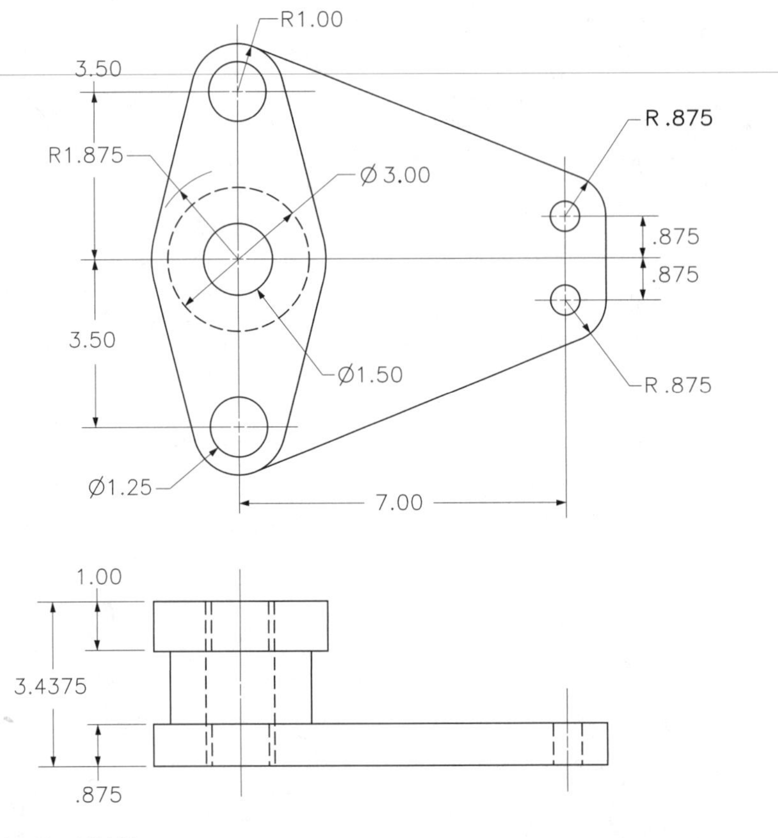

Problem 5.6 *Crank Weldment.*

Modeling with NURBS

CHAPTER OVERVIEW

NURBS (*Non-Uniform Rational B-Splines*) are mathematical descriptions of curves and surfaces. The surface itself is formed when displayed in the shaded view in the viewport or when rendered. The smoothness of the surface is determined both by the NURBS framework on which the surface rests as well as by the complexity of the *tessellation* (polygonal subdivision) assigned to the NURBS surface at rendering time. NURBS surfaces are geometrically stable, easily adjusted and edited, and economical in terms of storage requirements. Figure 6.1 shows both NURBS geometry and polygonal mesh for the same object. As you can see, the NURBS wireframe on the left is quite different, and geometrically less complex, than the polygonal mesh on the right.

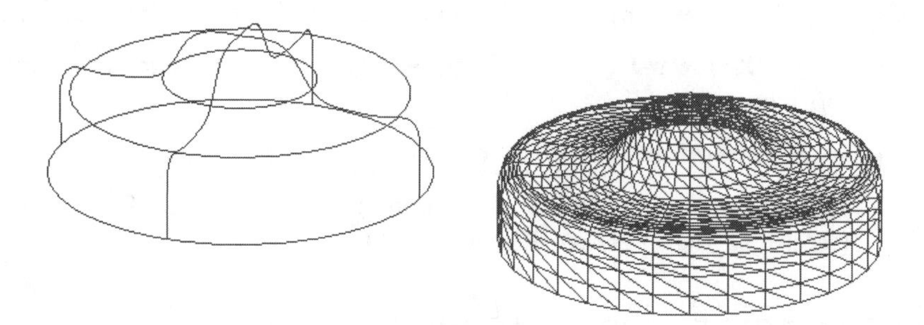

Figure 6.1 *NURBS and polygonal surface descriptions.*

NURBS are appropriate for curved surfaces, especially nonregular curved surfaces. In fact, it is difficult (though not impossible) to model sharp edges and straight sides using NURBS. This technique is also appropriate for curved surfaces that must be seamlessly blended together. However, to perform many operations (Booleans, for example), NURBS surfaces are converted to polygonal meshes. NURBS are less appropriate for linear, prismatic objects; NURBS are more appropriate for free-form surfaces with few, if any, sharp edges.

Many CAD systems make use of NURBS modeling engines. However, because NURBS give you complete creativity in pushing and pulling surfaces, you can often end up with models that would be either very expensive or impossible to manufacture. For that reason, this chapter will concentrate on those NURBS operations most applicable to manufacturing design.

There are, in fact, many NURBS operations and techniques not covered in this chapter. These are more appropriate for modeling organic, nonmechanical shapes—the shapes you would want were you modeling monsters, space aliens, or other creatures that populate games. If you are interested in this type of modeling, refer to any of the several excellent texts or Web sites listed under References for Further Study at the end of this book.

KEY COMMANDS AND TERMS

- **Base Object**—the first NURBS curve or surface in a model.
- **Control Vertex Curve/Surface**—curve or surface defined by a control lattice that surrounds but does not touch the object.
- **Datum**—a line (2D) or plane (3D) of reference from which measurements are made.
- **H Key Shortcut**—displays a list of parent and dependent object surfaces.
- **Host Surfaces**—the two surfaces between which a fillet is formed.
- **Plug-in Keyboard Shortcut Toggle**—a switch that facilitates single key shortcuts.
- **Point Curve/Surface**—curve or surface defined by points on the object.
- **Tessellation**—the polygonal subdivision of a NURBS surface. The higher the tessellation, the smoother the surface.

IMPORTANCE OF MODELING WITH NURBS

NURBS provide a range of modifications not available in 3D Studio with mesh models. Surfaces can be joined, blended, and filleted; surfaces can be formed from directrices and ruled elements; lofts, lathes, and extrusions are also available. Any of the 3D Studio standard geometric primitives can be converted into NURBS as can splines and loft objects. You are given the option for NURBS output when creating an extrusion or lathe. So you can start with a spline line (which may be easier to control than a NURBS curve) and easily convert the output to NURBS. With the options of NURBS you can choose the modeling method that best fits your own approach as well as the dictates of the model itself.

NURBS METHODS

Most technical specification for manufactured parts contain data establishing the object's edges, centers, limits, tangents, and so on; that is, points that are actually on or within the object. Note that 3D Studio provides two methods of creating NURBS curves and surfaces: *point surfaces* and *control vertex surfaces*. You will usually pick the point surface option based on the curve or surface information you have. Figure 6.2 shows these two methods of NURBS creation.

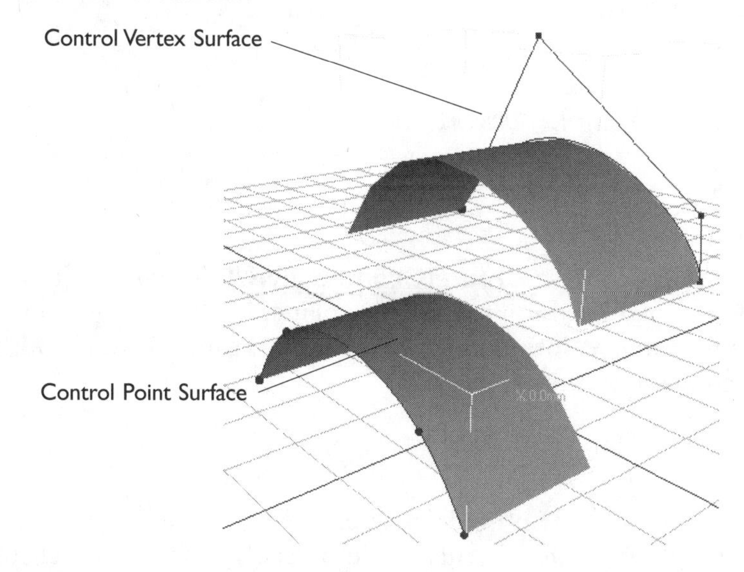

Figure 6.2 *A control vertex NURBS surface and a control point NURBS surface.*

NURBS surfaces are comprised of NURBS curves. Very often, you will begin a modeling task by creating a simple NURBS surface. This base object gets you into the **Modify** roll-out unique to NURBS. You must have a NURBS curve or surface selected to have access to NURBS modifiers. Consider Figure 6.3, a warped surface. (This is an irregular surface called a *cow's horn*, much like the pinched off end of a toothpaste tube.) It has always been a perplexing graphical and geometric task to represent warped surfaces such as a cow's horn and because warped shapes are difficult to manufacture, designs using them are scarce.

But as you will soon see, with NURBS, you can actually drape one end of a flexible plastic sheet (a NURBS surface) over a cylinder and create a warped surface as if you were vacuum forming the shape. Figure 6.3 shows views of a cow's horn. The fact that the surface blends smoothly from a circular cross section to a straight line suggests NURBS modeling.

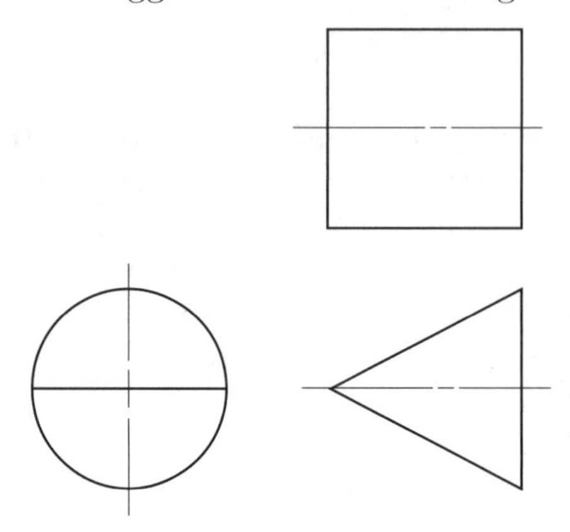

Figure 6.3 *A warped surface appropriate for NURBS modeling.*

STEP 1

Create a NURBS surface. Choose **Create|Geometry|NURBS Surface|Point Surf** and in the Top view draw out a rectangular plane. In the roll-out, enter **100** mm as its **Length** and **Width** and specify **Length Points=2** and **Width Points=4**. Move the plane to 0, 0, 0 world.

STEP 2

Create a circle template. In the Left Side view, create a 50 mm radius spline circle at 0, 0, 0 world. You will use this circle as the form over which to drape the surface (Figure 6.4).

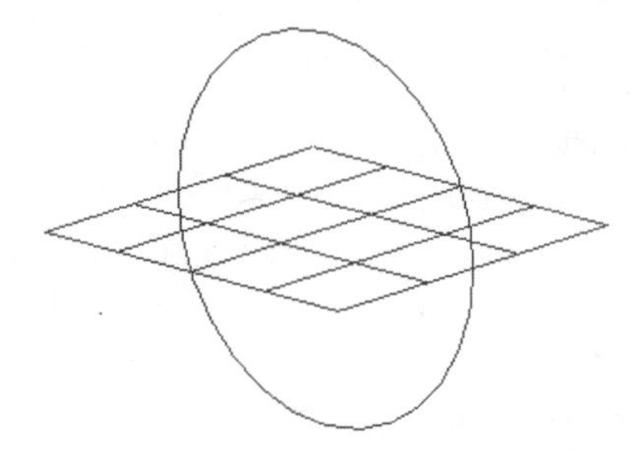

Figure 6.4 NURBS surface and circle template.

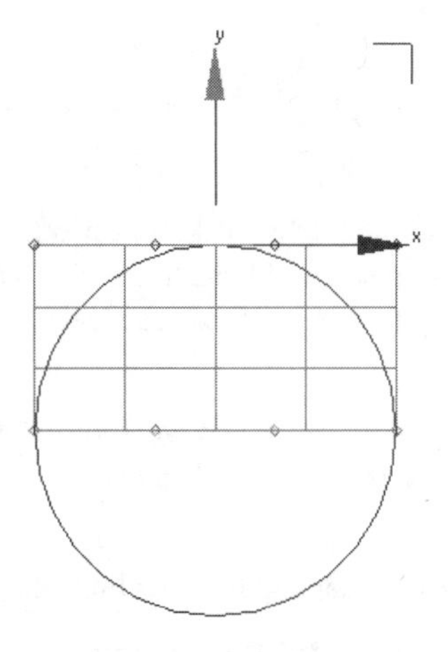

Figure 6.5 *NURBS surface and circle template.*

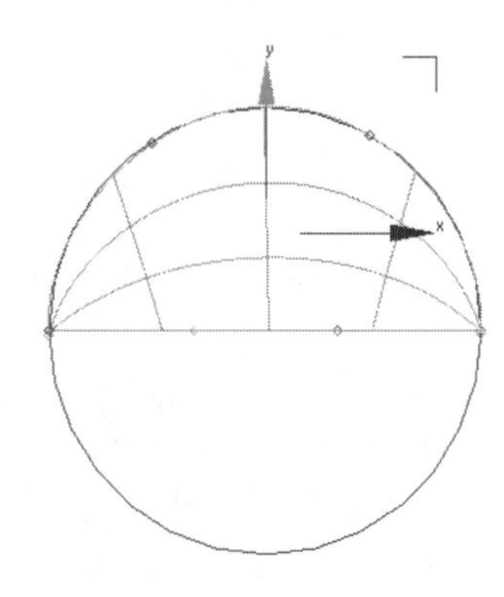

Figure 6.6 *Edited NURBS surface.*

STEP 3

Edit the NURBS surface. Switch to the Top view. Select the NURBS surface. Choose **Sub Object|Point** and select the four points on the left edge of the surface. Switch to the Left Side view.

Pull the points up until the left edge of the NURBS plane is even with the top of the circle template (Figure 6.5).

Individually, move the points until the left edge forms a semicircle concurrent with the circle template (Figure 6.6). The resulting surface is defined elegantly in wireframe view and smoothly blends from the circle to the straight line when rendered (Figure 6.7).

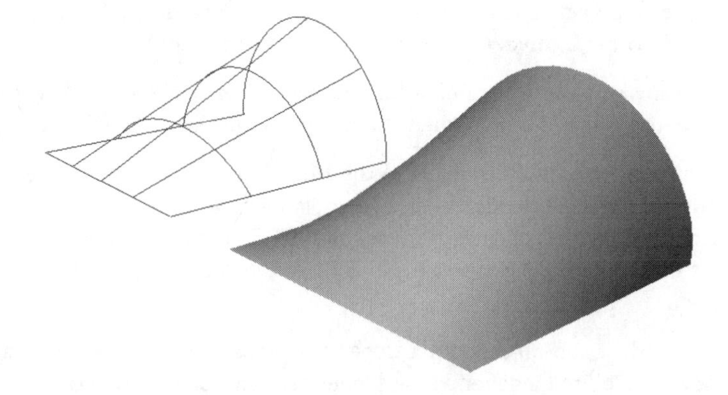

Figure 6.7 *The completed cow's horn.*

CONVERTING SHAPES TO NURBS

Any of the standard primitives, spline line shapes, lathes, and extrusions can be converted to NURBS by any of several methods.

- Immediately after creating a primitive, select and right–click. A pop-out menu appears; choose **Convert to NURBS**. Because the NURBS creation roll-out doesn't have primitives, this is the way to get basic shapes into a NURBS model.
- After scaling, rotating, moving, and so on, objects can be converted to NURBS by the preceding method or by clicking on the **Edit Stack** button in the **Modify|Modifier** roll-out.
- Some objects may have their modifier stacks collapsed back to NURBS.
- Before deselecting a lathe or extrusion, choose **NURBS Output** from the **Output** section of the **Modify** roll-out.

NURBS GUIDELINES

- Use NURBS to model surfaces that are not easily modeled using polygonal methods.
- Straight edges can be created with NURBS curves but require extra control points on point curves to control the curvature.
- Straight edges are difficult to impossible with control vertex (CV) curves.
- Non-NURBS objects assume NURBS properties when they are attached to NURBS curves or surfaces.
- In building a surface from NURBS sections, it is critical that all curves be drawn in the same direction and that the starting vertex be in the same position on each curve.
- Polygonal objects and spline curves can be converted to NURBS by selecting and choosing **Convert to NURBS** from the pop–out menu.
- You can always convert a NURBS object to a mesh. The underlying NURBS structure remains intact when the polygonal mesh is edited.
- While editing a NURBS object in the **Modify** panel you can create sub-objects (curves, surfaces, blends, fillets, sweeps, lathes, and extrusions) without going back to **Create**. This means for an all-NURBS model, once a starter surface has been created, you never have to leave the NURBS modifier roll-out and menu.
- Independent sub-objects don't depend on other sub-objects. They can be moved or edited independently. Dependent sub-objects rely on other sub-objects already in the model.

- Converted spline shapes that have been added, subtracted, or intersected by 2D Boolean operations become multiple NURBS curves, necessitating multiple NURBS surfaces.

GEOMETRIC ANALYSIS

When considering a NURBS approach to modeling, a careful analysis of the geometric characteristics of the object must be completed. From the previously enumerated guidelines, you now know that NURBS are not for every problem. The following sections outline those geometries most appropriate for NURBS: extrusions, lathed surfaces, lofted surfaces with varying cross sections, and filleted intersections.

EXTRUDED NURBS SURFACES

An extruded NURBS surface functions in much the same manner as the extrusion of a spline shape with the following exception:

- The extruded surface is interactively drawn from the cross–sectional shape.
- The cross–sectional shape can be extruded forward or backward but forward may produce uncapped ends and surface normal abnormalities.
- Once extruded and deselected, the distance in the shape **Amount** field does not affect the extrusion.

Consider the shape in Figure 6.8. Because the cross–sectional shape is consistent the length of the object, and the surfaces are perpendicular to the axis, an extrusion is suggested. To extrude this as a NURBS surface, follow these steps.

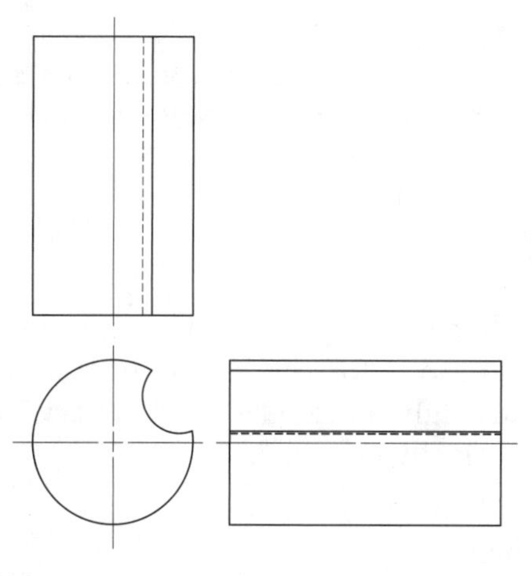

Figure 6.8 *Views of extruded shape.*

STEP 1

Model the shape. Choose **Create|Shapes|Circle** and create a 100 mm–diameter circle at 0,0,0 world. Create a 50 mm diameter circle at world X=40, Y=0, Z=40.

STEP 2

Edit the shapes. Select the larger circle and choose **Modify|Edit Spline|Attach** and click on the smaller circle, the shape that will be subtracted. Choose **Sub Object|Spline|Boolean** and click on the large circle. Click on **Boolean** again and the **Subtraction** icon (middle icon next to the **Boolean** button). Click on the smaller circle. The smaller circle is subtracted from the larger circle (Figure 6.9).

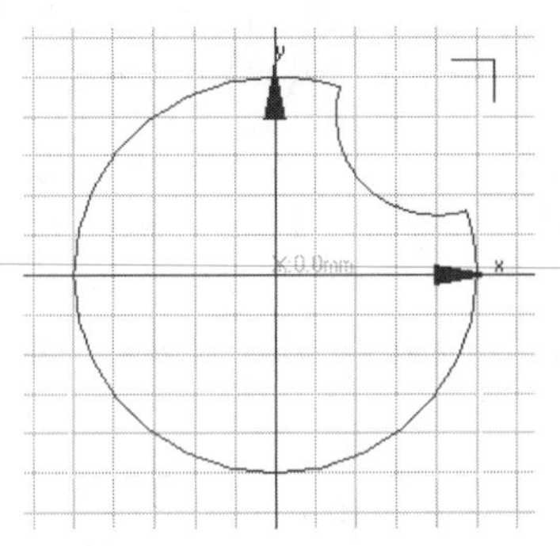

Figure 6.9 *Completed cross–sectional shape.*

Note: Spline shapes that have been Booleaned will appear as individual curves after conversion to NURBS. This means that extruded surfaces will be hollow. We'll fix this later.

STEP 3

Convert to the spline shape to NURBS. First, create two copies of the shape: one for the *front cap* and one for the *rear cap*. Hide these cap shapes. Select the visible shape and right–click. Choose **Convert To NURBS**. The **Nurbs Modify** roll-out automatically opens.

STEP 4

Extrude the surfaces. Choose **Modify|Create Surfaces|Extrud**e and pull the small curve in the negative local Z direction. Enter **–125** in the **Extrude Surface|Amount** field. Repeat this process for the larger curve. You may want to right–click on the view label and choose **Configure|Force 2-Sided** to see both sides of all surfaces. Remember, the nonclosed shapes don't generate capped ends (Figure 6.10).

Figure 6.10 *Completed hollow NURBS extrusion.*

Tip: Because you can extrude a NURBS surface in either direction, rotate the view slightly. Remember a positive extrusion is positive local Z and a negative extrusion is negative local Z.

STEP 5

Extrude the caps. Hide the NURBS surface and reveal the caps. Select the front cap and choose **Modify|Edit Spline|Extrude** and enter **–1** in the **Distance** field. Do the same for rear cap and enter **1** in the **Distance** field. (This extrudes one toward you and one away from you.) Position all three objects at 0,0,0. Move the rear cap to X=0, Y=125, Z=0 world. This aligns the two caps with the ends of the NURBS extrusion (Figure 6.11).

Figure 6.11 *Completed capped NURBS extrusion.*

Tip: Because a NURBS surface doesn't have surface polygons until it is rendered, your edges may at first appear rough. Select the NURBS surface and choose **Surface Approximation|Tessellation Presets** and click on the **Renderer** radio button and then **High**. This increases the complexity of the surface polygon mesh and renders smoother edges.

You probably wonder why we didn't just extrude the original spline object. NURBS have a unique surface creation option of **Fillet**, something of particular importance to manufacturing designs. Figure 6.12 shows the result of creating a fillet between the extruded NURBS surface and a NURBS cylinder. This is covered in detail later in this chapter under "Example: NURBS Fillets."

Figure 6.12 *Filleted NURBS surfaces.*

LATHED NURBS SURFACES

Lathed NURBS surfaces can be converted from spline lathes, or lathed directly from a NURBS curve. Figure 6.13 describes a wheel, appropriate for the lathe method. Because the cross–sectional shape contains straight line segments, a spline line curve will be drawn first and then converted to a NURBS curve for lathing.

STEP 1

Create the cross section. Create the underlying layout for the top portion of the wheel at the correct distance from the axis of the wheel. Use this layout to guide your construction with **Snap|Vertex** activated (Figure 6.14).

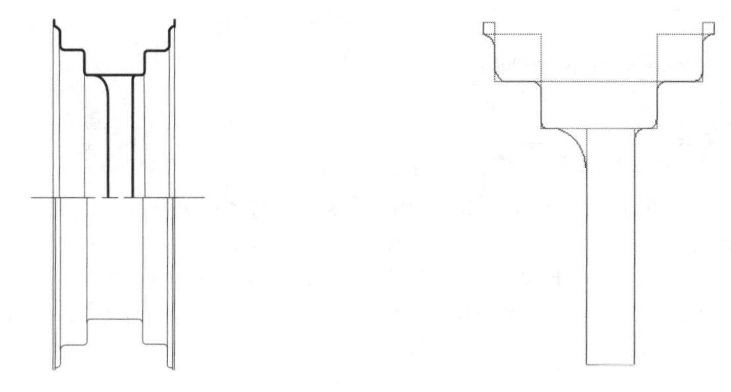

Figure 6.13 *Wheel for NURBS lathe operation.*

Figure 6.14 *Cross–sectional shape and layout.*

Tip: Don't be overly concerned if your first attempt to use the layout to guide your construction is only a crude approximation. Choose **Modify|Edit Spline|Sub Object|Vertex** and using **Refine** add points as necessary. These points have independent handles so one can be collapsed to the vertex itself, forming a straight line into the point, while the other handle creates a curve out of the point.

STEP 2

Lathe the surface (Figure 6.15). Select the curve and choose **Modify|Lathe**. Click on the **X AXIS** tab to form the surface about the X-axis. Choose **Sub Object|Axis** and right–click on the **Select and Move** tool. Enter **0** in the world Z-axis field. Finally, click the **Nurbs Output** button. This is now a NURBS lathe and can become part of a model.

Figure 6.15 shows the NURBS output in wireframe and rendered views. Notice that the NURBS wireframe has no polygonal surface definition.

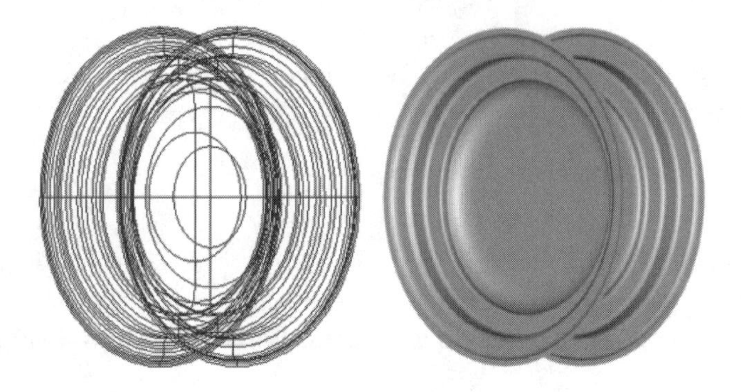

Figure 6.15 *Lathed spline converted to NURBS.*

LOFTED NURBS SURFACES

Many compound curves in manufacturing design (turbine blades, for example) are described by a series of sections taken at known intervals, each a distance from some datum. For example, Figure 6.16 shows detail views of an irregular feature defined by sections. Each shape is a known distance from a datum plane. Because the object is a transition between highly irregular curves, a NURBS solution is suggested.

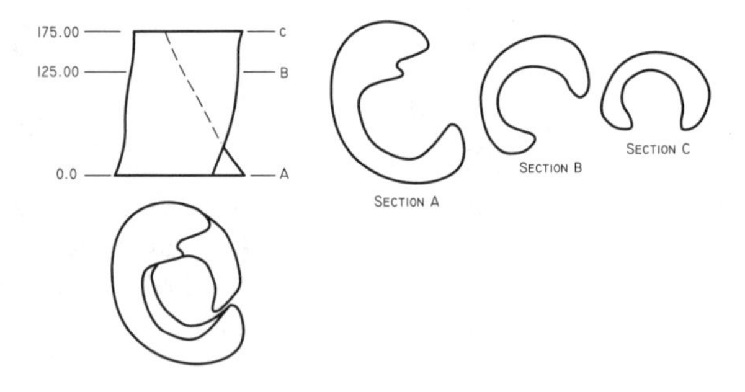

Figure 6.16 *Object defined by cross–sectional shapes.*

STEP 1

Create the sections. Create the shapes at scale and correct rotation in the Front View. (You can loft the surface and then rotate the shapes, but this may be less direct.) Call one *bottom*, one *middle*, and one *top*. This establishes the world X-Z plane as the datum=0 plane (Figure 6.17). The curves can be spline shapes that are converted to NURBS or NURBS curves.

STEP 2

Move sections using the datum dimensions from Figure 6.16. Select *middle* and right–click the **SELECT and MOVE** tool. Move *middle* to Y=125 mm. Move *top* to Y=175 mm. This is shown in Figure 6.18.

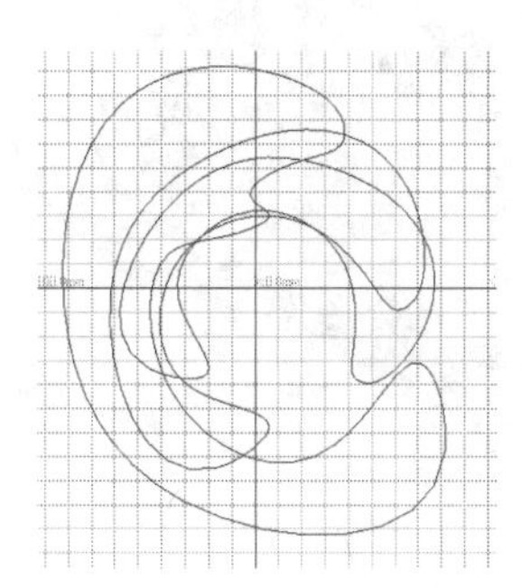

Figure 6.17 *Sections on the world X-Z plane.*

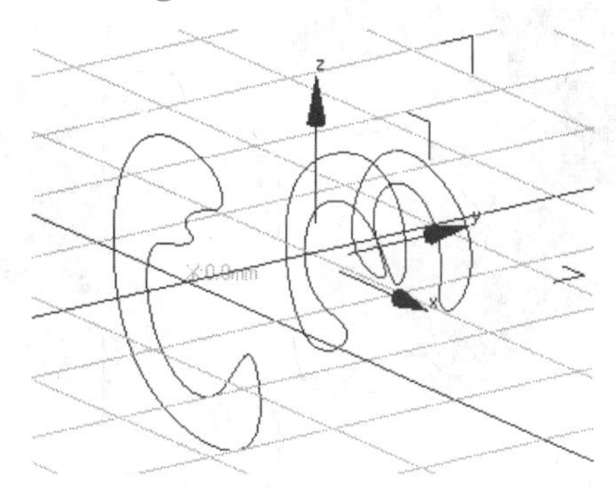

Figure 6.18 *Sections moved to datum distances.*

STEP 3

Attach the three curves. Select the bottom curve. From **Modify|General** choose **Attach** and select first the middle and the top curve. All three curves are now selected and named *bottom*, the base object. Change the name to *nurbs loft*.

STEP 4

Loft the surface between the sections. Choose **Create Surfaces|U Loft** from the **Modify** roll-out. As you move over the attached curves, they turn blue. Click the bottom curve and move the cursor to the middle curve. A dashed line connects the two curves as the cursor changes shape and the middle curve turns blue. Click the middle curve and repeat for the top curve.

Because you see both sides of the surfaces, some surface normals are pointing away from you, making the part of the surface transparent (Figure 6-19). Right–click on the view label and choose **Configure|Force 2-Sided**. This renders both sides of the surface in the viewport.

STEP 5

Cap the ends. Choose **Create Surfaces|Cap** and move the cursor over the open bottom end. The bottom curve turns blue. Click on the curve and the open end is capped. Repeat this procedure for the top end. The finished loft is displayed in Figure 6.20.

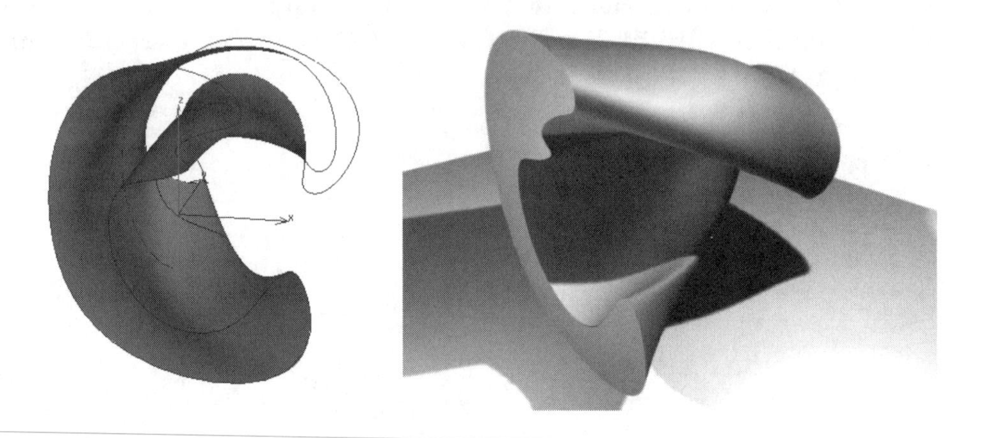

Figure 6.19 *Force 2-sided reveals both sides of the surface.*

Figure 6.20 *Finished loft from cross sections.*

Note: You can cap the ends because the NURBS curves that define the ends are closed. Were they not closed, you would be unable to cap the ends and would have to make explicit caps (see Figure 6.11).

EXAMPLE: NURBS FILLETS

As was demonstrated earlier, NURBS allow filleted surfaces to be formed between intersecting and attached NURBS surfaces. A fillet smoothly connects two host surfaces with a curved surface of consistent radius. Figure 6.21 shows a pressure vessel with filleted tube intersections. These tubes have filleted transitions that smoothly blend the cylindrical surfaces into the spherical surface. Additionally, the top surfaces of each flange as well as the flange-to-cylinder intersections receive fillets. Because of the need for fillets, a NURBS solution is suggested.

Note: 3D Studio calls all surfaces blended with a constant radius a fillet. Traditionally, rounded exterior corners are called rounds, while rounded interior corners are called fillets.

STEP 1

Create the shapes. Use standard sphere and tube primitives and the dimensions in Figure 6.21 to model the basic shapes. Create the tubes so that they extend into the sphere (Figure 6.22).

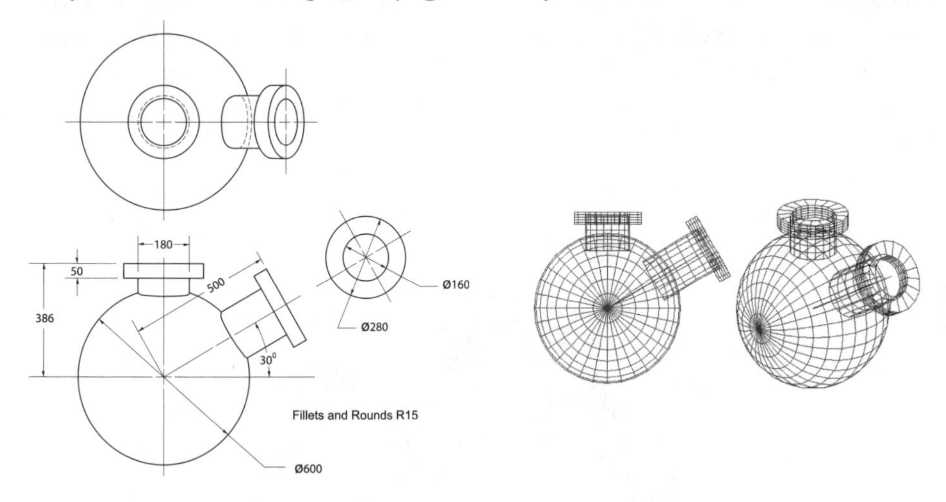

Figure 6.21 *Pressure vessel and connections.*

Figure 6.22 *Basic geometry in place.*

STEP 2

Convert to NURBS. In turn, select each of the primitives, right-click, and choose **Convert to NURBS**.

STEP 3

Attach the objects. Select the sphere. Choose **Attach Multiple** and select the two cylinders and flanges. A single NURBS objects called *sphere* is created. Rename this *pressure vessel*. Render the viewport. Your model looks like Figure 6.23.

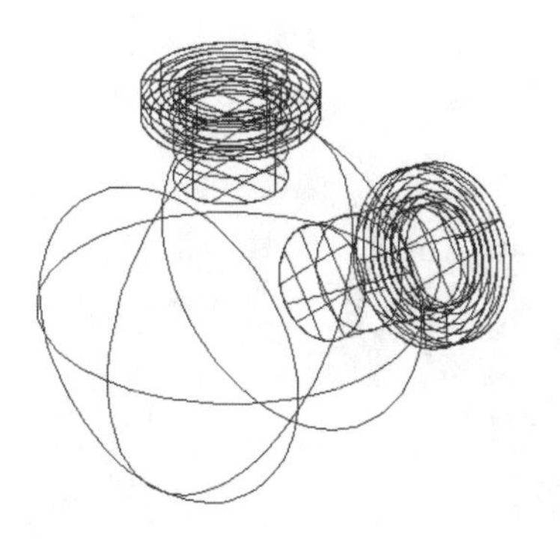

Figure 6.23 *NURBS model.*

STEP 4

Create the interior fillets. Each interior fillet will be created in the same manner:

Choose **Fillet** and in the roll-out enter seed values of **1**, **0**, **1**, **0**. Enter the fillet radius in the **Radius** field. Pass the cursor over the first host surface. It turns blue. Select this surface (Figure 6.24.). Pass the cursor over the other host surface. This surface also turns blue, signifying which two surfaces will be filleted (Figure 6.25).

Note: There are multiple solutions to most fillet constructions. Seed values determine which fillets are created.

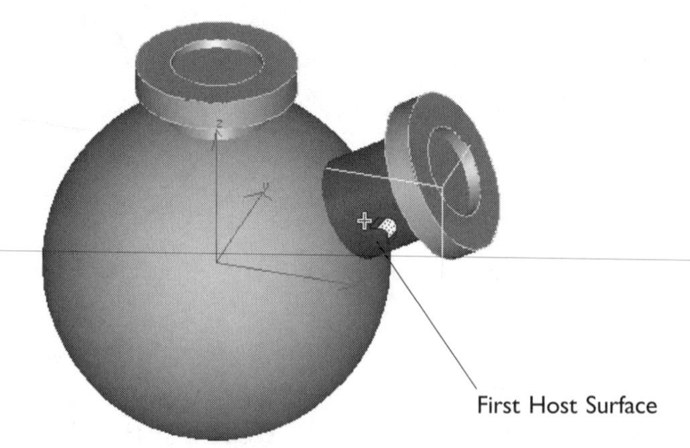

Figure 6.24 *First host surface is selected.*

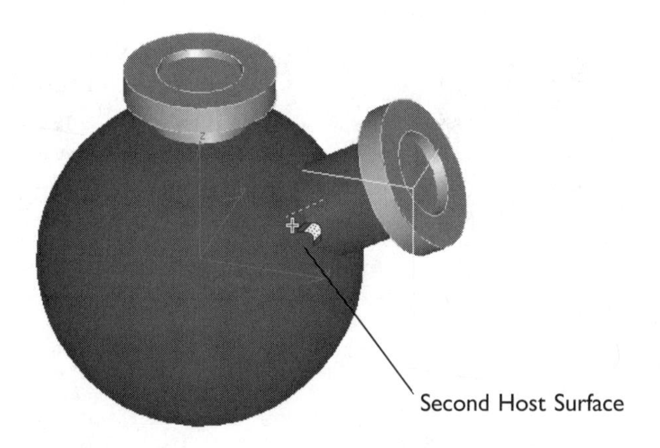

Figure 6.25 *Both host surfaces selected.*

Note: You may want to view the NURBS model in **Wireframe** display to better see if the correct fillet has been constructed. If you can't see the fillet in shaded display, go to the bottom of the roll-out and choose **Flip Normals**.

Notice that 3D Studio creates a new surface that blends the two host surfaces (Figure 6.26). It does not, however, trim the host surfaces back to the fillet. If the incorrect fillet has been formed, delete, change seed values, and repeat.

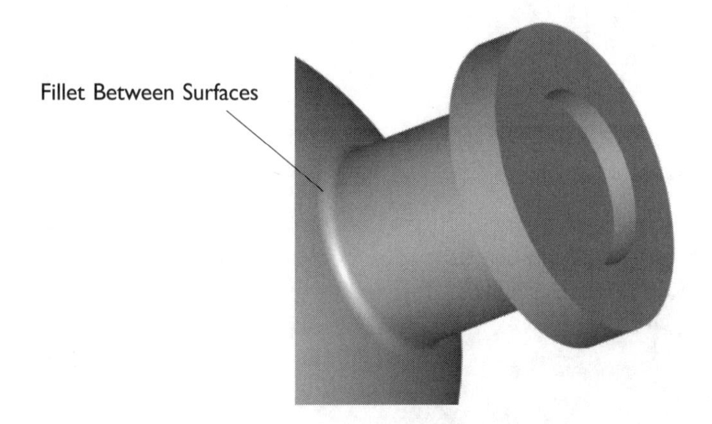

Figure 6.26 *Fillet created between host surfaces.*

If you create an incorrect fillet you can **Undo** or simply hit the delete key. If you have deselected the fillet, choose **Sub Object|Surface** and reselect the fillet (wireframe view), and delete.

STEP 5

Fillet (round) the exterior corners. The previous fillets were formed between interior corners and the untrimmed host surfaces disappeared behind the fillets. You did not have to trim the host surfaces to see the fillet. An exterior fillet (Figure 6.27) is formed under the host planes. These planes must be trimmed back to the fillet for the fillet to be visible.

Select the two surfaces as was done previously. Immediately after the second host surface is selected, choose **Trim Surface** under both **Trim Surfaces** options in the **Fillet** roll–out. This trims both host surfaces back to the fillet, revealing the curve (Figure 6.28). If either host surface is transparent, choose **Flip Normals** in the respective roll-out section. If the fillet itself is transparent, choose **Flip Normals** at the very bottom of the roll–out.

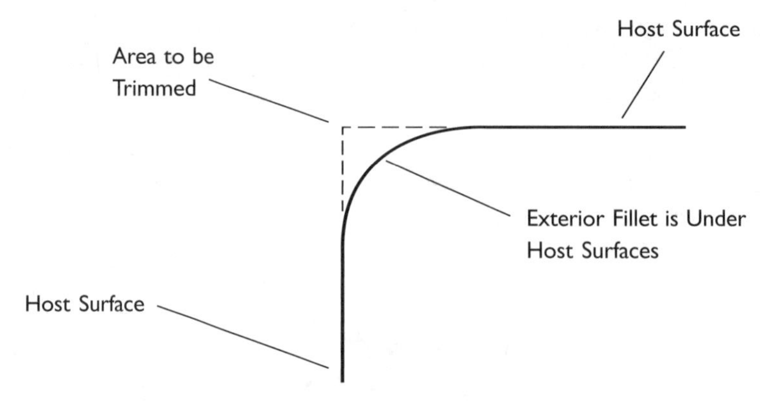

Figure 6.27 *Fillet on exterior corner is below host surfaces.*

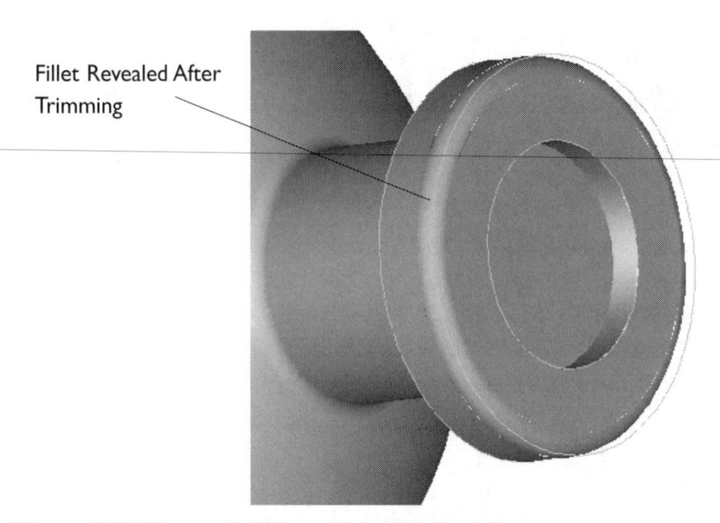

Figure 6.28 *Fillet revealed after surfaces are trimmed.*

Note: The host surfaces are not trimmed geometrically. Instead, no polygonal mesh is assigned to the surface beyond the fillet. If you choose **Edit Mesh** you can see how this external fillet is formed by the polygonal mesh. Choose **NURBS Surface** from the bottom of the **Modifier Stack** and see that the NURBS definition has not changed,

Tip: If you have trouble identifying the second host surface, activate the **Plug-in Keyboard Shortcut Toggle** and after identifying the first surface, press the **H** keyboard key. Move the selection window off to the side so when you select the various surfaces by name you can see them turn blue.

STEP 6

Smooth the curve. The NURBS curve has a surface only when rendered, either to the screen, or to the *Virtual Frame Buffer* (VFB). See Chapter 19, Still and Movie Output for a discussion of the VFB. Choose **Surface Approximation|Viewports|Tessellation Presets|High**. This increases the density of polygons on the NURBS object, smoothing the edges and surfaces in smoothly rendered viewports. Do this again with the **Renderer** option so that edges are smoothed when you actually render the file. The completed Pressure Vessel, with its interior and exterior fillets, is shown in Figure 6.29.

Figure 6.29 *Completed Pressure Vessel.*

PROBLEMS

Problems appropriate for solving with NURBS modeling are found on the following pages. Use the file *isogrid* found in the *tools* directory to plan your Boolean tree strategies. Assign appropriate names to all objects both on your sketches and models.

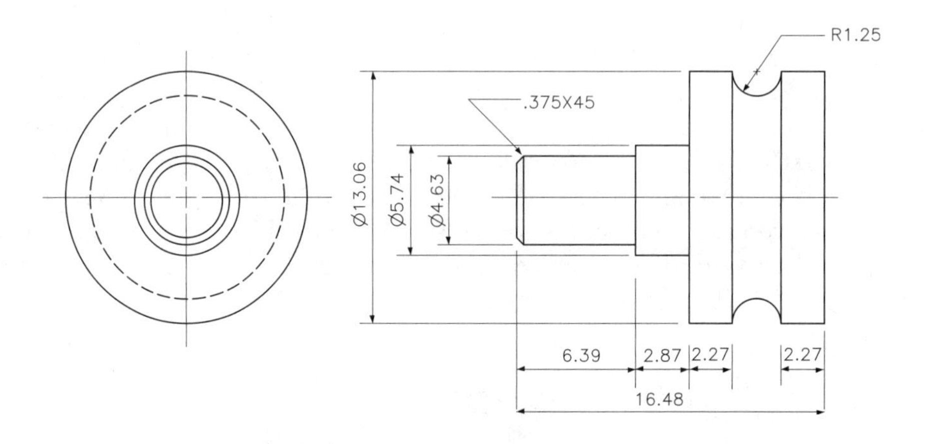

Problem 6.1 *Ring Retainer.*

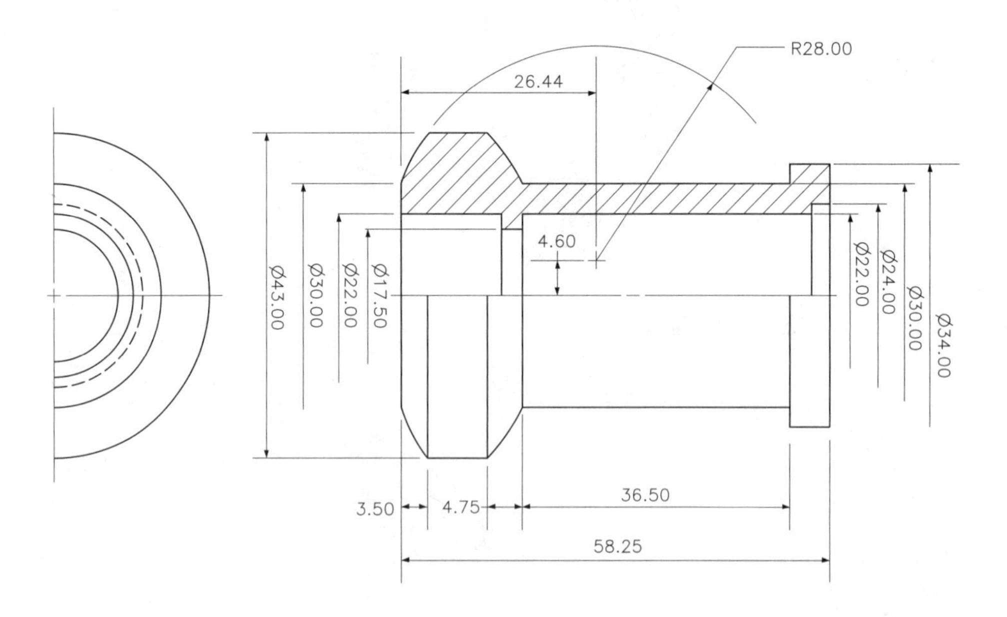

Problem 6.2 *Pressure Collar.*

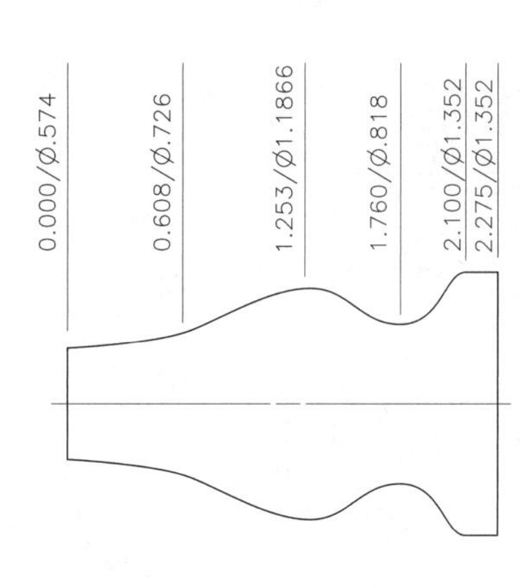

Problem 6.3 *Weaving Spindle.*

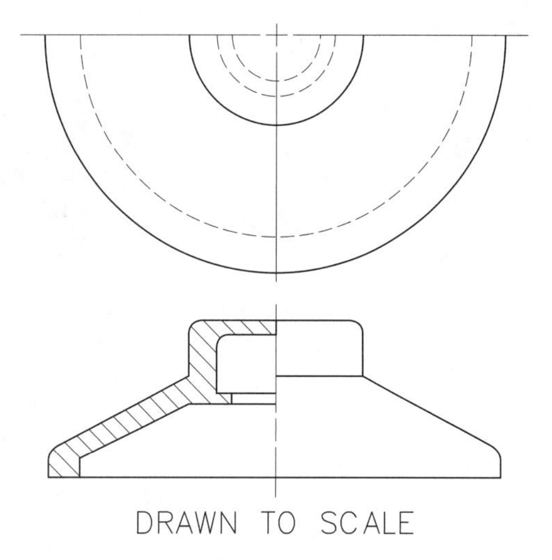

Problem 6.6 *Bearing Housing.*

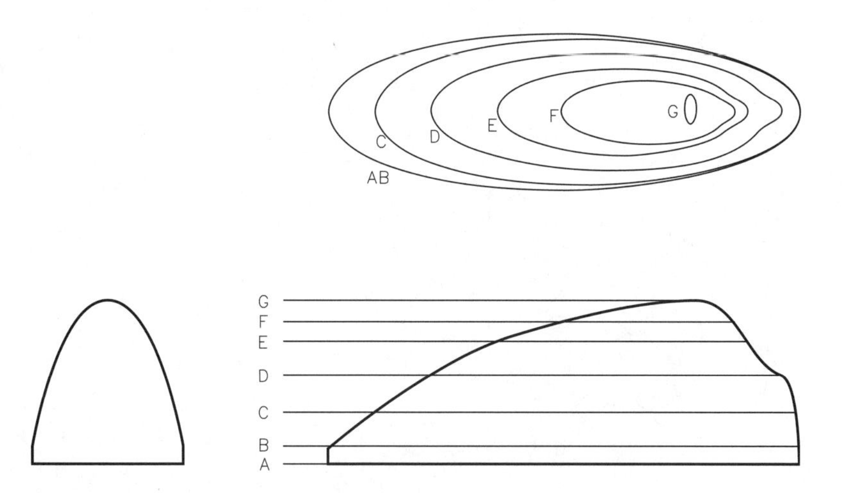

Problem 6.4 *Canopy. NURBS profiles can be found on the CD-ROM in chapters\chapter_6\canopy.max.*

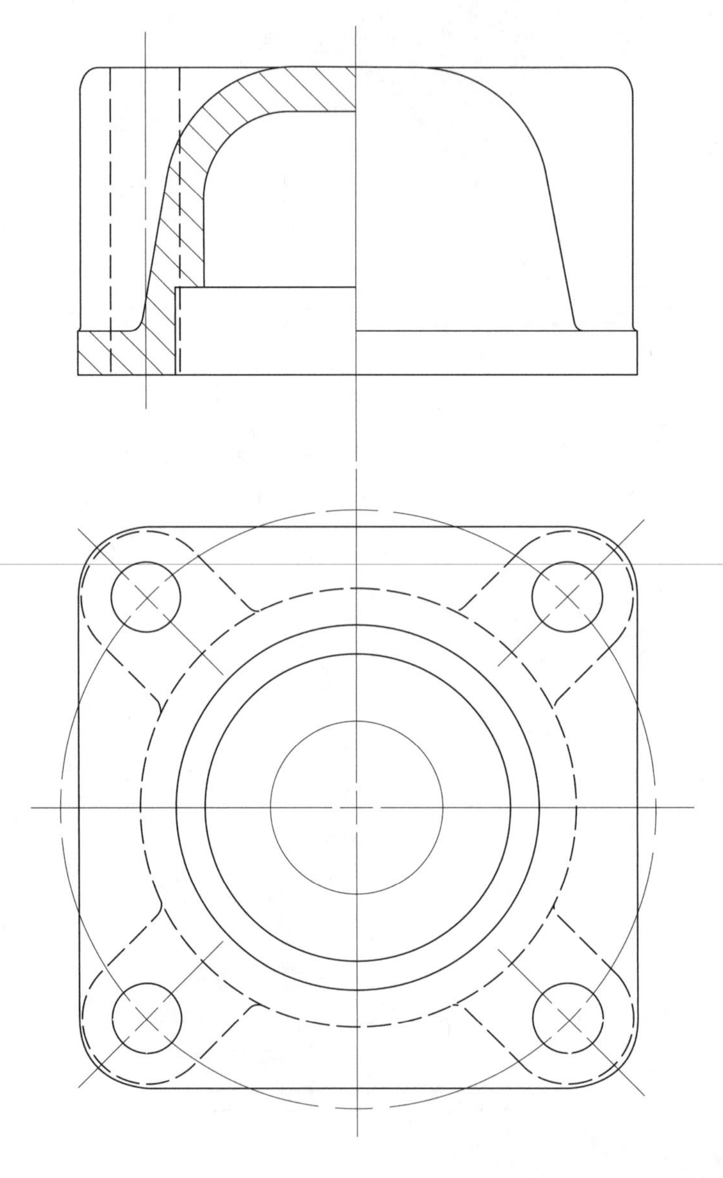

Problem 6.5 *Hub Cover.*

CHAPTER 7

Modeling with Modifiers

CHAPTER OVERVIEW

Using modifiers is a completely different way to look at modeling. Rather than constructing geometry by assembling precision components, generalized shapes are edited to arrive at final geometry. To effectively use modifiers, you need to predict the effect of the modifier before starting the modeling process. It is a waste of time to apply a modifier to geometry that simply won't respond in the way you anticipate, or to create geometry that won't work with any modifier.

Modifiers act directly on the geometry in *object space*. This is different from *transforms* that act on geometry in *world space*. We will discuss the **Scale** transform in this chapter because it directly relates to modeling with modifiers. Modifiers are recorded in the **Modifier Stack** where they can be revisited, reordered, and adjusted. The default VIZ/MAX interface displays a panel of ten modifiers; however, custom button sets can easily be configured so that you have access to related modifiers at various points in the modeling process.

KEY COMMANDS AND TERMS

- **Axis Lock**—a toggle at the top of the 3D Studio interface that restricts transformations to the selected axis.
- **Collapsing the Stack**—locking the effect of modifiers into an editable mesh. A collapsed stack cannot be uncollapsed.
- **Custom Button Sets**—collections of modifiers that can be assigned to the ten-button modifier panel.
- **Editable Mesh**—the polygonal surface description of an object where vertices, edges, faces, polygons, and elements can be selected.

- **Gizmo**—the proxy that applies modifier settings to a selected object.
- **Measuring Plane**—a plane set at an exact location used to guide freehand transforms.
- **Mesh-**—he polygonal description of a surface.
- **Modifier Stack**—the history of all modifications that have been applied to an object. The stack can be rearranged, modifiers deleted, copied to other objects, or collapsed.
- **Named Selection Set**—selections that are saved so they can be easily reselected by assigned name.
- **Object/Space Modifiers**—modifiers based on the object's local axis system and are evaluated before transforms and world/space modifiers.
- **Parametric Modifiers**—modifiers whose parametric settings can be readjusted without impacting the surface topology.
- **Pinning the Stack**—freezes the stack to the current object and stays available even though other objects may be selected.
- **Topology**—the manner in which the surface is described.
- **Transforms**—world space modifications (scale, move,rotate).

IMPORTANCE OF MODIFIERS

Modifiers can be applied to an entire object or any portion of the object you can select. By applying modifiers you can arrive at geometry that would be impossible with extrusions, lathes, lofts, Booleans, or NURBS alone. Because modifiers impact all subsequently applied modifiers, the effect is cumulative. Bending and then twisting produces different results than does twisting and then bending.

The weakness of modifiers in engineering and technology are several fold. First, modifications, though parametric, are generally applied to change how an object *looks*. The parameters controlling the modifier are not easily discerned from standard engineering documentation. Also, because modifiers give you complete freedom in adjusting any part of the object (vertices, edges, polygons, or faces), it is easy to arrive at "visually pleasing" though impossible to manufacture designs.

For these reasons, it may be advantageous to insert engineering views into a scene to guide these modifiers. This way the views can be used as a template of sorts, with modifiers applied to edit beginning shapes and objects into final form.

MODIFIER GUIDELINES

- It is paramount to assign easily identifiable names to objects when using modifiers.
- Selection subsets are vital in applying modifiers to portions of object geometry.
- A modifier affects all modifiers applied after it because the modifier stack is evaluated bottom to top.
- Object-space modifiers operate on topology.
- Topology-dependent modifiers require the same polygonal mesh to produce consistent results. Change the topology and the result of subsequent modifiers changes.
- Parametric modifiers don't affect polygonal typology and can be independently adjusted.
- Collapse the modifier stack when you are satisfied with the results. This "locks" the results of the modifications into an editable mesh and reduces the amount of memory required.
- Some modifiers act directly on the selected object. Some modifiers act on a proxy, called a gizmo, and the gizmo transfers the changes to the object.
- A modification applied to a sub-object selection is preceded with an asterisk (*) in the modifier stack.
- The order in the modifier stack is base object followed by a single line followed by object space modifiers followed by a double line followed by world space modifiers. The double line represents transformations (move, scale, rotate).
- If the modifier stack is "pinned" it is locked to the object though other objects may be selected.
- Because affecting complex geometry with modifiers may be computationally intensive, you may wish to turn off the effect of individual modifiers, the effect up to the currently selected modifier, or their entire display.
- Modifiers are repositioned by choosing edit stack. cut the modifier from the stack list. Select the position above which you want the modifier and paste.

USING SUB–OBJECT MODIFIERS

Modifiers can be applied to entire objects or portions can be selected and acted upon. Depending on the type of object selected, different **Sub–Object** modifiers appear in the **Modify** drop–down menu. This sub–object menu is slightly different, depending on the modifier chosen, and is arrived at by various methods depending on object characteristics.

- With standard and extended primitives you must select the object and choose **Modify|Edit Mesh|Sub Object**.
- Splines require selection and choosing **Modify|Edit Mesh|Sub Object**.
- Splines that are lathed are immediately in the **Modify** menu and **Sub Object** is automatically available.
- Extrude doesn't have any sub–objects.
- Any surface converted to a mesh will have sub objects. Select and choose **Modify|Edit Mesh|Sub Object**.

Assume a **Box** primitive with four segments each in length, width, and height. This produces a matrix suitable for the **Edit Mesh** modifier. The box is comprised of vertices, edges, faces, polygons, and elements. The five available sub–objects are illustrated (Figure 7.1) in how the box is modified when the sub–objects are moved along the world Y-axis.

- **Sub–Object|Vertex** selects only the vertex (a). When the vertex is transformed, edges at the vertex are also transformed.
- **Sub–Object|Edge** selects individual edges connecting vertices (b). When the edge is transformed, it pulls connected edges along by common vertices.
- **Sub–Object|Face** selects a triangular polygon supported by three edges (c). When the face is transformed, it pulls connected edges by common vertices.
- **Sub–Object|Polygon** selects the entire polygon (d). When the polygon is transformed, it pulls connected edges by common vertices.
- **Sub–Object|Element** selects all elements on the object (e). When the elements are transformed, the entire object is moved.

Tip: When selecting sub–objects, use the single click method to select the visible sub–object. Use a region selection when you want to select all sub–objects within the region at any screen Z-axis depth. The region selection actually creates a selection solid, extending infinitely in the screen Z-axis direction.

MODIFYING WITH SCALE AND MOVE TRANSFORMS

Even though the SCALE and MOVE transforms are not actually modifiers in 3D Studio parlance, they are what you often do at the very beginning to alter the shape of primitives, lathes, or extrusions. Then, when actual modifiers are applied, these transforms may again be used to change the design of surfaces.

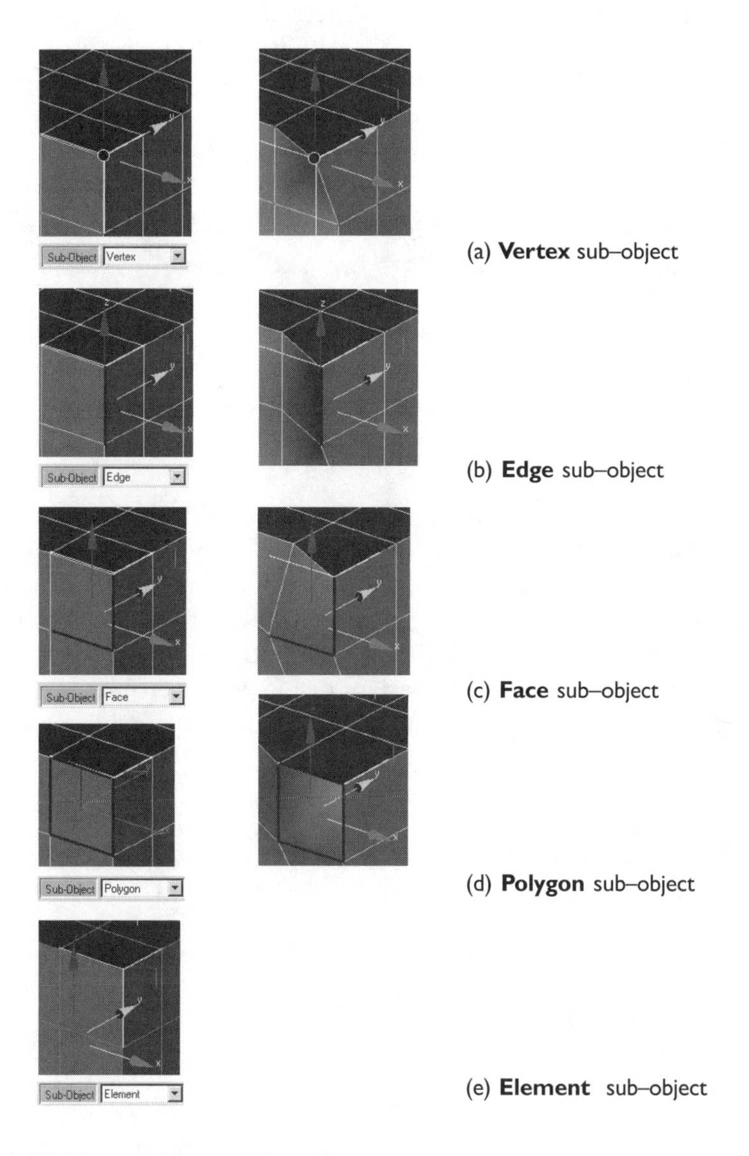

Figure 7.1 *Sub–object modifiers and their effect.*

The **Scale Transform** quickly reshapes objects relative to world, view, or local axes. Figure 7.2 shows the progression of a sphere (a) into an oblate spheroid by scaling unequally in the world Z (Front view) and Y (Right Side view) directions (b). By choosing **Modify|Edit Mesh|Sub Object|Vertex** the vertices on the left side can be selected and moved (transformed) in the negative X-world direction. With these vertices still selected, the **Select and**

Uniform Scale transform (c) can move them inward or outward about the selection origin (d). So in this case the modifier (**Edit Mesh**) uses two transforms (**Move and Scale**) to alter the geometry of a transformed (**Scale**) sphere.

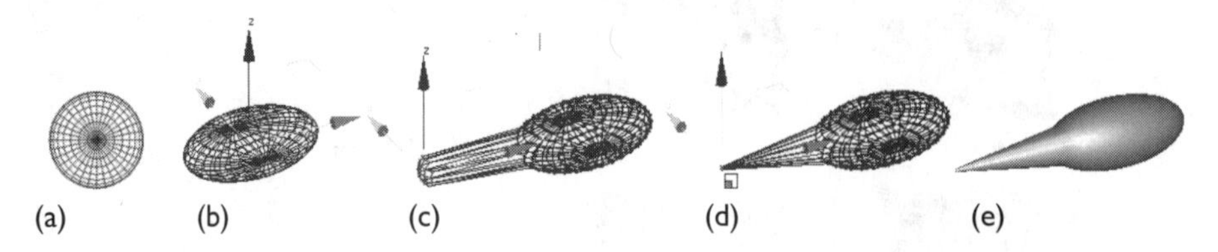

Figure 7.2 *Transforms work with modifiers.*

THE MODIFIER STACK

The *modifier stack* keeps track of an object's history. In our previous example first we created a sphere, then we transformed it by scaling, and then we edited its mesh-first by moving some of its vertices, and then by scaling them to a point. The modifier stack in Figure 7.3 shows this history. Note that transforms that are part of the **Edit Mesh** modifier don't explicitly appear in the stack. Transforms can't be revisited and adjusted parametrically as can modifiers. Because the sphere is the base object, we can revisit its creation parameters, with caution.

Note: You will receive a warning when returning to topology that modifiers depend on. If you change the underlying topology, it will impact (possibly negatively) the action of later modifiers.

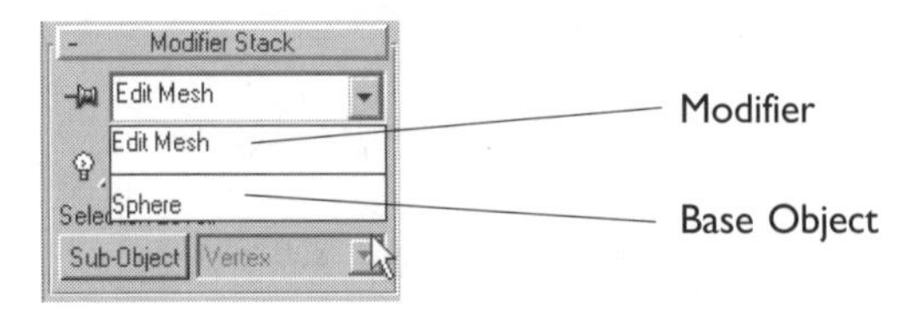

Figure 7.3 *Modifier stack.*

If we are satisfied with the results of our modifications we can *collapse the stack* to the underlying mesh. Once collapsed, access to the modifications is lost. To collapse the stack, follow these steps.

STEP 1

Click on the **Edit Stack** icon at the right of the **Modifier Stack** roll-out.

STEP 2

Choose Collapse All from the options. This permanently applies the actions in the **Edit Mesh** modifier to the sphere. The **Modifier Stack** now shows an **Editable Mesh** as the base object.

Note: You can capture modifier actions down to a position in the modifier stack by choosing **Edit Stack** and the modifier you want on the top of the stack. Choose **Collapse To**.

SUB–OBJECT SELECTION SETS

Because you will often apply modifiers to the same geometry at different times, it is helpful to create selection sets. Figure 7.4 shows a cylinder that was modified first by transforming a group of vertices. Later, those vertices were modified by applying the **Stretch** modifier. Here's how the **Named Selection Set** was created and the modifications made.

Figure 7.4 *Cylinder modeled with transforms and modifier.*

STEP 1

Create the cylinder. In the Right Side view, create a cylinder of any reasonable length and radius with **10 Height Segments**, **8 Cap Segments**, and **24 Sides**.

STEP 2

Identify the sub–objects. In the Front view, select the **Rectangular Region** tool, select the faces as shown in Figure 7.5. All faces contained within the region selection, on both sides of the cylinder, are selected.

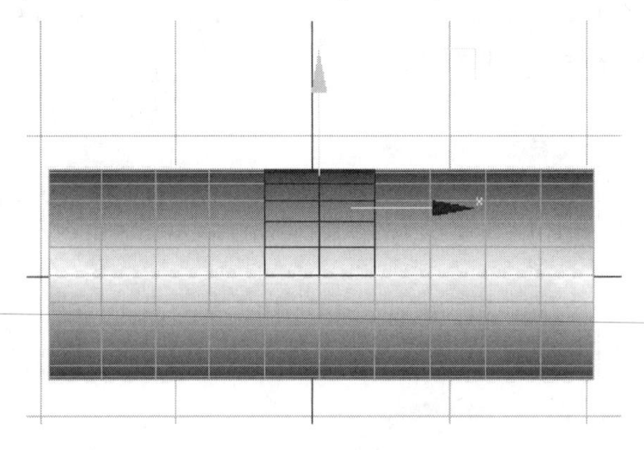

Figure 7.5 *Faces selected.*

STEP 3

Create the selection set. With the faces selected, click in the **Named Selection Sets** field in the toolbar at the top of the 3D Studio interface. Type *rib* in this field and press the return key. These faces will be available when you return to the **Edit Mesh** modifier in the **Modifier Stack** and choose **Sub Object|Faces**.

STEP 4

Transform these faces upward, forming the rib (Figure 7.6). You can see by the Left Side view that the faces have been transformed straight upward.

Figure 7.6 *Rib is transformed along world Z-axis.*

Note: To transform along an axis either move the selected geometry by grabbing the desired axis pivot vector or toggle the appropriate **Axis Lock** and transform by grabbing anywhere.

STEP 5

Reselect the faces using the selection set. Select the cylinder. In the **Modifier Stack**, **Edit Mesh** is selected. On a taller stack, you may have to select this modifier. This is the modification in which you identified the faces as a selection set. Choose **Sub Object|Faces**. Because you created a face sub–object selection set, you must choose the same option for the rib to appear in the drop-down (Figure 7.7).

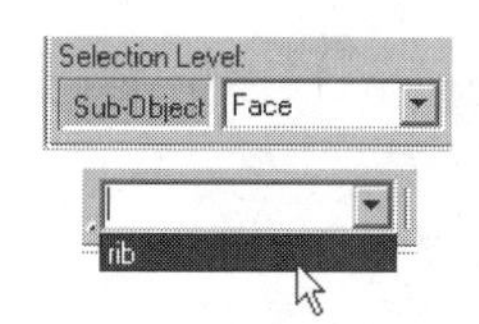

Figure 7.7 *Selection set.*

STEP 6

Apply the **Stretch** modifier. Switch to the Left Side view. Choose **More** from the **Modify** panel and select **Stretch** from the modifier list. In the **Stretch** roll-out enter **2** in the **Stretch** field and select **Stretch Axis: X**. The rib is reshaped by the **Stretch** modifier (Figure 7.8).

Note: In this example, as well as most examples in this book, transforms are assumed to be in relation to *view axes*. When *world* or *local axes* are used, they are explicitly referenced.

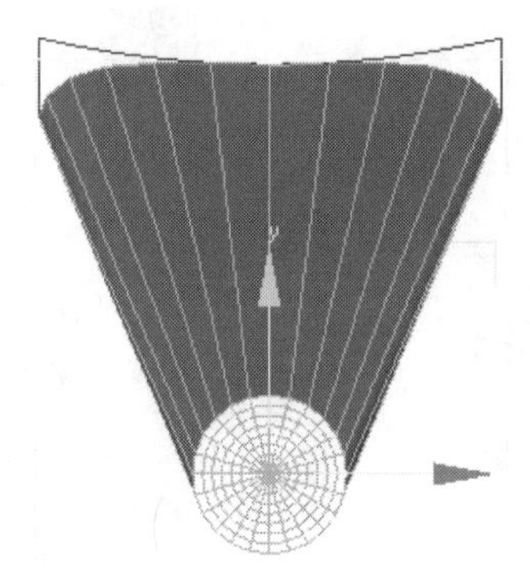

Figure 7.8 *Effect of **Stretch** on the transformed faces.*

STEP 7

Apply the **Mesh Smooth** modifier. Rotate the view into a pictorial User View. Choose **More** from the **Modify** panel and select **Mesh Smooth**. In the **Sub Division Amount** section enter **2** in the **Iterations** field. Note that 3D Studio creates a finer polygonal mesh to define the surface. Figure 7.9 shows the original mesh after **Stretch** (a) and after **Mesh Smooth** (b). The modifier stack (c) is shown, recording the history of modifications performed.

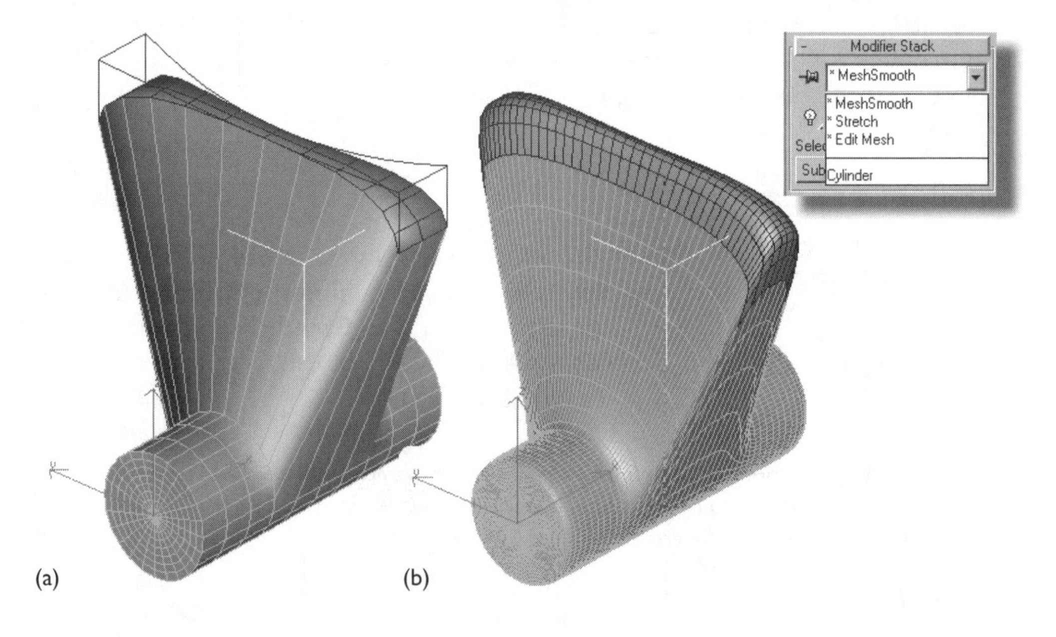

Figure 7.9 *Effect of **Mesh Smooth** on the transformed faces.*

MODIFYING PRIMITIVES

Almost every 3D Studio book demonstrates how a box can be turned into an airplane by the application of modifiers. The reason is that this simple example demonstrates so well the power and flexibility of modifiers that every 3D Studio student should practice the procedure at least once. Not to leave this technique uncovered, here's how to turn a box into an airplane.

Figure 7.10 shows views of a hypersonic Re-entry Vehicle. The irregular cross–sectional shape might at first suggest a NURBS loft solution. However, by carefully modifying a primitive box the compound curve shape is attainable. The underlying 3 x 3 x 3 **Box** is superimposed on the views for reference.

STEP 1

Create the base box. **Choose Create|Standard Primitives|Box** and draw out a box in the Front view whose height and width are proportional to the height and width of the plane in Figure 7.10). Enter **3** in each of the **Segments** fields in the **Box** roll–out (Figure 7.11).

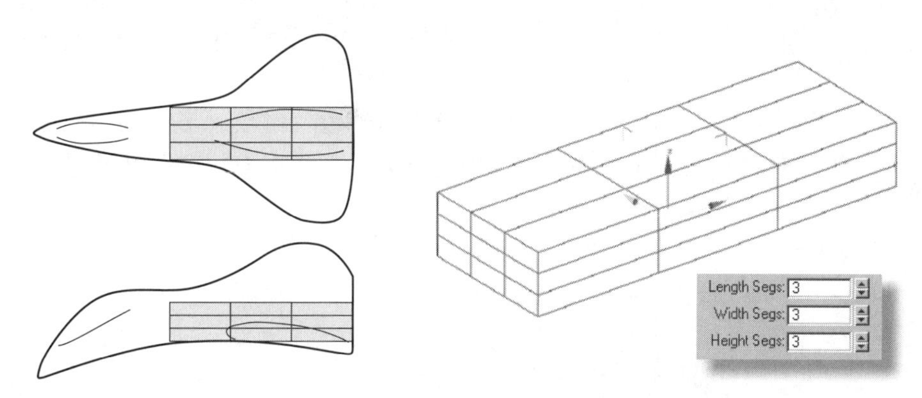

Figure 7.10 *Views of plane.*

Figure 7.11 *Base **Box** for plane construction.*

STEP 2

Select the wing polygon. In the Front view, select the box and choose **Modify|Edit Mesh|Sub Object|Polygon**. Select the middle polygon on the right (Figure 7.12) with a single click.

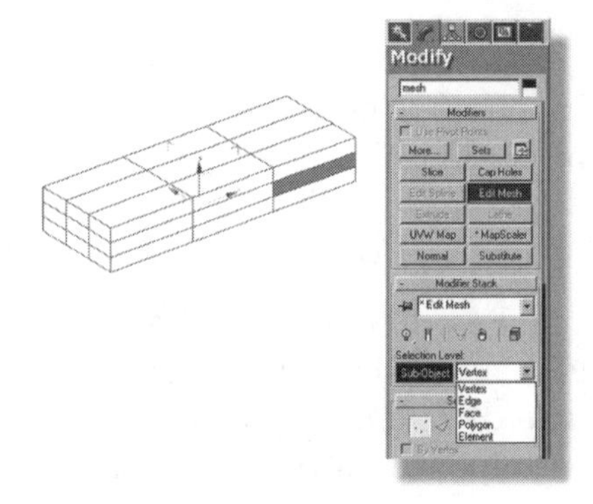

Figure 7.12 *Wing polygon selected.*

STEP 3

Transform the wing polygon. Use measuring planes (see p. 144) to control the amount that each wing is pulled from the box. Work in the Top view. With the **Select and Move** tool, grab the Y-axis and pull the polygon out to the length of the wing (Figure 7.13).

Switch to the Back view. Select the corresponding polygon on the back of the box. Switch to the Top view and pull the polygon to its measuring plane (Figure 7.14).

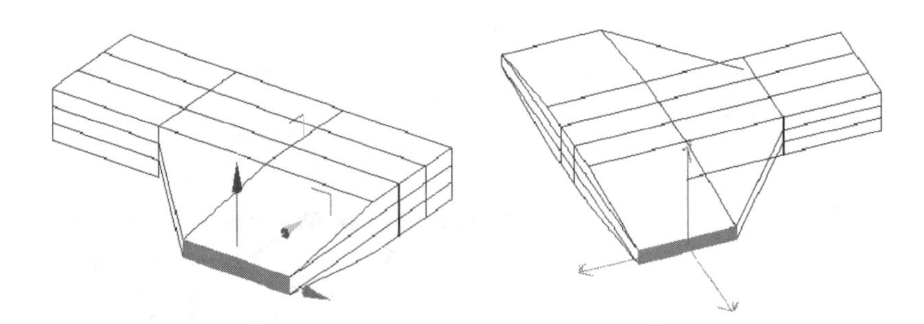

Figure 7.13 *First wing is pulled from the box.*

Figure 7.14 *Second wing is pulled from the box.*

STEP 4

Transform the tail. In the Top view, select the tail polygon. Switch to the Front view. Using a measuring plane, pull the tail polygon upward along the Y–axis (Figure 7.15).

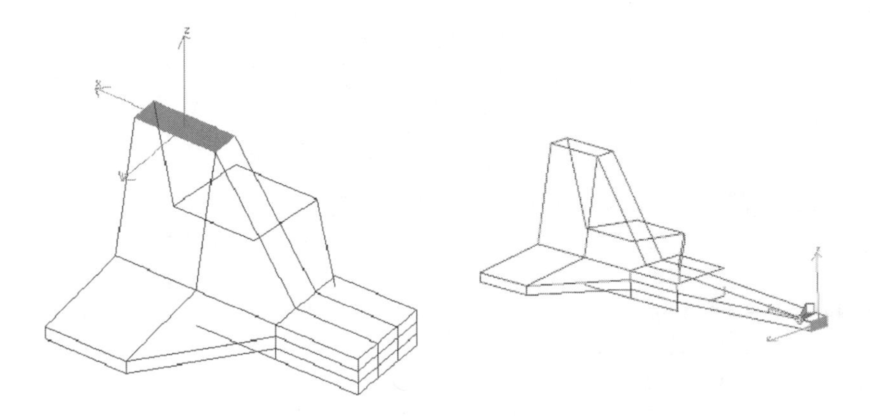

Figure 7.15 *Tail is pulled upward from the box.* **Figure 7.16** *Nose is transformed from the box.*

STEP 5

Transform the nose. Switch to the Left Side view. Select the middle polygon. Switch to the Front view. Using a measuring plane, pull the nose out along the negative X–axis (Figure 7.16).

Note: When polygon sub–objects were manually transformed, attached edges were pulled along by common vertices. If polygons are extruded, only the polygon is transformed, and not the connected edges.

STEP 6

Extrude the canopy. In the Top view, select the top polygon of the extended nose. Switch to **Local Axis** space. In the **Extrude** roll–out, enter a positive amount equal to the height of the canopy (Figure 7.17).

Note: Multiple extrusions will add elemental subdivisions, something that may be helpful when transforming details.

STEP 7

Reshape the canopy. Switch to **Sub Object|Vertex** and in the Front View, select the two vertices on the right side of the canopy with the **Select and Move** tool and the **Region** option. Reposition these vertices and those on the left to the position shown in Figure 7.18.

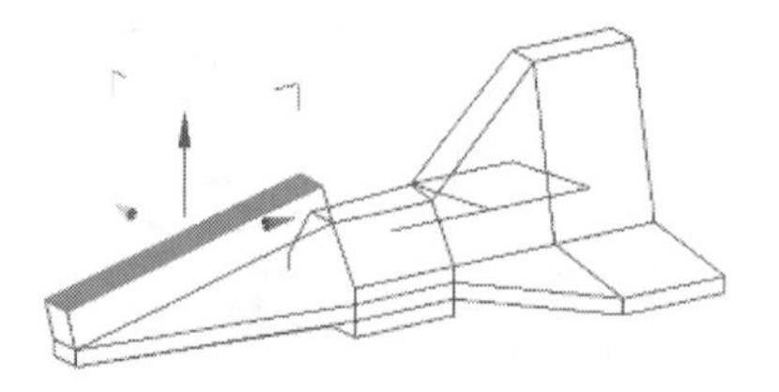

Figure 7.17 *Canopy is extruded.*

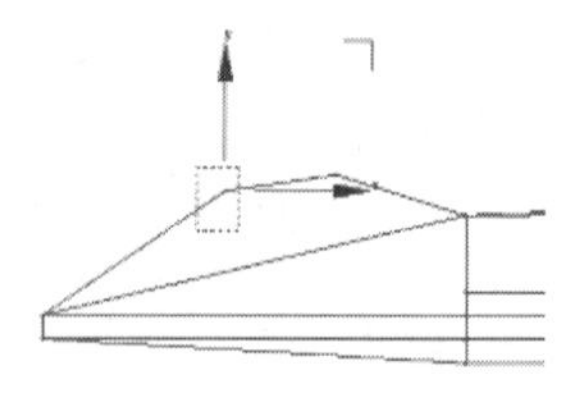

Figure 7.18 *Canopy is transformed.*

Change to **Smooth+Highlights** and rotate the view into a pictorial User view. You can see that the basic shape of the re–entry vehicle is progressing nicely (Figure 7.19).

Figure 7.19 *Re–entry Vehicle in basic form.*

STEP 8

Smooth the mesh. Select the object and choose **More** from the **Modify Panel**. Choose **Mesh Smooth** from the displayed list. In the roll-out enter **2** in the **Iterations** field. 3D Studio increases the tessellation of the surface, smoothing sharp intersections (Figure 7.20).

Tip: The **Bend Gizmo** deflects geometry *along* the selected axis. The position of the gizmo *about* the axis determines the direction of the bend. Rotate the gizmo in 90–degree increments about the Front view's X–axis to select the correct orientation.

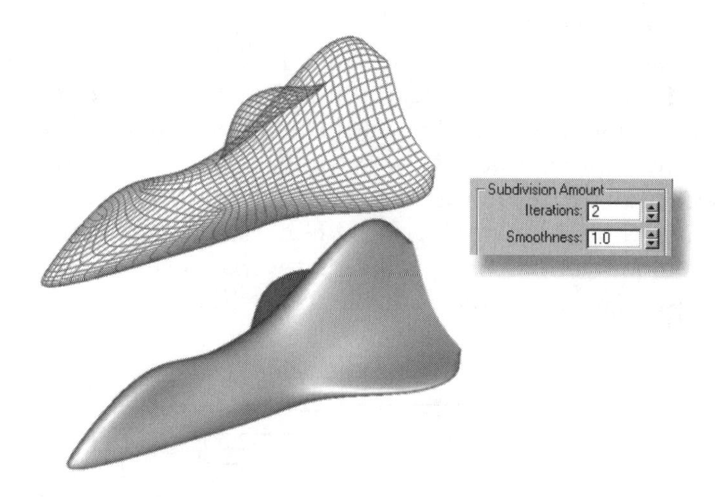

Figure 7.20 *Effect of **Mesh Smooth** on the transformed **Box** primitive.*

STEP 9

Apply a **Bend** modifier. In the Front view, select the object and choose **More** from the modifier buttons. Choose **Bend** from the displayed list. In the **BEND** roll–out, choose **Sub Object|Gizmo**, enter **30** in the **Angle** field, and check **Bend Axis: X Axis**.

The **Gizmo** may not be in the correct position to transfer the angle to the selected axis. In Figure 7.21 the **Gizmo** was rotated 90 degrees about the X–axis and the bend value changed to **–30**. This produces the desired effect. The completed re–entry vehicle is shown in Figure 7.22.

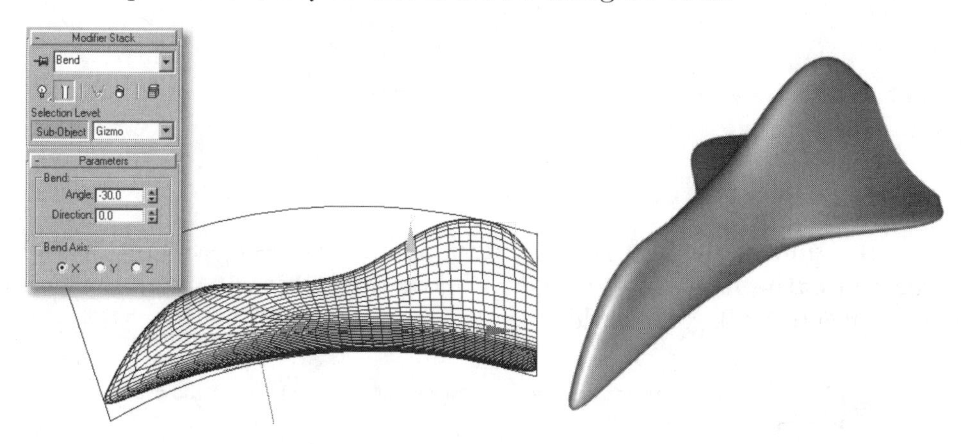

Figure 7.21 *Application of **Bend** modifier.*

Figure 7.22 *Completed re–entry vehicle.*

EXAMPLE: AUTOMOTIVE WHEEL

An effective example of modeling by using modifiers is the automotive wheel shown in Figure 7.23. A geometric analysis of the wheel immediately reveals that the rim is a logical lathed profile. The hub is not symmetrical about an axis so a lathe approach would not be productive. The star-shape with holes might suggest an extrusion with attached holes but the sectional view shows counter–bored holes, impossible with an extrusion. From our exploration of sub–object modifications, it appears that were a cylinder transformed and modified just right, the interior of the wheel (the *hub*) could be accurately represented.

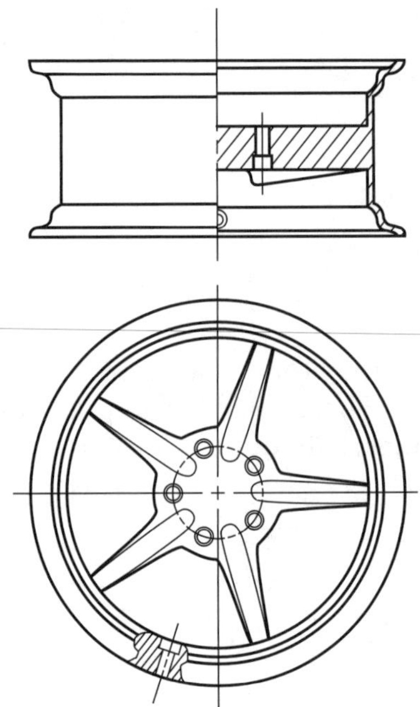

Figure 7.23 *Automotive wheel for modeling with transforms and modifiers.*

STEP 1

Create the rim. In the Top view create a spline shape describing the rim profile and **Lathe** it about the Y-axis. Choose **Sub Object|Axis** and reposition the axis to Y=0, X=0 world.

Note: Use a region selection to assure that all vertices along the cylinder's depth are selected.

STEP 2

Construct the hub as a cylinder. In the Front view, construct a cylinder at 0, 0, 0 world the diameter and height of the hub and specify a number of **Cap Segments** and **Sides** divisible by 5 (try 10 and 35). This assigns the same number of vertices to each of the spokes and gives you geometry to work with to raise the spoke's rib. The rim and base hub are shown in Figure 7.24.

STEP 3

Create circle templates. In the Front view, create circle templates representing the width of the horizontal spoke at both the hub and the rim. Relocate the **Pivot** of each to 0, 0, 0 world and create arrays of five copies over 360 degrees (Figure 7.25). Group these ten circles and move them straight forward, in front of the wheel geometry.

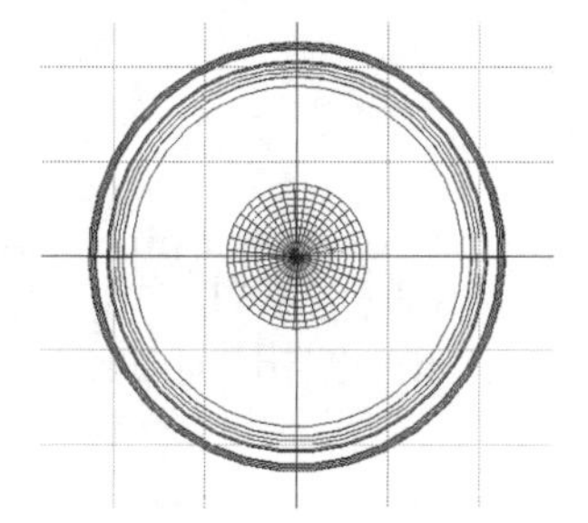

Figure 7.24 *Base rim and cylinder.*

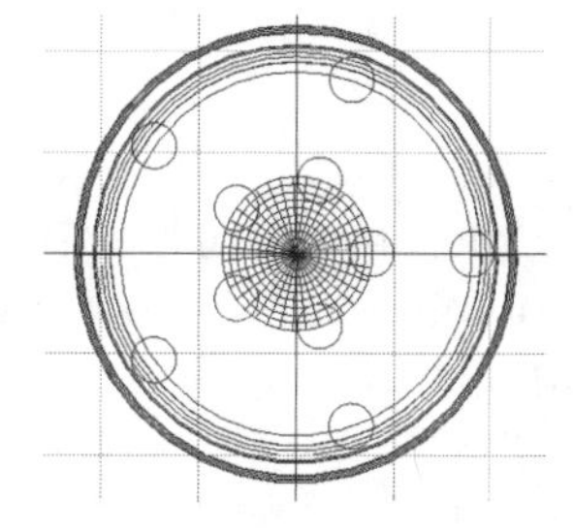

Figure 7.25 *Circle templates for spokes.*

Note: When you modify the vertices contained within the circle templates a natural fillet will be formed between the hub and the spoke as the vertices and connected edges are moved.

STEP 4

Select and move spoke vertices. **Lock** the circle templates. Select the *hub* and choose **Modify|Edit Mesh|Sub Object|Vertex** and with the **Select and Move** tool select all vertices within the hub circle template. Move the vertices using the X-axis vector to the rim circle template (Figure 7.26). The spoke is extended slightly into the *rim*, assuring an intersection.

STEP 5

Create the remaining spokes. In turn, select and move identical spoke vertices to their corresponding rim templates (Figure 4.27).

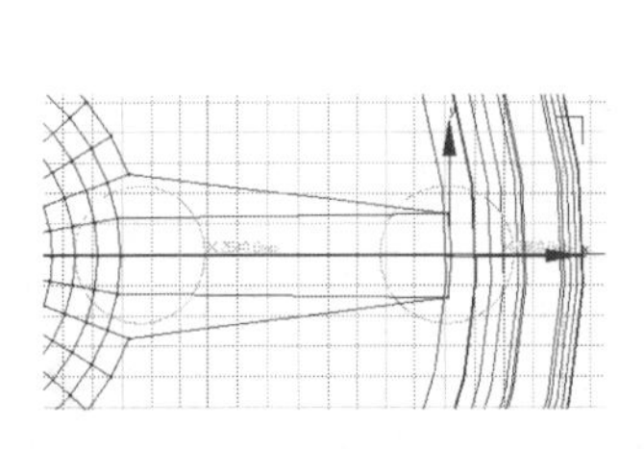

Figure 7.26 *First spoke is created.*

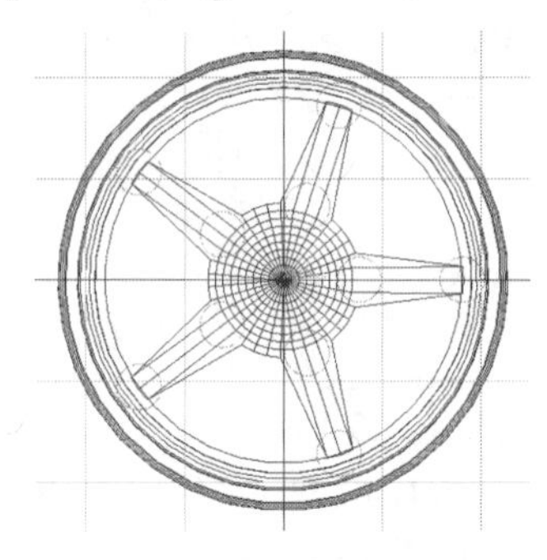

Figure 7.27 *Spokes correctly modified.*

STEP 6

Establish the measuring plane. In the Front view, create a circle at 0, 0, 0 world large enough to cover roughly half of the spokes. In the Top view, move this circle (appearing as an edge), to a position signifying the forward position of each spoke's rib (Figure 7.28).

STEP 7

Pull the rib vertices. Work in the Front view. Select the *hub*. Choose **Modify|Edit Mesh|Sub Object|Vertex** and select the same vertices for each rib (Figure 7.29). Use a single click because you want to select only the vertices on the front side of each spoke.

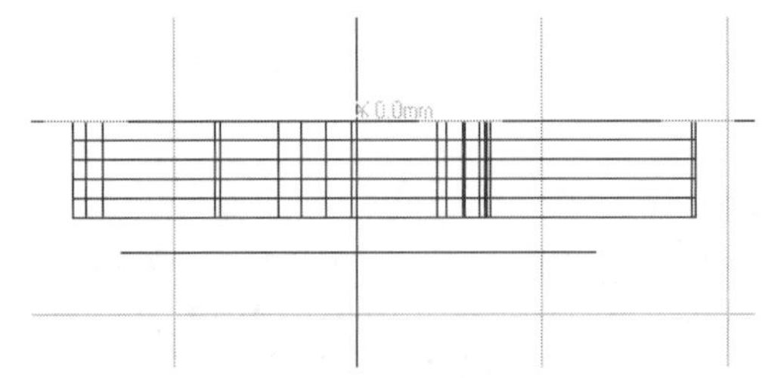

Figure 7.28 *Circle used as measuring plane.*

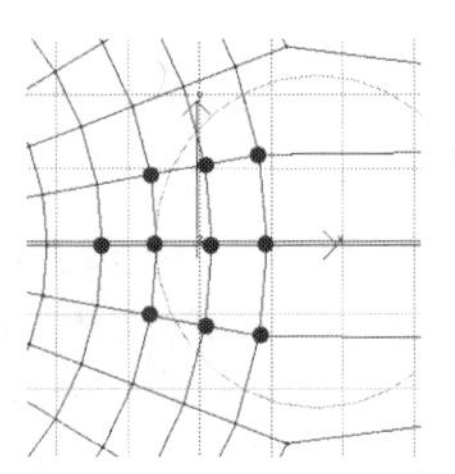

Figure 7.29 *Vertices selected for rib.*

When all vertices have been selected (use the **CTL** key to add to selection), press the **Space Bar** to lock the selection and switch to the Top view. Using the Y–axis vector, move the vertices forward until they are aligned with the measuring plane (Figure 7.30).

STEP 8

Smooth the mesh. The sharp polygonal edges of the *hub* need to be smoothed, forming filleted and rounded edges. Select the *hub* and choose **Modify|More** and select **Mesh Smooth** from the list of available modifiers. In the **Mesh Smooth** roll-out, select **NURMS|Apply to Whole Mesh** and enter **3** in the **Iterations** field and **1** in the **Smoothness** field. Figure 7.31 shows the difference between the unsmoothed (a) and smoothed (b) hubs.

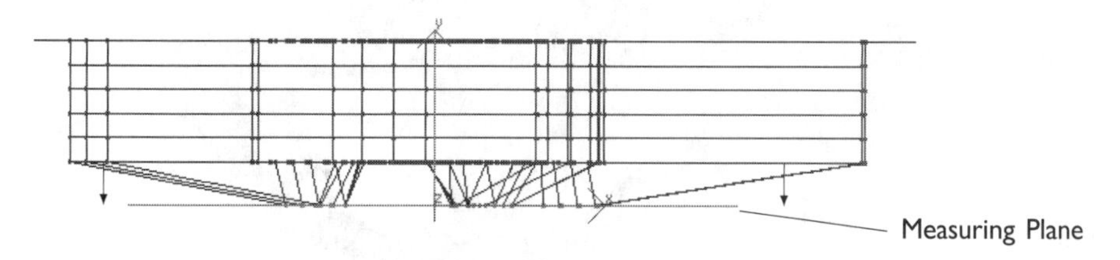

Figure 7.30 *Vertices moved to measuring plane.*

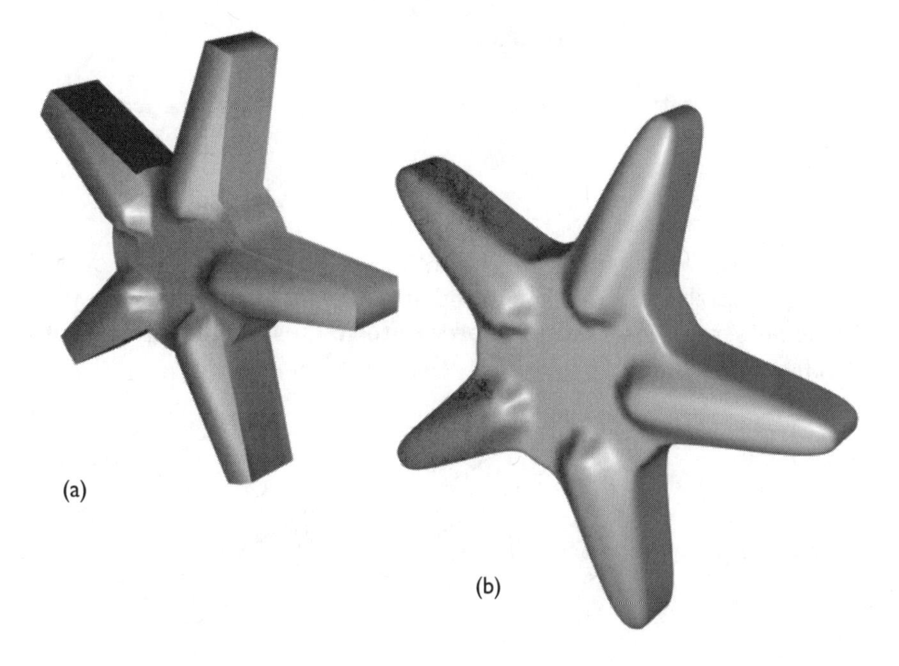

Figure 7.31 ***Mesh Smooth*** *is used to form fillets.*

When combined with the *rim*, and after counter bored lug and valve stem holes have been Booleaned away, the finished automotive wheel is a realistic and proportionally correct rendition (Figure 7.32).

Figure 7.32 *Completed automotive wheel.*

PROBLEMS

Problems appropriate for solving using modifiers modeling are found on the following pages. Use the file *isogrid* found in the *tools* directory to plan your modification strategies. Assign appropriate names to all objects both on your sketches and models.

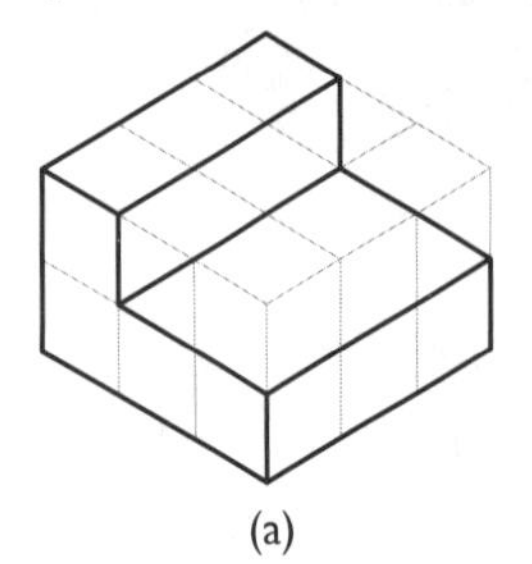

(a)

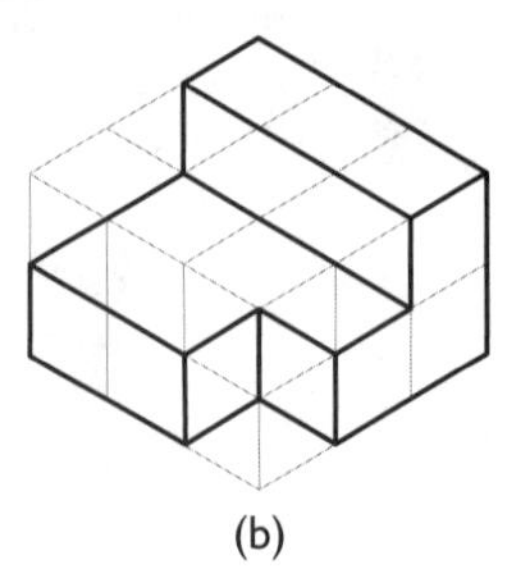

(b)

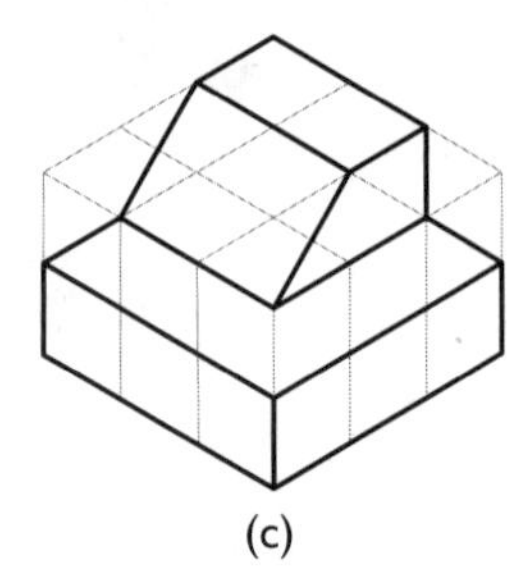

(c)

Problem 7.1 *Start with a 3 x 3 x 2* **Box** *primitive and use sub–object modifiers to arrive at the final geometry shown.*

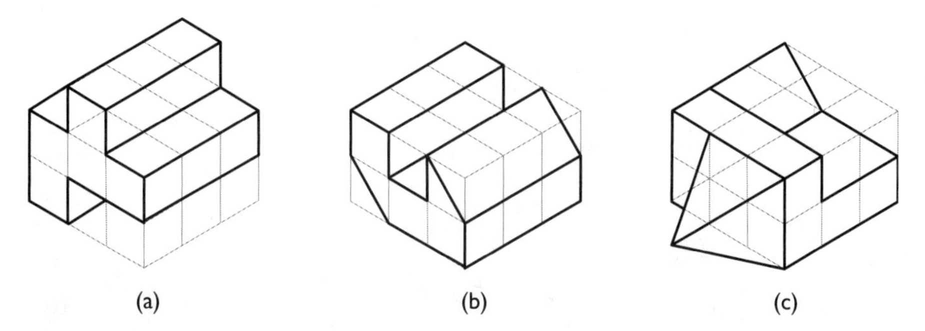

Problem 7.2 *Start with a 3 x 3 x 2* **Box** *primitive and use sub–object modifiers to arrive at the final geometry shown.*

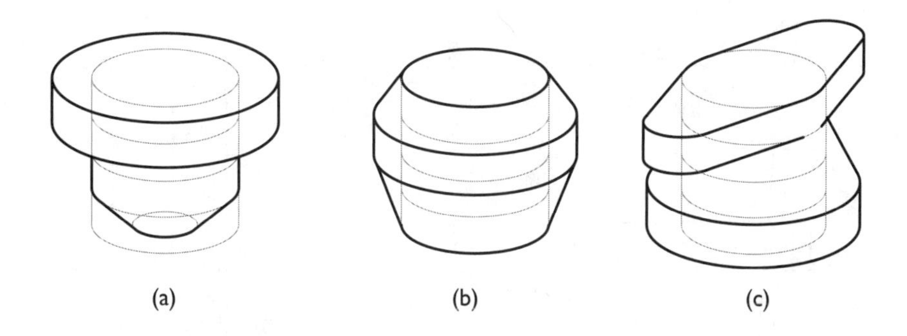

Problem 7.3 *Start with a four-segment, eighteen-sided* **Cylinder** *primitive and use sub–object modifiers to arrive at the final geometry shown.*

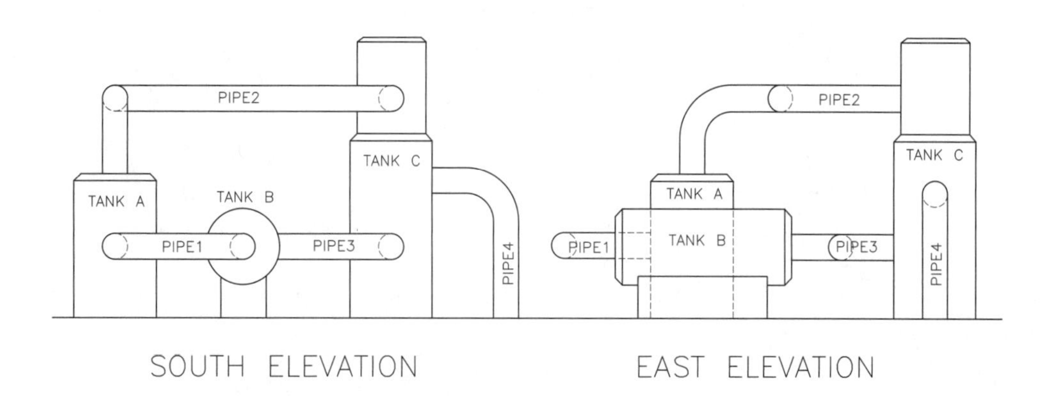

Problem 7.4 *Storage Tanks. Using* **Cylinder** *primitives and modifiers, model the piping layout above. A single lofted pipe can be cloned and modified for all positions.*

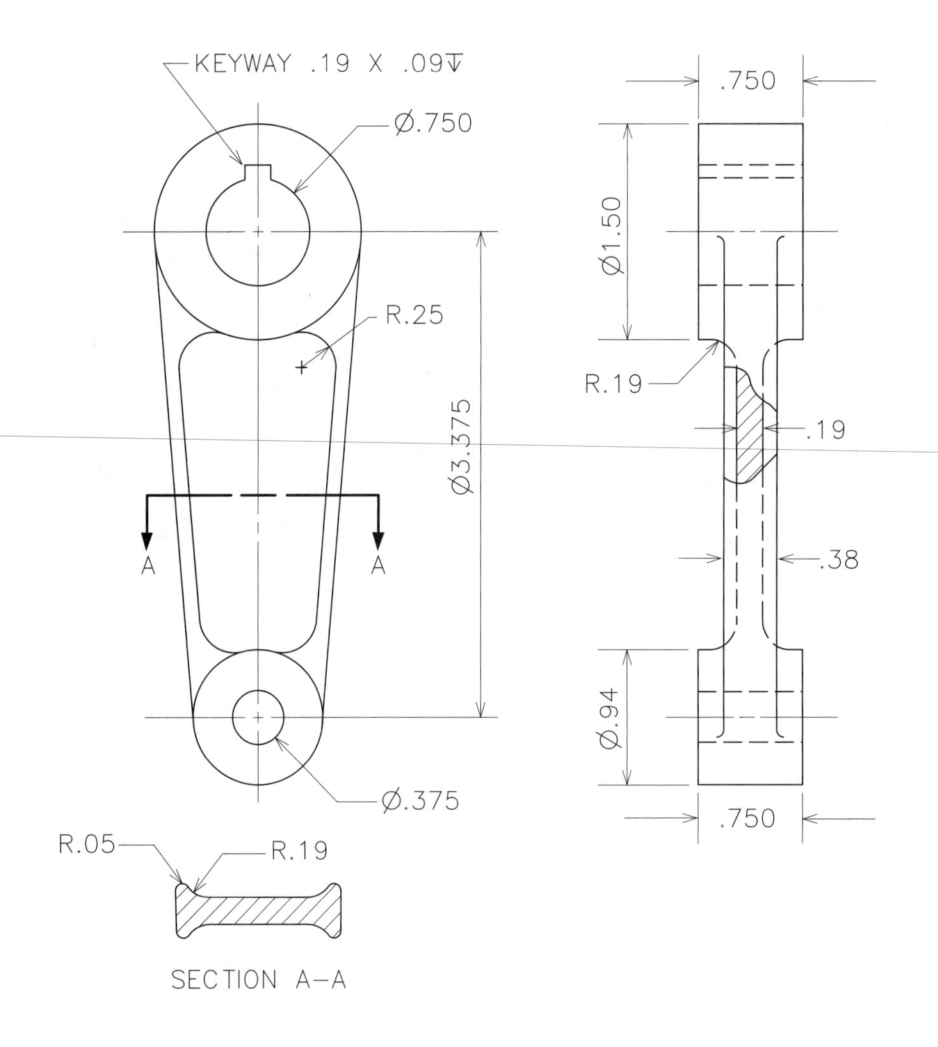

Problem 7.5 *Connecting Link. Begin this problem by modeling the rough shape with primitives.* ***Boolean Union*** *and collapse the stack to an* ***Editable Mesh****. Complete the fine detail making use of modifiers.*

Modeling Assemblies

CHAPTER OVERVIEW

Most industrial products are comprised of groups of parts in an assembly. How these parts are constructed and positioned for fit is the subject of this chapter. In subsequent chapters you will learn how to increase the realism in the appearance of parts in assemblies and animate how the parts are brought together to make a final product.

You now have experience with many of the modeling techniques available in 3D Studio: primitives, extrusions, lathes, NURBS, and modifiers. You will find the need to employ more than one of these modeling techniques in an assembly—that's why assemblies are so interesting graphically. Assemblies also have parts with various materials and that's what makes them interesting visually.

Some assemblies have only a few parts while others have tens, even hundreds, or thousands of parts. You can see the need to plan object names that both make sense and are unique. This is especially true when parts are merged together.

KEY COMMANDS AND TERMS

- **Assembly**—a group of individual parts that collectively comprise a product shown in functioning position.
- **Assembly Axis**—the axis along which a part in an assembly will be moved during its installation.
- **Assembly Drawing**—a 2D representation of a product that identifies individual parts and their position.

- **Bump Map**—a technique whereby fine detail (threads, knurls, lettering) can be engraved or embossed on a surface at rendering time by evaluating the relative brightness of a raster map applied to the surface.
- **Central Part**—the part in an assembly that controls the placement of other components; the central part is modeled and positioned first.
- **Detail Drawing**—a 2D representation of an individual part in an assembly with the information necessary for its manufacture.
- **Exploded Assembly**—a group of individual parts that collectively comprise a product shown in removed spatial position.
- **Import/Insert**—command that brings in non-3D Studio part geometry.
- **Merge**—the command that brings objects (geometry, lights, cameras) from previously saved 3D Studio files into the current scene.

IMPORTANCE OF ASSEMBLIES

Assemblies consist of stationary and moving parts, custom parts, standard parts (like fasteners), packaging (like covers), electrical parts, mechanical parts, hydraulic parts, pneumatic parts, hoses, lenses, switches, and controls. The inclusion of these parts, accurately modeled and then material mapped, is key in creating effective computer mock-ups, simulations, and presentations.

Assemblies can be shown with components in either unassembled or assembled positions. We call the first an *exploded assembly* and the second an *assembly*. When assembled, many of the interior parts will be hidden. When exploded, the position of the parts is critical in understanding spatial relationships (Figure 8.1).

Assemblies also make use of parts from other assemblies. It's how companies can leverage design and development costs on one product to include many others. For this reason it's important to model in *real world units*. If you model at full scale you are assured that components of one assembly will easily fit in another.

Tip: It is common that details of parts making up an assembly may be drawn at different scales when the CAD geometry is intended for traditional engineering drawings. Small parts need to be detailed at an enlarged scale; large parts need to detailed at a reduced scale. It may be convenient to put all parts into the same scale *before* merging them into a single 3D Studio file.

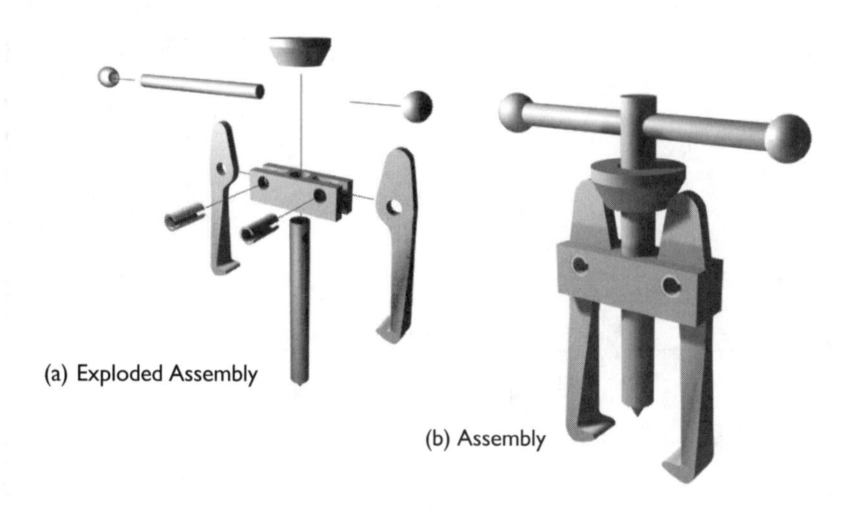

Figure 8.1 *An exploded assembly and an assembly.*

A SIMPLE ASSEMBLY

Two types of engineering documents are used in modeling an assembly: *assembly drawings* and *detail drawings*. Figure 8.2 shows an assembly drawing for a small Electric Motor Assembly. The parts in the assembly are identified by balloons and numbers and recorded in a parts list. This drawing demonstrates how the parts are assembled. Figure 8.3 shows detail drawings for the parts necessary to complete the assembly. Note that the mounting screws and pinion gear are not detailed; they are standard parts. The motor is detailed because the design of the bracket depends on it.

STEP 1

Establish the assembly axis and datum. In the case of the Motor Assembly, the assembly axis is parallel to the world X–axis in the Front view. The datum is shown in the assembly drawing and establishes a reference for the placement of the parts; 0, 0, 0 world is marked by a black balloon.

STEP 2

Determine the central feature. It is evident that the Mounting Screws are not central in this assembly. Likewise, the Electric Motor is mounted to the bracket. We can decide then that the Mounting Bracket is the central feature and begin with that. The Mounting Bracket is created in a position that matches its relation to 0, 0, 0 world.

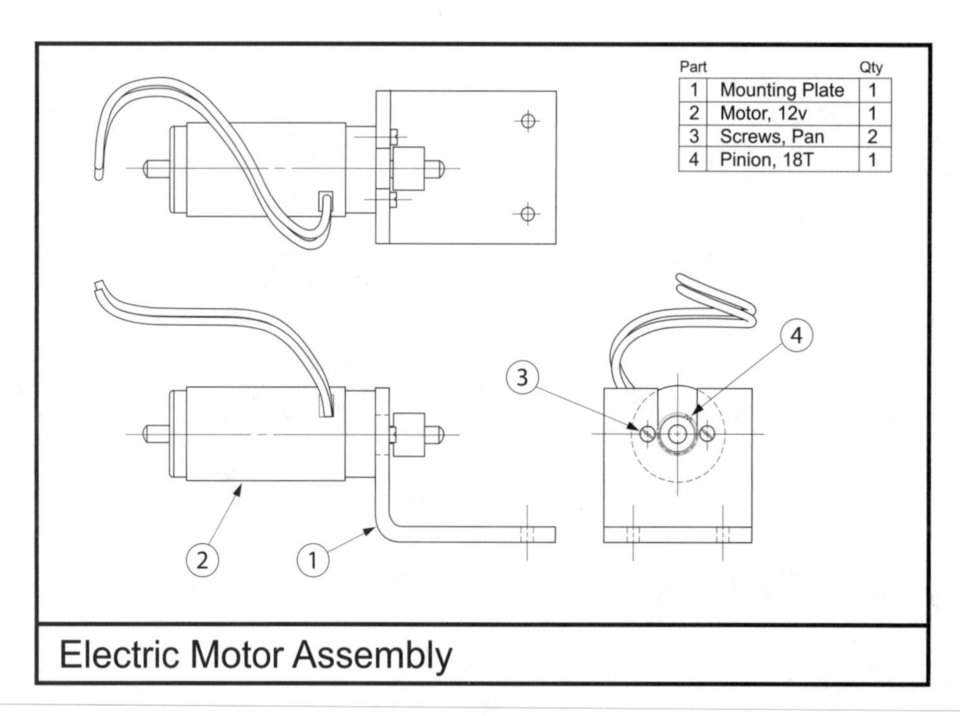

Figure 8.2 *Assembly drawing of Electric Motor Assembly.*

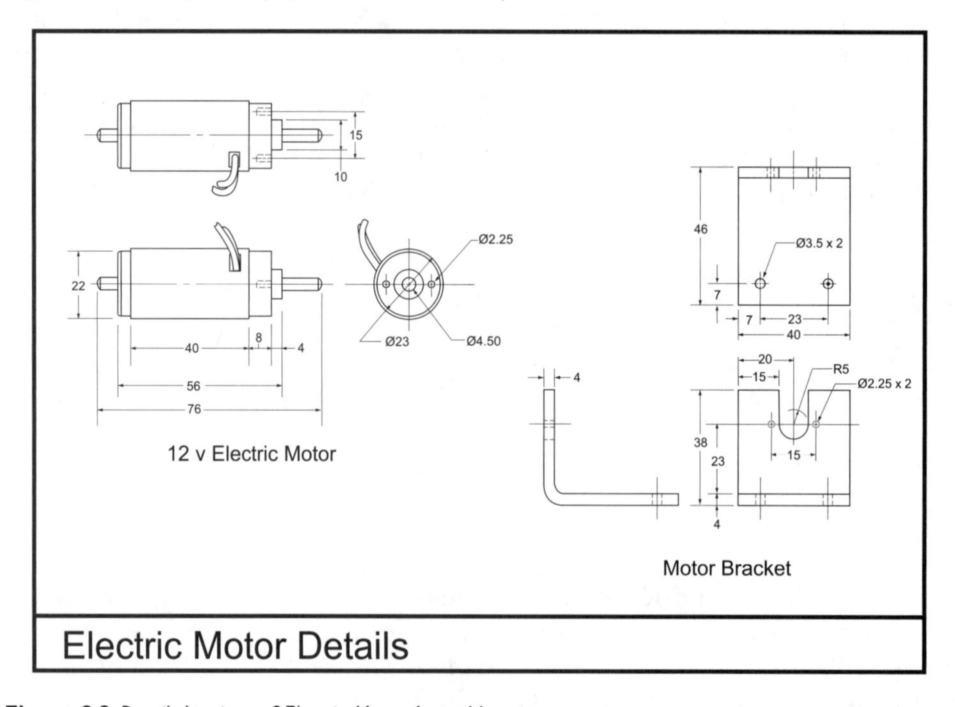

Figure 8.3 *Detail drawings of Electric Motor Assembly parts.*

A geometric analysis reveals that extrusion and then Boolean subtraction is an efficient modeling strategy. Begin with a profile of the bracket and extrude this profile to the bracket's depth. Create drill tools for the holes and a slot tool for the boss on the gear end of the motor. Figure 8.4 shows the bracket with tools in place and the Booleaned result.

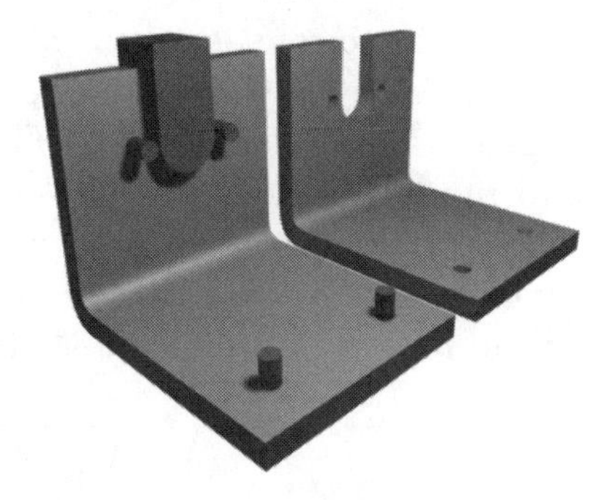

Figure 8.4 *Bracket before and after Boolean subtraction.*

STEP 3

Create the Electric Motor. Hide the bracket. Using the location of 0, 0, 0 world as a reference point, create the upper half of the motor profile as a spline shape. **Lathe** this shape about the **X Axis** with **32** steps. Relocate the **Pivot** to the right center of the lathe and the entire part to 0, 0, 0, world. If you have correctly aligned the parts, they should appear as in Figure 8.5.

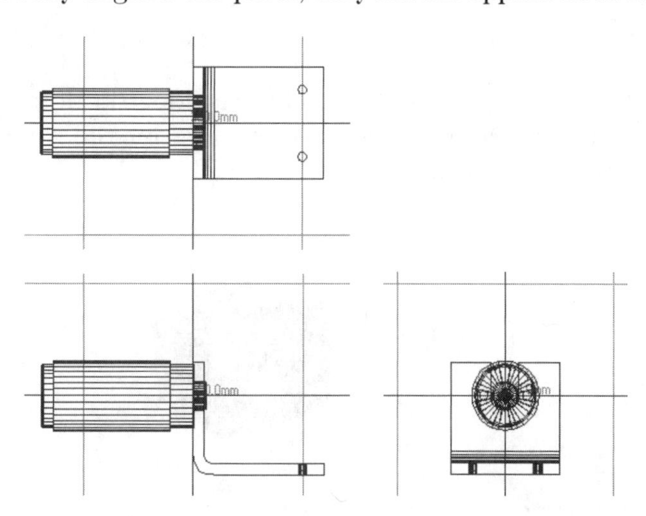

Figure 8.5 *Motor and bracket in position.*

STEP 4

Detail the Electric Motor. Begin by creating a **Standard Primitive|Oil Tank** to represent the Motor Shaft. Align the shaft with Z=0, Y=0 world. Make sure that the shaft extends sufficiently to hold the Pinion Gear.

Create the 18-tooth pinion Gear by extruding a gear profile (Figure 8.6). In the Right Side view create a spline circle equal to the outer tooth diameter. Create a gear root profile and array this around the circle eighteen times (a). Create a circle (the gear's hole) to insert the Gear on the Shaft. Attach the root profiles and mounting hole to the Gear circle and perform Boolean subtractions of each profile until you arrive at a complete gear shape (b). Leave the hole for extrusion. Extrude this profile a distance equal to the thickness of the gear and move out along the shaft to position (c).

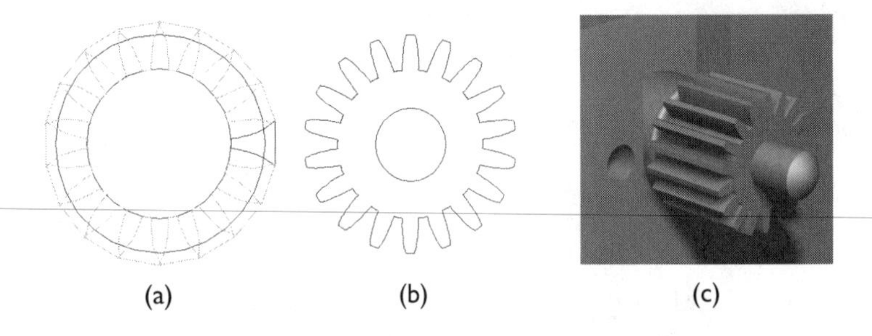

Figure 8.6 *Pinion Gear development.*

STEP 5

Create the commutator opening. Hide all but the Electric Motor. In the Front View create a cutting tool to provide the opening for the wires. Select the motor and subtract the cutting tool (Figure 8.7).

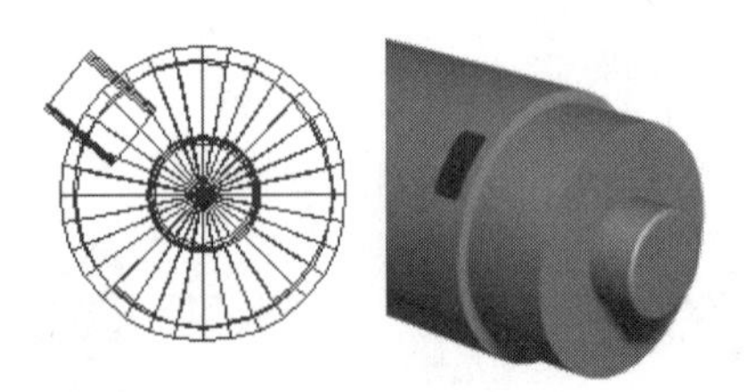

Figure 8.7 *Opening in motor case.*

Note: Because the case was modeled with a single spline it has no actual thickness. Were it important to show this motor case thickness you might first entertain adding a thin surface to the opening. This way you wouldn't have to go back and create a new double sided motor case profile.

STEP 6

Create the wires. Working between the Front, Top, and Side views, create a spline path representing the centerline of one wire. Give the wire a natural twist as it comes out of the commutator opening. Create a circle the diameter of the wire. Choose **Create|Compound Objects|Loft** and click **Get Shape**. A lofted wire is formed. Enter **3** in **Shape Steps**, **20** in **Path Steps**, and turn on smoothing options. Move this first wire to the bottom part of the opening.

Create a clone copy of the wire and move it slightly above the first wire, making sure that it emerges from the opening in the motor. By making a copy of the first wire they appear naturally in tandem (Figure 8.8).

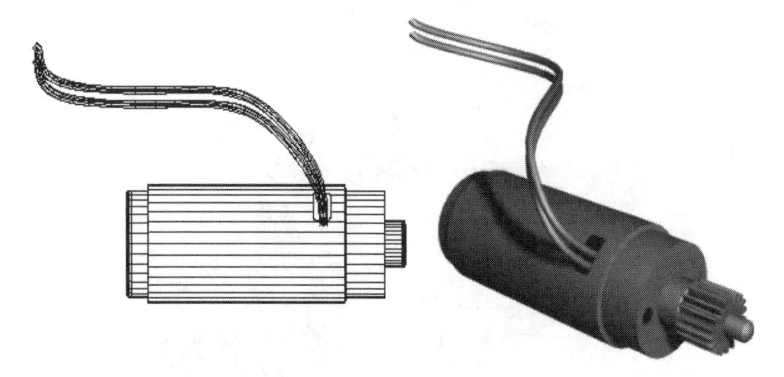

Figure 8.8 *Motor with wires.*

STEP 7

Create fasteners. Ideally, you should have access to a library of standard fasteners, but let's assume you don't. You need a pan head self-tapping screw. You won't actually model the threads. Quite often the view and scale will make showing the threads a moot point: you can't see them anyway. If you do need threads, you'll *bump map* rather than model them (see Chapter 12, Modeling With Bump Maps).

First, create the mounting holes in the motor end by placing a drill tool through the mounting holes in the bracket (acting as a jig). Subtract these tools from the motor (Figure 8.9).

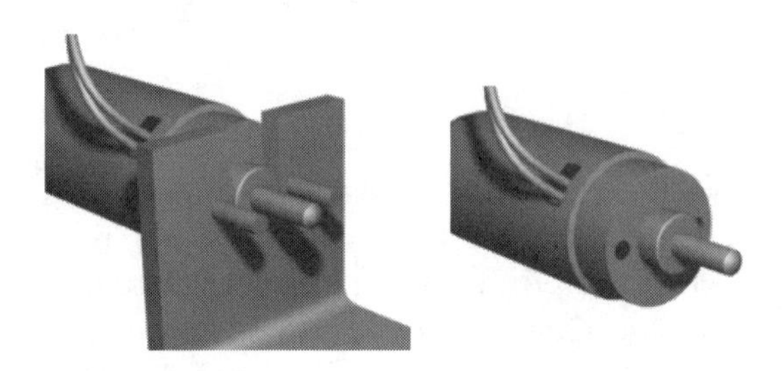

Figure 8.9 *Drill tools use bracket as a jig.*

Create the self-tapping pan head screw as a lathed profile using the bracket as a guide. Lathe the profile and subtract a small box to create the slot. The final model is shown in Figure 8.10 as an assembly and an exploded assembly.

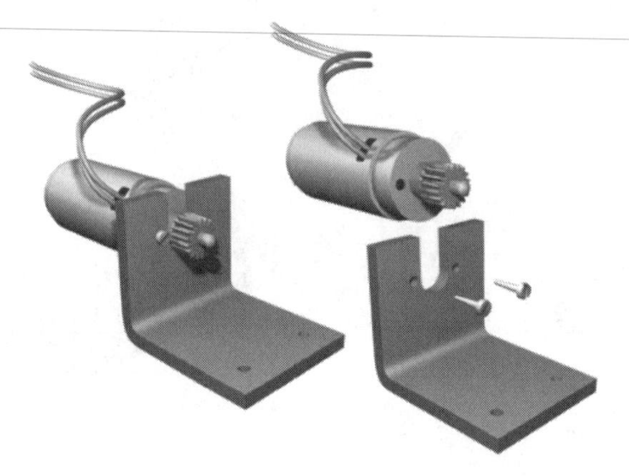

Figure 8.10 *Assembly and exploded assembly.*

ASSEMBLY GUIDELINES

- It is often more efficient to model parts in individual files and assemble them later by merging.
- Create all parts of an assembly at the same scale so that merging the parts is made easier.
- Establish a datum from which all measurements can be made. That way, when parts are merged, they are placed into world space at precisely the right spatial location.

- Because parts often mate in an assembly, it may be more efficient to make one part from another, assuring that sizes are identical.
- It is important to create instances of repeating parts so that when the design changes, all instances of the design change also.

USING DRAWING DATA

Assembles are often comprised of standard parts for which you may already have drawings or digital part files and as you would imagine, it's almost always more efficient to use existing geometry than create it from scratch. But of course you have to ascertain that the data is accurate and current.

- CAD data can be brought directly into 3D Studio through VIZ's **Insert** function and MAX's **Import** command. These operate essentially in the same manner. However, CAD data often contains more information that you want or need to build models in 3D Studio. You may need to edit CAD drawings down to what you need; otherwise, be prepared to do some massive editing in 3D Studio.
- CAD data may not be in a form usable in 3D Studio. This is especially true in the case of spline curves that may be represented as numerous chords or elements. These shape are not continuous and are problematic when making lathes or extrusions. In these cases it may be more effective to **Insert/Import** as a single object and use the CAD data as a template over which you create valid geometry.
- In the case of paper drawings, you have three options: use the information to create necessary geometry in a drawing application (CAD or illustration), scan the drawing and use as a template in an illustration application, or scan and use it as a texture map in 3D Studio, tracing over the image with 3D studio spline or NURBS line tools. This last technique is covered in Chapter 10, Raster Material Maps.

MERGING PARTS TOGETHER

You **Merge** geometry from a 3D Studio file into the currently open 3D Studio file. When the files are at different scales, you are given the option to adopt either file's scale. You can merge any object: geometry, lights, cameras, helper objects, and paths. Several provisions make merging geometry more effective.

- Large projects can be broken into subprojects, often done by different people at diverse locations, and brought together. Smaller part files open, save, zoom, pan, and render more rapidly.
- Objects in the merged files with identical names will trigger a warning. Objects in the same scene can have the same name but this really isn't a good idea. Determine unique part names so that later, when the files are merged, you can easily select, hide, and lock parts by name.

- Objects will be merged in world space. By establishing common datum planes objects can be merged with little or no repositioning. This is critical when determining fit and clearance.

Note: The portable .3ds file format gives you a way to move geometry back and forth from different versions of VIZ and MAX. However, some features unique to later versions cannot be passed to earlier versions.

The Gear Puller Assembly in Figure 8.11 is an excellent example of how establishing datum relationships can ease this process of merging objects from separate files into an aligned assembly. (These files are on the accompanying CD-ROM in *problems/chapter_8*.) By establishing 0, 0, 0, at the center of the assembly, getting the parts to align as they are merged is a snap.

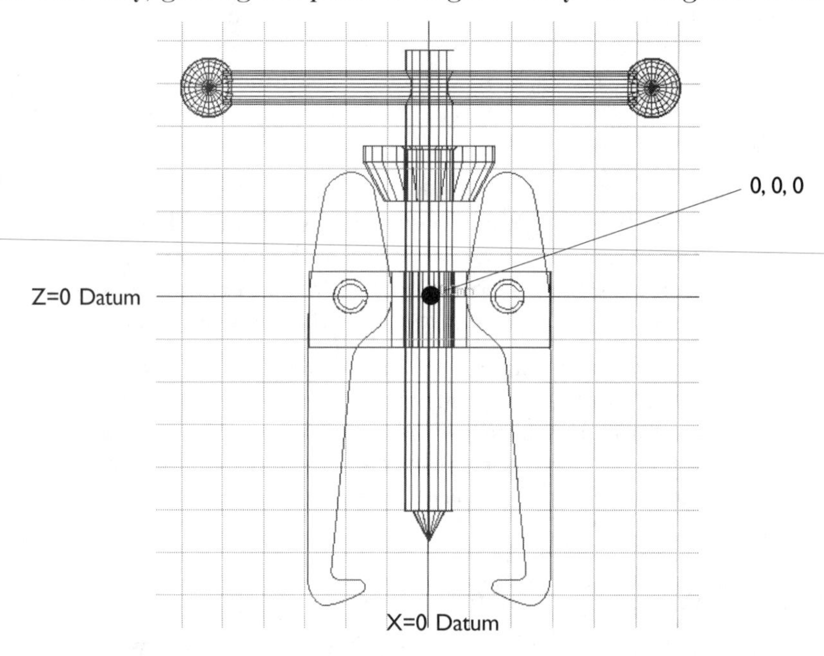

Figure 8.11 *Assembly aligned to datum planes.*

STEP 1

Determine the central part. The part identified as the Yoke controls the alignment of all other parts. Open the file *yoke.max* from *chapters\chapter_8*. This becomes the base file into which the other parts will be merged. Display the Front view as a single viewport. Note that the Yoke is aligned to the datum planes (Figure 8.12).

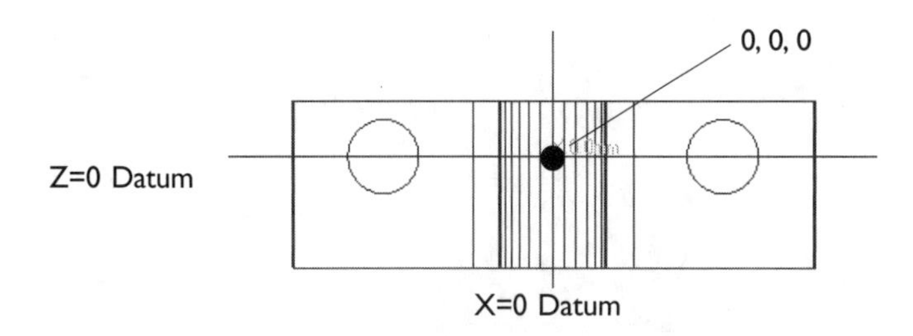

Figure 8.12 *Yoke aligned to datum planes.*

STEP 2

Merge the arms. Choose **File|Merge** and browse to the CD-ROM and select arm.max. There are two objects in this file: *arm_l* and *arm_r*, the left and right arms. Select and click **OK**. The arms are merged with the current scene at the place in space where they were constructed and align perfectly with the Yoke (Figure 8.13.)

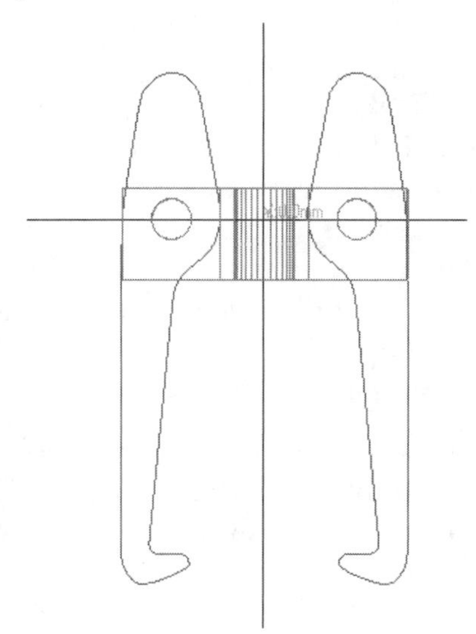

Figure 8.13 *Arms aligned with Yoke.*

STEP 3

Continue merging parts. In order, merge objects from the following files: *spring_pin.max*, *spread_nut.max*, *screw.max*, and *handle.max*. When the parts are all merged, the Gear Puller Assembly is complete (Figure 8.14).

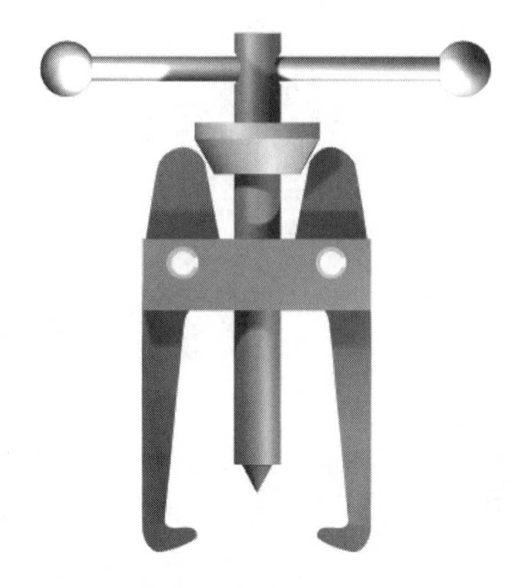

Figure 8.14 *All parts of Gear Puller Assembly merged.*

EXAMPLE: FAN ASSEMBLY

Figure 8.15 displays yet another type of assembly document, a *sectional assembly drawing* of a Fan Assembly. Parts with interior detail are shown sectioned; parts with no interior detail (nuts, bolts, shafts, bearings) are shown full and unsectioned. The parts are identified and specified by notes. The sectional assembly is drawn at full scale so measurements can be taken directly from the drawing. The surface to which the pulley is mounted is not specified and will not be modeled in this assignment.

This Fan Assembly is an excellent candidate for modeling in 3D Studio. Most of the parts are profiles suitable for lathe modeling. The assembly contains a number of standard parts that may exist in libraries or if not, can be modeled once and cloned as instances.

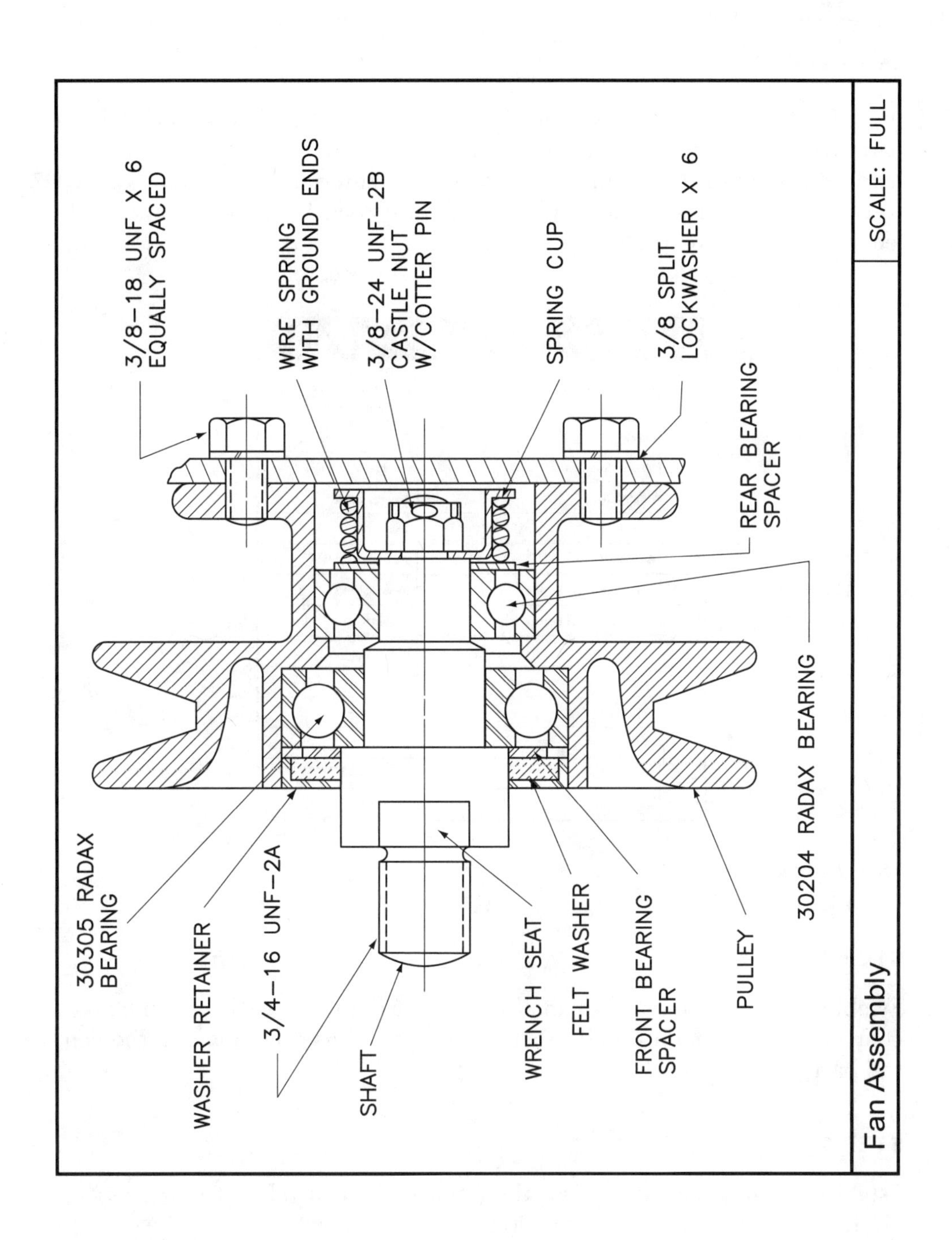

Figure 8.15 *Sectional assembly drawing of a Fan Assembly (Luzadder and Duff:* Fundamentals of Engineering Drawing*).*

STEP 1

Identify all parts in the assembly. The sectional assembly does not have a parts list as did the Electric Motor Assembly in Figure 8.2. Table 8.1 is a list of all parts in the assembly. Additionally, materials have been assigned to each of the parts so that provisional colors can be assigned to the objects as they are created in 3D Studio.

Fan Assembly Parts List		
Qty	Part Description	Material
1	Shaft	Machined Steel
1	Pulley	Cast Steel
1	Washer Retainer	Anodized Aluminum
1	Felt Washer	Absorbant Felt
1	Front Bearing Spacer	Brass
1	Rear Bearing Spacer	Brass
1	30305 Radax Ball Bearing	Treated Steel
1	30204 Radax Ball Bearing	Treated Steel
1	Wire Spring	Steel
1	Spring Cup	Pressed Steel
1	Castle Nut	Nickel Plate
1	Cotter Pin	Steel
6	Hex Cap Screws	Nickel Plate
6	Spring Lockwashers	Nickel Plate

Table 8.1 *Parts in Fan Assembly.*

STEP 2

Establish the central part. An analysis of the assembly reveals that the Shaft controls the placement of all other parts. We will use the Shaft as the central part.

STEP 3

Establish datum planes. Because the Shaft is the central part in the assembly, its centerline will lie at Z=0 along the X world axis. From an inspection of Figure 8.15, the Front Bearing establishes both the position of the Pulley relative to the Shaft. The position of the assembly in relation to 0, 0, 0 world is then established (Figure 8.16).

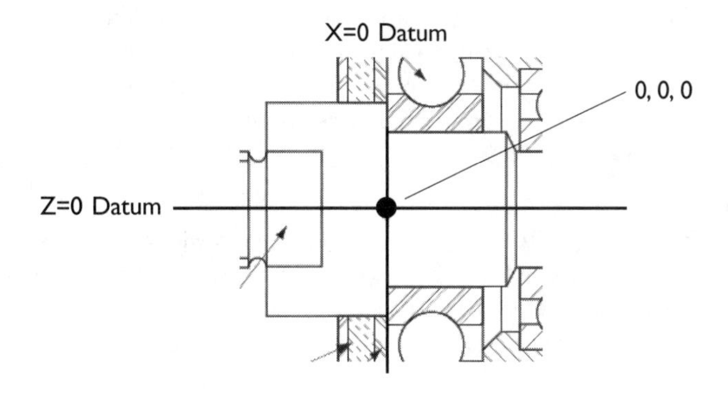

Figure 8.16 *Datum position of Fan Assembly.*

STEP 4

Create the profiles. Because the Fan Assembly exists as a paper engineering drawing, it is scanned and placed into a drawing application such as CorelDRAW or Adobe Illustrator. There, after resizing to assure full scale, it serves as a template for creating precision profiles using inch units (Figure 8.17). Notice that not all geometry becomes a profile. Do not create profiles for the spring, castle nut, cap screws, or washers. These will be created individually. Create only that portion of each shape on contact with the Z-X world plane.

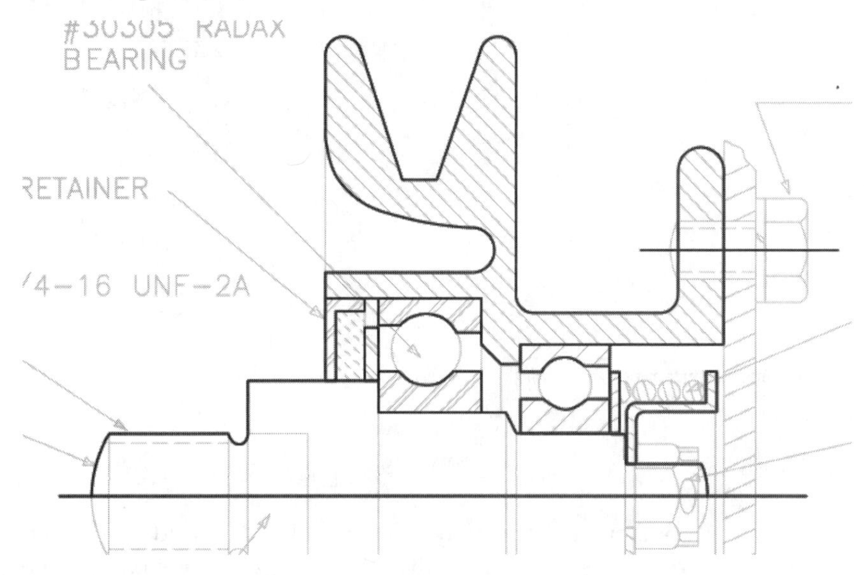

Figure 8.17 *Profiles are generated from the scan.*

Tip: When using a scanned engineering drawing as a template, make use of the drawing application's snaps and guides to control spline placement. Use the fewest spline control points necessary as 3D Studio will use those control points as vertices. Assure that profiles are single contiguous curves so they can be lathed and extruded in 3D Studio.

Tip: It is helpful to bring centerline geometry in with the profiles. In the Fan Assembly, the $^{3}/_{8}$ Cap Screw centerline will establish the position of drill tools, for creating the holes in the Pulley, and the Cap Screws that will eventually be arrayed to fit the holes.

STEP 5

Import/Insert the profiles. In 3D studio, set units to inches and choose **File|Import** in MAX or **Insert|Other** in VIZ. Choose the correct import filter and place the profiles as separate objects in 3D Studio's Top view. Move these profiles to the datum origin (Figure 8.18).

Note: It is important to merge multiple part profiles as separate objects so they can be used individually for lathes and extrusions.

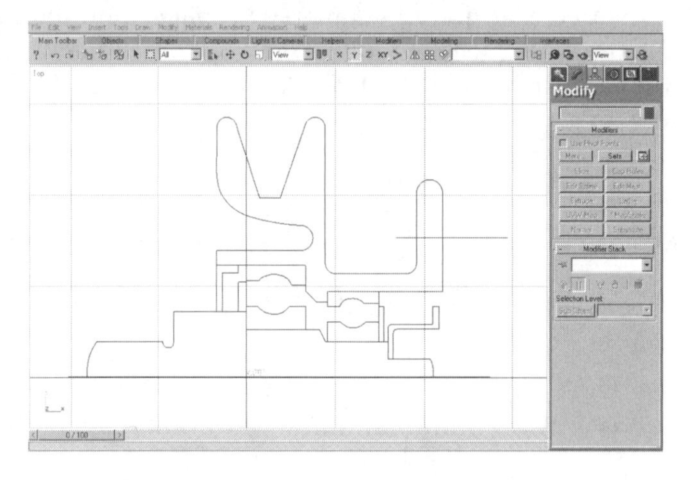

Figure 8.18 *Profiles inserted into 3D Studio.*

STEP 6

Lathe the shapes. Begin with the parts list in Table 8.1 and work your way down the list. Figure 8.19 shows the result of lathing the first part, the Shaft.

After lathing each part and relocating each **Axis** (**Pivot**) to X=0, Z=0 world, name the part and hide it. When you have completed the parts, save the file as *fan_assy.max*.

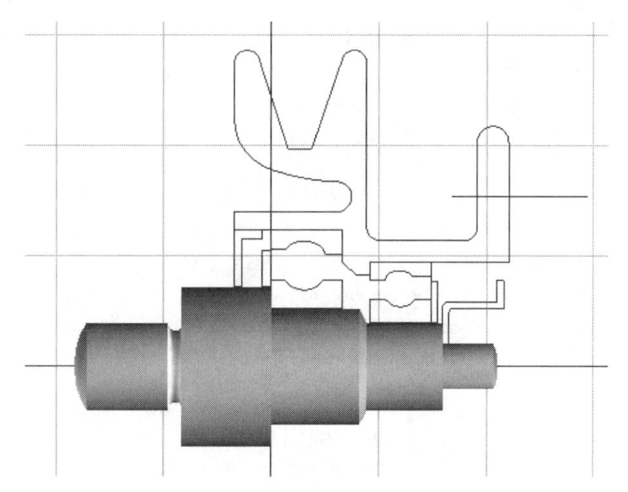

Figure 8.19 *Shaft profile lathed and AXIS moved to Z=0 world.*

Tip: Always keep a version of the assembly with all parts in correct alignment. Once moved, it is laborious to accurately align them again.

Note: The bearings are comprised of outer and inner races. These are lathed. The balls in the bearings are created later as primitive spheres and arrayed.

Save the file again as *explode_fan_assy.max*. In this file, pull the parts along the X world axis and you can begin to see how the assembly is aligned (Figure 8.20).

Figure 8.20 *Fan Assembly with parts exploded.*

STEP 7

Detail the Shaft. Work in *fan_assy.max* and display the Shaft in the Front view. Create a **Box** tool that will cut the wrench seat from each side. (You can measure the width of the seat from the sectional assembly and then in the shaft's Left Side view, determine how far into the Shaft the tool must be placed to arrive at the desired height.)

Create the cotter pin hole by positioning a .375 diameter drill tool .125" from the right end of the shaft. The finished Shaft is shown in Figure 8.21.

Tip: The Boolean engine in VIZ is sensitive to the tessellation of the surface being acted upon. The engine in MAX is less so. You may want to choose **Modify|Edit Mesh|More|Tessellation** and increase the complexity of the surface if Boolean results are incorrect.

Note: The threads on each end of the Shaft are not modeled. They are *bump mapped.* This keeps the complexity of the Shaft's geometry to a minimum and allows changes in the material and surface without having to model the Shaft again. Bump mapping is covered in Chapter 12, Modeling with Bump Maps.

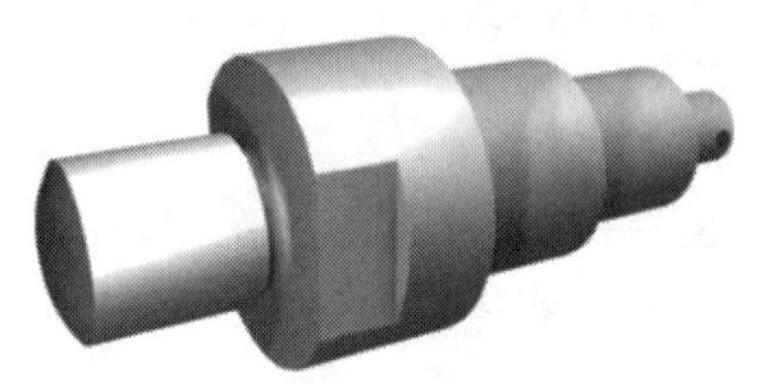

Figure 8.21 *Finished Shaft with wrench seats.*

STEP 8

Detail the bearings. Hide all parts other than the outer and inner races of the Front Bearing. View the bearing in wireframe display. In the Front View create a **Sphere** primitive that fits between the inner and outer races at the top of the bearing. (Because the bearing is at Y=0 world, the ball will automatically be in the middle of the races.) Relocate this sphere's **Pivot** to Z=0 world and in the Left Side view array twelve times about 360 degrees as *instances* (Figure 8.22). Group both races and balls into front bearing. Repeat this operation for the Rear Bearing with eleven balls and group into *rear bearing*. When you are finished, the bearings are complete and in correct position.

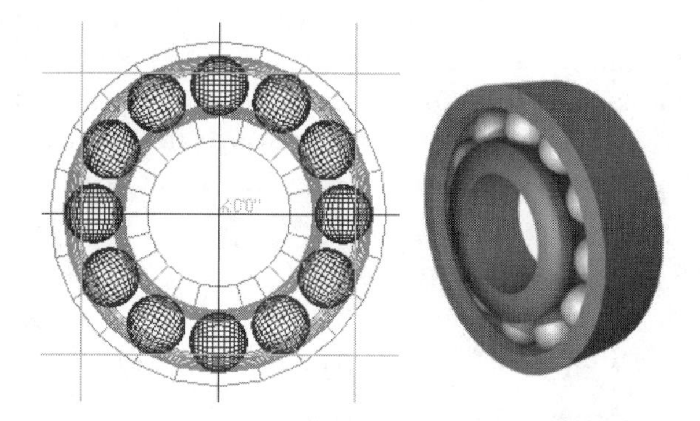

Figure 8.22 *Arrayed balls in bearing.*

STEP 9

Complete the Pulley. In the Right Side view create drill tools for the $^3/_8$ hexagonal fasteners. You should still have the centerline available from the original profiles. If not, bring in the profiles, move to datum origin, and delete all but the centerline. Array the $^3/_8$ drill tool six times about the Z=0 world axis. Perform Boolean subtractions, deselecting and reselecting after each operation. The Pulley is shown in Figure 8.23, both with tools in place and after the Boolean subtractions.

Note: Don't be surprised if you lose the centerline. Remember that the line is two-dimensional and will be visible with the profiles only in the Top view.

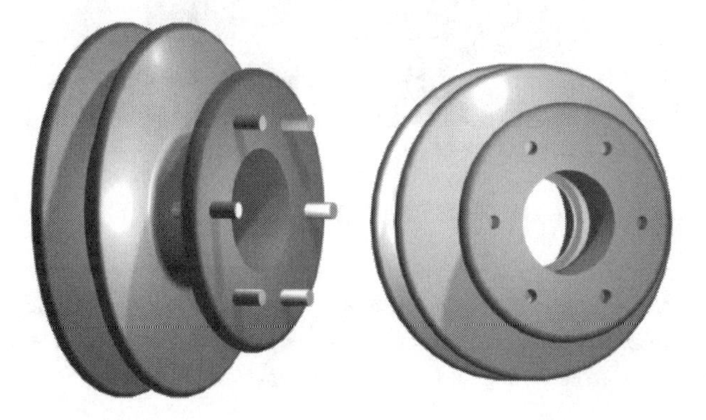

Figure 8.23 *Pulley with drill tools and completed.*

STEP 10

Create the Spring. An inspection of the sectional assembly reveals the length of the Spring (.5"), the diameter of the wire (.125"), the radius of the helix (.625) and the number of turns ($3^{1}/_{2}$). In the Right Side view choose **Create|Spline|Helix** and draw out the radius (.625") and push the length (.50"). Enter these numbers in the appropriate fields in the roll-out. Note that **Radius 1** and **Radius 2** are both **.625"**. This is the path for a lofted shape.

In the Front view create a **Circle** radius **.0625**. This is the cross–sectional shape for the Spring. The shape's **Pivot** will be matched to the beginning of the helix. Select the helix path and choose **Create|Compound Objects|Loft|Pick Shape** and select the circle. A rough Spring is lofted.

In the **Skin Parameters** roll–out enter **10** in the **Shape Steps** field and **32** in the **Path Steps** field. Make sure **Smooth Length** and **Smooth Width** are checked.

Notice that the ends of the Spring have been ground flat to provide a bearing surface between the Spring and the Spring Cup on one end and the Rear Bearing Spacer on the other. Display these two parts and lock them. Create a **Box** tool on either end to remove the material from the Spring and provide these flat surfaces. The completed Spring is shown in Figure 8.24 with its ground ends and between the Spring Cup and Rear Bearing Spacer.

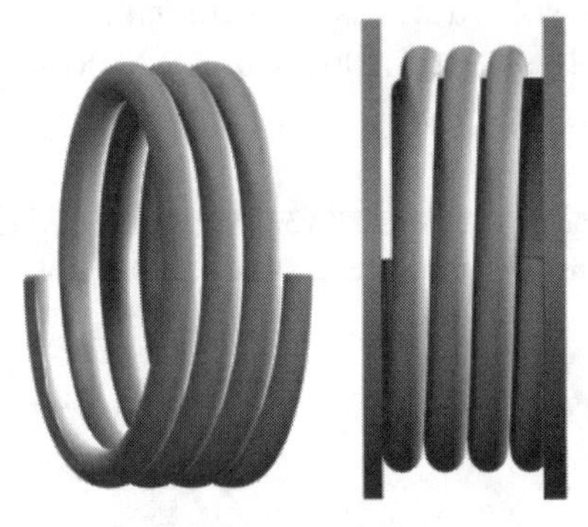

Figure 8.24 *Spring with ground ends and in position.*

Note: This is the spring in the compressed state. We really don't know what the unsprung length is. Were it critical, we could determine this length and **Morph** between unsprung and compressed lengths during an assembly animation.

STEP 11

Merge standard parts. On the accompanying CD-ROM you'll find the four standard parts required for this assembly. In the *tools/ directory* you'll find the file *parts.max*. Merge *castle_nut*, *cotter_pin*, *hex_head* and *washer* into the Fan Assembly. These parts are at full scale and aligned along the bottom of the scene. Move and rotate the castle nut and cotter pin into position. Move and rotate the hex cap screw and washer into position, relocate the **Pivot** of each to Z=0 world and **Array** six copies. These should be in alignment with the Pulley's mounting holes.

Note: Standard parts are proportionally correct but may have to be scaled to fit specific applications. For example, scale the shaft of the hex cap screw to fit the $^3/_8$" holes. Scale the castle nut to fit the $^3/_8$" shaft end.

STEP 12

Assign colors to the assembly. Display the parts from the inside out and assign representative colors. Select the Pulley. Choose the **Materials|Material Editor**. The upper left material sphere is selected. In the **Anisotropic Basic Parameters** roll–out, enter **25** in the **Opacity** field (you can't see the interior parts through the Pulley so we will make it translucent). Click the **Assign Material to Selection** button (third from left under the samples). Render the front view with a white environment (Figure 8.25).

Figure 8.25 *Fan Assembly with Pulley transparency.*

Note: You'll add some basic colors here so that adjacent parts can been discriminated. In Chapter 9, Material Basics, you will learn how basic color materials add realism. In Chapter 10, Raster Material Maps, you will add realistic materials maps.

FLOW LINES IN EXPLODED ASSEMBLIES

Save your assembly and save again with the file name *exp_fan_assy.max* and go back and set the **Opacity** to **100**. Pull the parts out along the X world axis until all parts clear the Pulley. Rotate the view into a pictorial User view.

It is helpful to add lines to an exploded assembly that represent the axes along which parts will be assembled. Spline lines generally are not rendered unless you specifically instruct 3D Studio to do so. Because all parts of the Fan Assembly lie along the X world axis, your flow lines will be parallel to this axis.

STEP 1

Create the axes. Turn on **Point Snap**. In the Front view, create flow lines (straight spline lines) from parts to their targets. Assign black color to these lines.

STEP 2

Make the assembly flow lines renderable. Select all the flow lines, choose **Modify|General|Rendering** click **Renderable**, and specify **.01"** as a starting thickness. Render the scene and check the line thickness. Adjust as necessary. This is an *exploded assembly* of the Fan Assembly (Figure 8.26).

Tip: You don't have to be overly careful about precisely starting and stopping flow lines. In 3D, flow lines will disappear inside the parts, making it appear that they emanate directly from the part.

Figure 8.26 *Exploded assembly with flow lines.*

PROBLEMS

Problems appropriate for modeling as assemblies are found on the following pages. Use the file *isogrid* found in the *tools*\ directory to plan your modification strategies. Assign appropriate names to all objects both on your sketches and models.

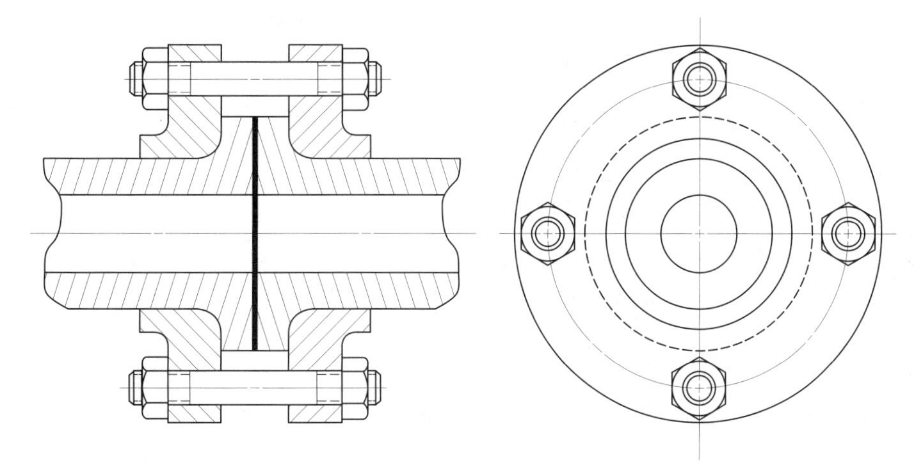

Problem 8.1 *Flange Assembly.*

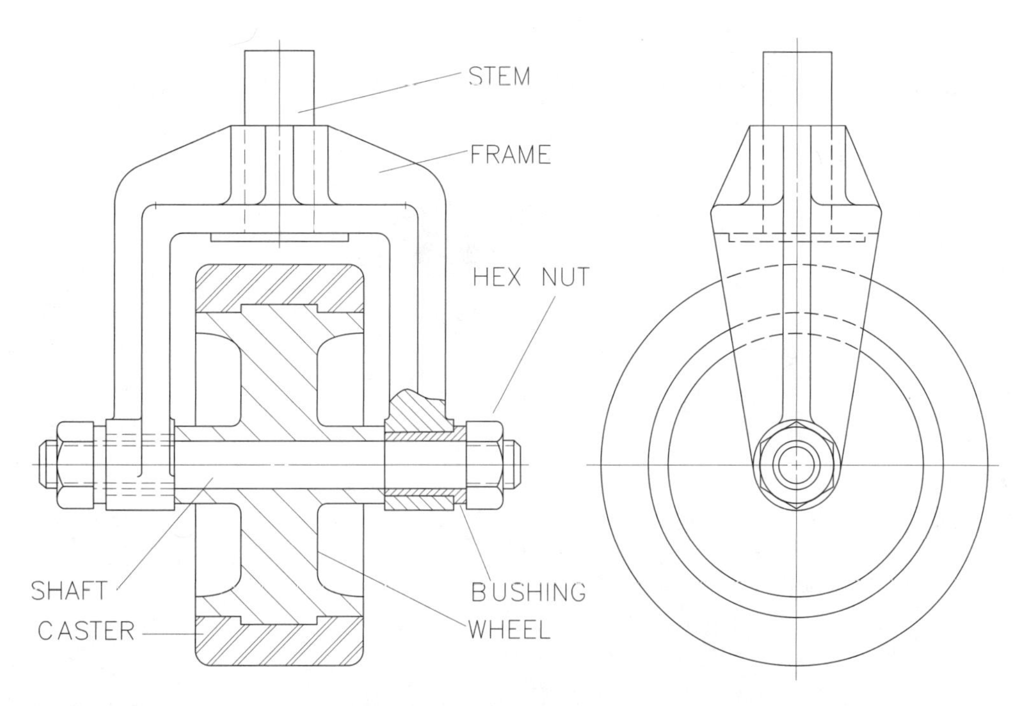

Problem 8.2 *Caster Assembly.*

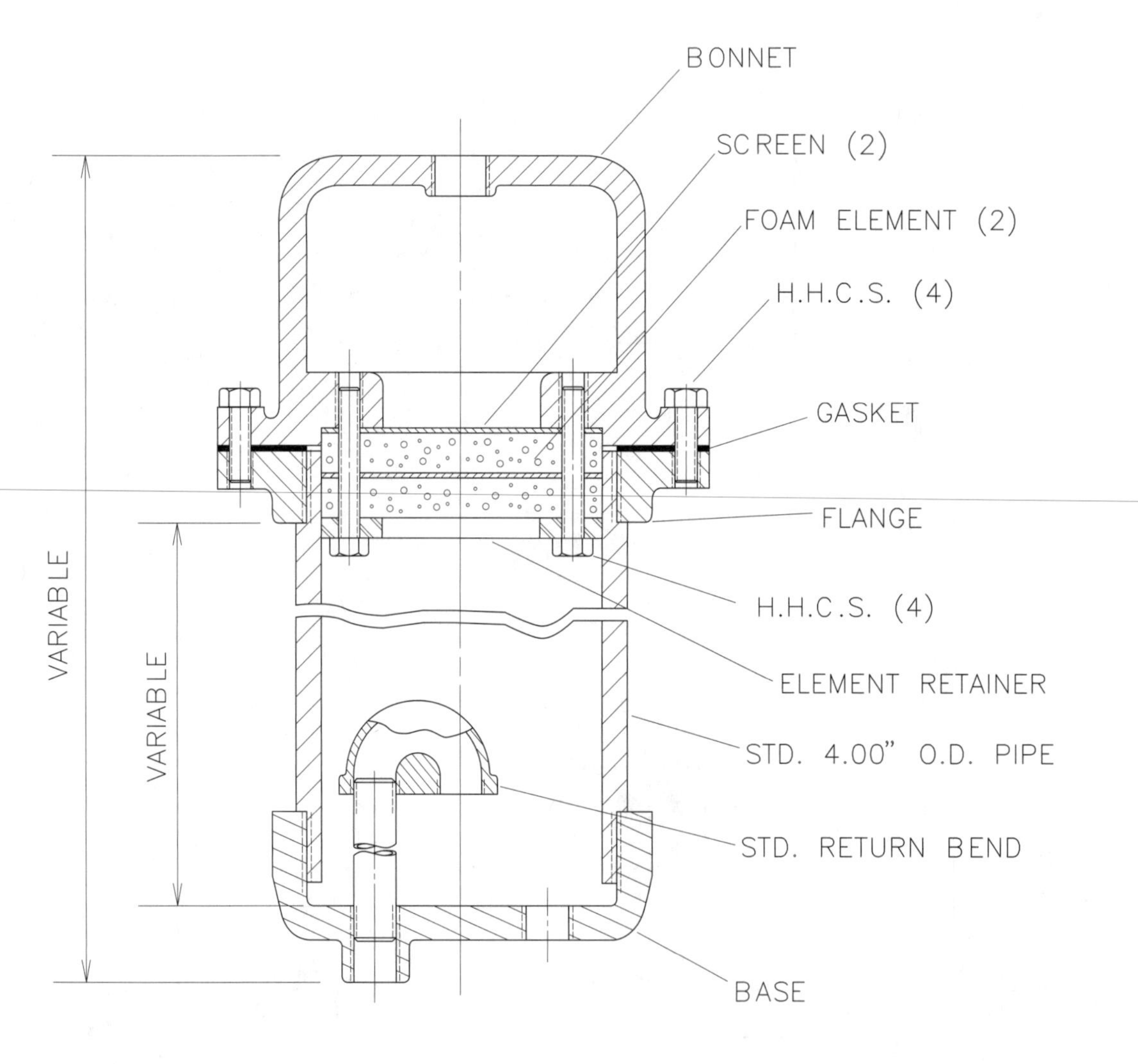

Problem 8.3 *Air Cleaner Assembly.*

Material Basics

CHAPTER OVERVIEW

Designers in engineering and technology may never need to concern themselves with anything other than the geometric accuracy of their models. For them, how a part *functions* is much more important than how it *looks*. Still, many technical decisions hinge on being able to discriminate between parts in an assembly, between a part and its background, or from one surface to another. Take Figure 9.1 for example. In (a) parts are the same color (the same hue, saturation, and brightness). Compare this to (b) where the components are assigned colors having varying brightness. You can see the effect of value contrast in being able to perceive different parts.

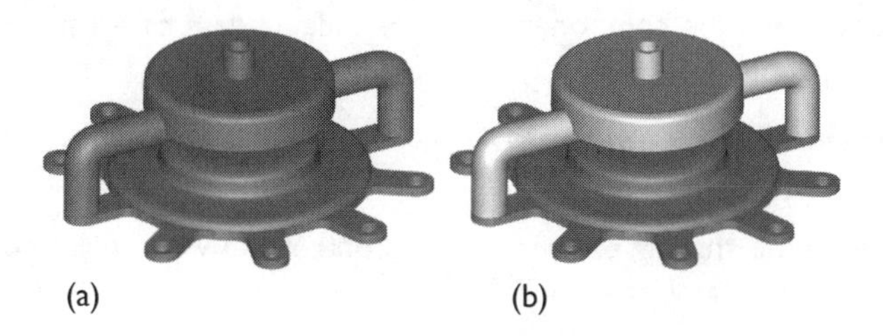

Figure 9.1 *Effect of color brightness change.*

KEY COMMANDS AND TERMS

- **Ambient Color**—the color not directly illuminated; the color in the shade.
- **24-Bit Color**—color data having 8 bits of red, green, and blue.
- **Ambient Light**—the general environmental light before any specific light objects are added to the scene.

- **Anisotropic Shading**—a shading method appropriate for flat reflective metal surfaces.
- **Base Color**—a color from which other colors are derived.
- **Brightness**—the component that describes the lightness or darkness independent of hue or saturation.
- **Blinn Shading**—a shading method appropriate for most reflective surfaces.
- **Color Palette**—the display of colors available for assignment to objects.
- **Diffuse Color**—the color evenly illuminated, neither in shade nor highlight.
- **Glossiness**—a setting that controls the dispersion of specular light.
- **HSV**—a model that describes color in terms of hue, saturation, and value (brightness).
- **Hue**—the component of color that describes the pure color itself, independent of saturation or brightness.
- **Material Editor**—the portion of 3D Studio where the appearance of surfaces is controlled.
- **Name and Color Roll-out**—the area in the 3D Studio menu where object name and color assignment can be changed.
- **Oren-Nayer Blinn Shading**—a shading method appropriate for non reflective surfaces.
- **RGB**—the model of color space that uses red, green, and blue primaries.
- **Saturation**—the component of color, independent of brightness, that describes the amount (purity) of a given hue.
- **Selection Set**—a method of associating objects so that they may be easily identified.
- **Soften**—the shading parameter that controls how specular and diffuse colors blend into the ambient color.
- **Specular Color**—the color that blends from intense highlight to the diffuse color.
- **Specular Level**—a shading parameter that controls how much light a surface will reflect.
- **Tint**—the environmental illumination setting that impacts the color of all light objects in a scene.

IMPORTANCE OF BASIC MATERIALS

A *basic material* represents an object's material by color and surface finish. It is the first step in creating effective surface representation and for many computer models, is sufficient. Assigning colors to objects also provides a way to organize your work. By assigning colors, it becomes easier to quickly distinguish alignment, placement, and orientation. In fact, you may want to use a consistent color scheme to keep track of object function and condition. For example, you may want to assign light blue to underlying rough geometry. Reference planes are easily recognized as such if colored brilliant red; preliminary features may be green; final model geometry may be a color representing the material. An assigned color is reflected in both wireframe and smooth shaded modes and when rendered; 3D Studio provides several methods of assigning colors to parts.

The easiest method is to let 3D Studio assign a random color to each object as it is created. This is an easy way to keep different objects separated. You can always go back and explicitly change the color by assigning a color from the **Create|Name and Color** roll-out. Note that 3D Studio has two color palettes: the 3D Studio Palette and the AutoCAD ACI palette. Both of these address 24-bit RGB color space and provide access to 16.7 million colors. The AutoCAD ACI palette displays a wider choice of standard colors; the 3D Studio palette provides empty slots in which custom colors can be stored.

Note: VIZ and MAX should assign random colors to objects by default. This is controlled in VIZ by **Tools|Options|3D Studio VIZ** with **Default to Layers for New Nodes** deselected.

Creating *selection sets* plays an important role in assigning colors to families of objects (Figure 9.2). It may be helpful to be able to select all the centerlines at once, for example, or all washers, or bolts.

STEP 1

At the root level of the **Create** menu, select the centerlines using the **Select** tool. Hold down the **CTL** key to add to the selection.

STEP 2

With the centerlines selected, type the title **centerlines** in the selection field at the top of the menu bar. Any time you want to select all the centerlines at the root level of **Create** (if you want to change their color, for example) all you have to do is go to the selection set field and choose **centerlines**.

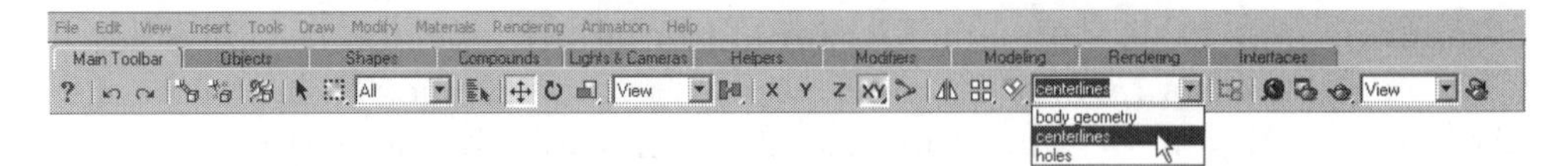

Figure 9.2 *Selection sets facilitate color changes.*

Note: Assigning colors to objects can be problematic depending on your viewport background color. A black background is horrible to look at, but few objects are black so contrast is good. A middle gray background is easy to look at but provides poor contrast with either light or dark objects (let alone a medium background). A white background is better than a black background, but there are more light objects (chrome, glass, white plastic, etc.), so contrast may be poor.

DEFAULT ILLUMINATION

The 3D Studio program illuminates its scene or design with a default ambient (allover) light so you don't have to set lights to see the basic material colors assigned to objects. These basic colors are, however, impacted by changes in this ambient light. Say, for example, you are modeling machinery that will operate in a steel mill. The ambient light while pouring molten steel (orange-yellow) would be entirely different than the ambient fluorescent light in a laboratory (blue-green). To change the ambient light (and thereby the appearance of any color you apply), do the following.

STEP 1

Choose **Rendering|Environment**. The **Ambient** color broadcasts a color evenly throughout the scene. A change in the ambient color will affect everything in the scene.

STEP 2

The **Tint** impacts all lighting other than the ambient light. This is like a gel placed over light sources.

STEP 3

The **Level** impacts the general level of illumination in the scene. Enter **2** in this field to approximate bright sunshine. Enter **.5** to approximate dusk.

Tip: It's critical to push object colors apart in value (brightness). If you have a question as to how a model will appear were it printed in black and white, print a color-rendered file on a black and white printer.

COMPONENTS OF COLOR

There are several models or methods for specifying color. Because it is displayed on RGB (Red, Green, Blue) monitors, 3D Studio uses RGB color space. However, humans don't perceive color in RGB color components. We say a color is a certain *hue* (such as yellowish green, or bluish green); a *saturation* (such as pale, pastel, or really intense); and a certain *value* (such as bright or dark). This HSV (Hue, Saturation, Value) system lets you assure that objects next to one another have sufficient value difference, irrespective of their color. Study Figure 9.3. In (a) the shaft and collar are two different colors, but as shown in the HSV fields, their value settings are identical. Stripped of color, they appear identical. In (b) the Hue and Saturation remain unchanged but with the value of the shaft increased dramatically. This allows the shaft to be more easily distinguished from the collar.

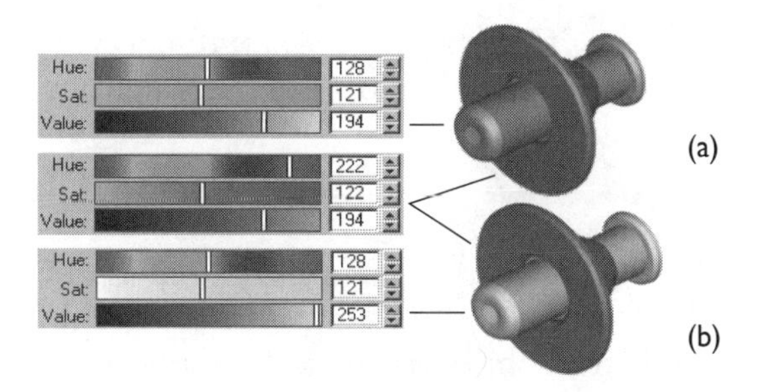

Figure 9.3 *Color contrast in HSV.*

It is interesting that the human eye is weighted toward brightness. Given two samples of different color, the eye will first distinguish not the color difference, but any difference in value (brightness). This makes material decisions in 3D Studio all the more important because colors, though different in hue, may be difficult to distinguish without adequate value change.

MATERIAL BY COLOR ASSIGNMENT

The colors contained in either 3D Studio palette respond to lighting in smooth, matte to semigloss finish. As mentioned previously, three individual color parameters can be adjusted: hue, saturation, and value. By changing the values of these three components you can assign colors that represent many materials:

- Cast metal: H=0, S=0, V=132
- Machined metal: H=128, S=32, V=255
- Aluminum: H=128, S=27, V=201
- Brass: H=36, S=144, V=255
- Copper: H=18, S=173, V=218
- Plated metal: H=122, S=18, V=255
- Wrought iron: H=170, S=6, V=66

Assume you want to assign a machined metal color to the shaft shown in Figure 9.3. A color assignment is the lowest level of material representation available in 3D Studio and is sufficient to distinguish machined metal from another material such as cast metal.

STEP 1

Select the target object. If you have let 3D studio cycle through random colors, this part will already have a color assignment from its palette of basic colors.

STEP 2

In the **Name and Color** roll-out, click on the color chip to the right of the name. This is the selected object's current color.

STEP 3

Notice that 3D Studio displays its default color palette. Click on the first open position of available **Custom Colors**. A black frame surrounds the box. Click on the **Add Custom Colors** button and the **Color Selector** is displayed. In the **Hue**, **Saturation**, and **Value** fields, enter **128**, **32**, and **255**. **Close** the **Color Selector**. The machined metal color is added to the **Custom Color** palette. Click **OK**.

STEP 4

The custom color is assigned to the object. Figure 9.4 shows the turned object in default gray (a) and machined metal (b) with default ambient environment lighting.

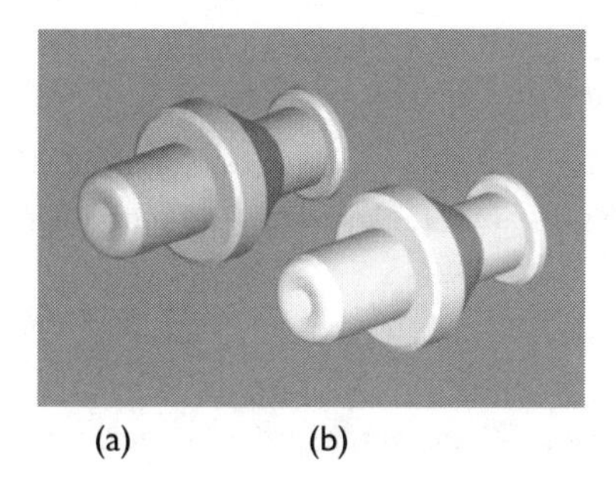

Figure 9.4 *Impact of custom color.*

Tip: You may not want to assign representative colors until all geometric construction is completed. It is difficult to identify machined metal or aluminum parts on a gray viewport background. If you choose to work in wireframe on a white background, assign black to objects not being worked on. Then, assign red to the active object and replace this with a representative color assignment when completed.

IMPACT OF LIGHTING ON BASIC MATERIALS

Although the default ambient lighting in 3D Studio illuminates a scene sufficiently to distinguish various basic material colors and reveal shape distinctions, the addition of basic lights (see Chapter 14, Camera and Light Basics) can have a dramatic effect both on the basic material color and the perception of shape. Material color with default lighting in Figure 9.5(a) is improved dramatically with the addition of basic lighting (b) from Chapter 14.

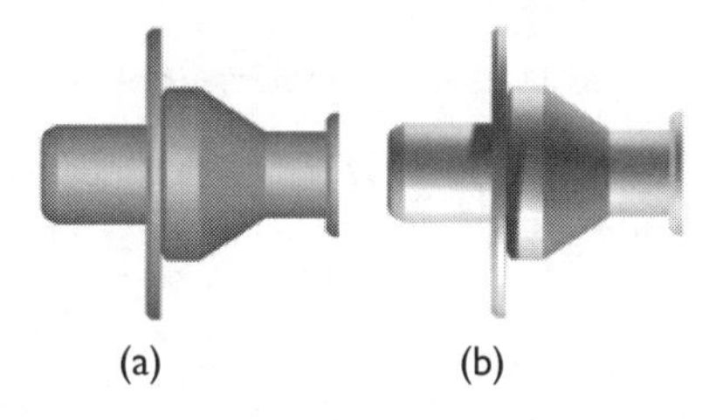

Figure 9.5 *Impact of basic lighting.*

BASIC MATERIAL GUIDELINES

- If you use black for inactive objects, choose a color you would never assign to an object (fuchsia?) to locked objects.

- Be prepared to change viewport background color to contrast with material colors. If your modeling materials are metallic, a gray background may not be productive.
- Using 3D Studio's default ambient environmental lighting will yield only marginal results.
- If you model mostly metallic parts and assemblies, you may want to fill the sixteen custom color slots in the 3D Studio palette with black on one end, white on the other, and the remaining fourteen slots with even steps between 1 and 254.
- Depending on the complexity of the scene and capabilities of the workstation, rendering viewport windows while working may slow screen display considerably.

THE MATERIAL EDITOR

The second level of material fidelity involves 3D Studio's **Material Editor**. In this chapter you will learn how just a few adjustments to basic color can add considerable material realism.

In VIZ choose **Rendering|Material Editor** or in MAX, **Tools|Material Editor**. The **Material Editor Window** appears. This window can be left open on top of the 3D Studio interface while you work or you can minimize and expand it as needed. Figure 9.6 shows the top portion of this window, the part we will concern ourselves with in this chapter. In the next chapter we will delve into the lower portion of the **Material Editor Roll-out** and material maps.

The top portion of the **Material Editor** contains:

1. **Sample**—a sphere or cube that displays material settings.
2. **Toolbar**—commands that access material options.
3. **Material Name**—a field that carries default names for materials. You are encouraged to enter an appropriate name for your model.
4. **2-Sided**—applies material to both sides of surfaces with no thickness.
5. **Shading Modes**—these options determine how light will interact with surfaces. For our purposes you will use **Blinn** shading for everything but flat metal and fabric. Use **Anisotropic** for flat metal and **Oren-Nayer Blinn** for objects that have matte or rough surfaces.
6. **Material Type**—this should be **Standard** for everything you do in this chapter.

⑦ **Object Color**—the base color separated into ambient, diffuse, and specular components. The RGB/HSV **Color Selector** is available by clicking on each chip.

⑧ **Opacity**—**100** signifies a solid object; a value of **0** creates an invisible object. **Opacity** is impacted by object color. Surface characteristics determine the level of **Specular Highlights**.

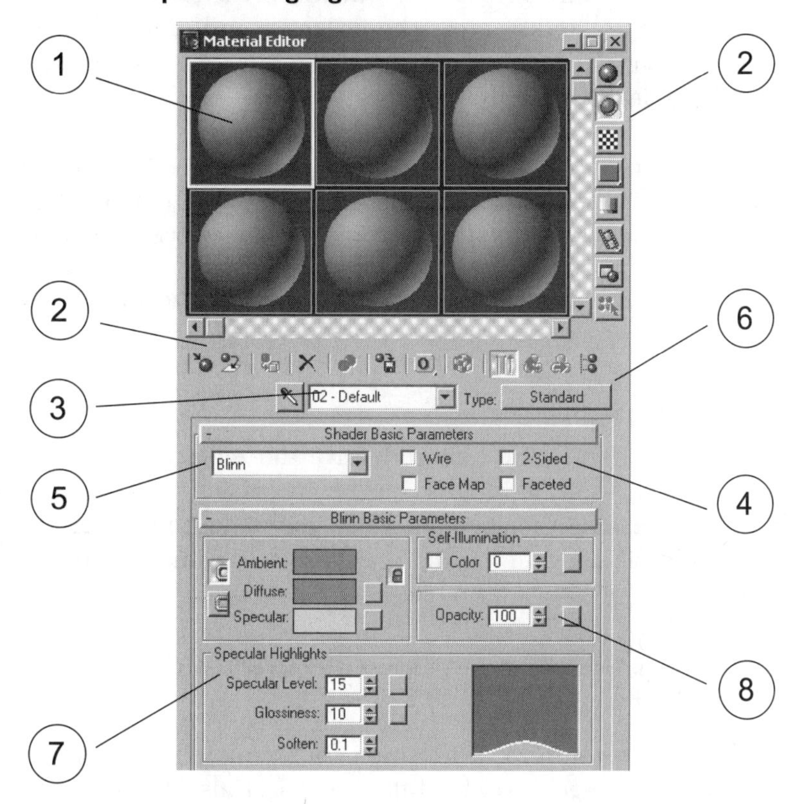

Figure 9.6 *The basic material portion of the Material Editor.*

AMBIENT, DIFFUSE, AND SPECULAR COLOR

Earlier in this chapter we assigned a single color to an object and using either 3D Studio's ambient lighting environment or the basic lights from Chapter 14, let the VIZ or MAX renderer decide how to represent the colors. In the Matrerial Editor, that color is further broken into three components.

Diffuse color is the "base" color or the object as it appears in light. We call it the base color because it is evenly illuminated with no highlights or shadows. The entire object would be this color.

Ambient color is the base color with less brightness or the object in the umbra formed by an interrupted light source. As lighting levels decrease, the ambient color mixes with the diffuse color, darkening the object. If the ambient color is a different hue from the diffuse color, the impression of reflected light is achieved.

Specular color mixes with diffuse color in the presence of strong lighting. It forms the highlights and displays the hardness and reflectivity of the surface. If the specular color is different from the diffuse color, the impression of a colored light source is achieved.

Tip: The easiest way to create realistic materials is to start with the diffuse color. Drag that color chip on top of the ambient chip and click **Copy**. Open the ambient chip and subtract at least 100 from the value reading. Copy diffuse to specular and enter 255 (white) in the **Value** field for shiny, hard surfaces. Lower this number to reduce reflectivity. Leave diffuse and specular colors the same as the ambient color for nonreflective or matte surfaces.

SURFACE HARDNESS

These two controls determine how reflective the material will be. Soft materials have broad and soft highlights. Hard materials have tighter and sharper highlights. By increasing the specular level, you make the material more receptive to illumination. By increasing the glossiness value, you make the surface harder.

Note: Specular level and glossiness work together. A low specular level lowers glossiness.

The best way to see the interrelationship of these material settings is to test render representative geometry as different materials (Figure 9.7).

STEP 1

Start with a basic color with basic lights. Begin with an object that has a variety of surfaces that demonstrate material settings and apply a basic material color in the **Name and Color** roll-out. We use HSV values of **36**, **64**, **100** and arrive at a brownish-gray (a). This will form the diffuse color in the Material Editor.

STEP 2

Establish colors in the **Material Editor**. Open the **Material Editor** and select **Blinn** shading. Click the color locks off and click on the **Diffuse** color chip. Enter **36** in **Hue** field, **64** in the **Saturation** field, and **100** in the **Value** field.

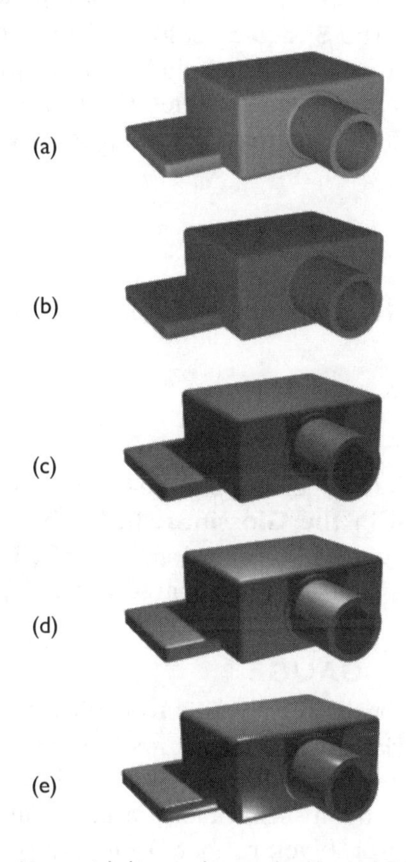

Figure 9.7 *The effect of basic material changes done in the Material Editor.*

Drag this color to the **Ambient** and **Specular** chips and **Copy**. All three colors are now identical.

Enter **0** in **Specular Level**, **Glossiness**, and **Soften**. If the target object is selected, click the **Assign Material to Selection** button in the **Material Editor** toolbar. Or, drag the sample from the Materials Editor to the scene and drop it on the target. (If you wondered why we take the time to name all our objects, here's where you find out!) There should be a slight change between basic color assignment and flat colors in the **Material Editor** (b).

Note: Once you have assigned a material to an object using the Material Editor, subsequent changes in that material will automatically be reflected in the scene.

STEP 3

Spread the values. Open the **Ambient** color and lower the **Value** number by a maximum of 100. Open the **Specular** color and raise the value by a maximum of 100. If you don't raise the specular level or glossiness, the change in ambient and specular color won't produce the desired effect. Enter **30** in **Specular Level** and **10** in **Glossiness**. The object takes on moderate surface reflectivity (c).

STEP 4

Dial up the shine. Enter **90** in **Specular Level**. The object reflects considerably more light (d).

STEP 5

Dial up the hardness. Push the specular color even lighter by increasing specular color **Value**. Enter **50** in the **Glossiness** field. You will immediately see that the object has less specular illumination. Enter **200** in the specular level field. The object takes on a hard, epoxy-coated finish (e).

EXAMPLE: PRESSURE GAUGE

The best way to practice basic materials is to work with a product that contains widely varying materials. Figure 9.8 shows the views of a Pressure Gauge. The Right Side view is shown in broken section so that interior parts can be observed. The gauge contains a satin black Case, chrome Bezel, glass Lens, white Face Plate, red Pointer, black rubber Hose, and plated Nut. The gauge can be found on the CD-ROM in *chapters\chapter_9*.

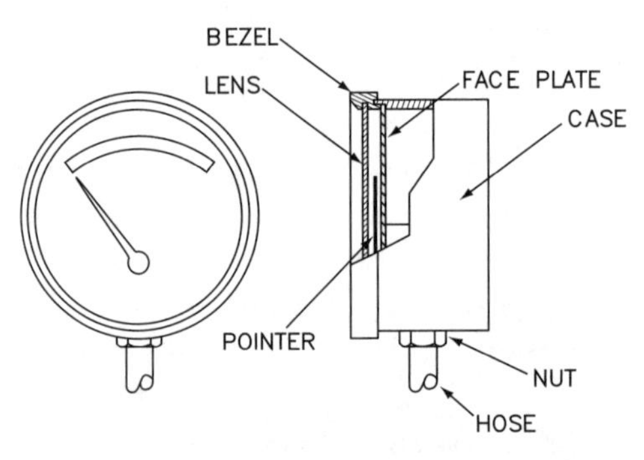

Figure 9.8 *Pressure gauge.*

STEP 1

Open *chapters\chapter_9\gauge.max*. Basic lights and a gray color have been established for you (Figure 9.9). The glass lens has also received a gray color and obscures the Face Plate and Pointer. Hide all but the bezel. Open the **Material Editor**.

Figure 9.9 *Pressure gauge in single color material.*

STEP 2

Create a basic chrome material. Chrome is essentially colorless, picking up the color of its surrounding environment. Pick an empty sample, name the sample chrome, and establish:

shader = Anisotropic

ambient = 0, 0, 0

diffuse = 128, 10, 255

specular = 0, 0, 255

spercular leve l= 80

glossiness = 50

STEP 3

Create a glass material. Pick the next empty sample. Name the sample glass and click on the checkerboard background tool. This lets you see the effect of your opacity setting. Establish:

shader = Blinn

ambient = 0, 0, 138

diffuse = 0, 0, 138

specular = 0, 0, 255

specular level = 90

glossiness = 5

opacity = 30

Note: Just because you make a surface transparent doesn't mean that lights don't know it's still there. Transparent surfaces will cast opaque shadows so you have to exclude these surfaces from certain lights so what's behind them can be seen.

STEP 4

The Pointer is relatively unimportant, as long as it's red, really red. Name this color pointer and establish:

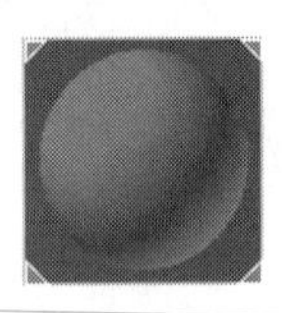

shader = Blinn

ambient = 255, 255, 255

diffuse = 255, 255, 255

specular =2 55, 255, 255

specular level = 10

glossiness = 10

STEP 5

Make the Face Plate a dull, nonreflecting white. Name this color face plate and establish:

shader = Oren-Nayer Blinn

ambient = 255, 255, 255

diffuse = 255, 255, 255

specular = 255, 255, 255

specular level = 0

glossiness = 0

STEP 6

Make the Case a satin black plastic. Name this color black plastic and establish:

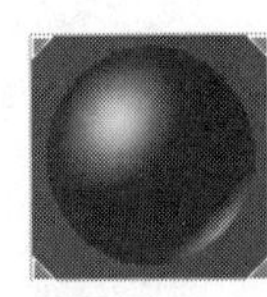

shader = Blinn

ambient = 0, 0, 0

diffuse = 0, 0, 0

specular = 255, 255, 255

specular level = 90

glossiness = 10

soften = 0.8

STEP 7

Use the chrome color for the Nut and the black plastic for the rubber Hose. (Or you can drag copy these to unused samples, rename, and tweak to get slightly different results.) The final rendered Pressure Gauge is shown in Figure 9.10. Compare the results with that in Figure 9.9 and you will see that the effort to create materials by simple color assignment is well worth the effort.

Figure 9.10 *Pressure gauge in basic material colors.*

PROBLEMS

Use the following basic material assignments for problems in Chapters 1 through 8. Create the basic lighting configuration from Chapter 14 and produce an 800 x 600 rendered User view showing the assignment(s) in favorable position. Be prepared to document your solutions as you did on pages 185–186 of this chapter.

Problem 9.1 *Adjustment tool basic materials. Create the following basic materials for the Adjustment Tool, Problem 1.1.*

Handle: Hard Yellow Plastic

Grip: Smooth Black Rubber

Tool: Polished Steel

Problem 9.2 *Guide Plate basic materials. Create different versions of basic materials for the Guide Plate, Problem 2.6.*

Dull Black Plastic

Green Anodized Aluminum

Dull Metal

Painted White Enamel

Problem 9.3 *Support Base basic materials. Create different versions of basic materials for the Support Base, Problem 3.5.*

Shiny Brass

Turned Aluminum

Cast Aluminum

Chrome

Problem 9.4 *Fan Assembly basic materials. Create the following basic materials for the Fan Assembly, Figure 8.15.*

Shaft: Machined Metal

Washer Retainer: Red Anodized Aluminum

Felt Washer: Dull Dark Brown

Spacers: Brass

Pulley: Gray Epoxy Painted Metal

Spring Cup: Bright Pressed Metal

Bearing Races: Dull Metal

Bearing Balls: Chrome

Spring: Dull Metal

Raster Material Maps

CHAPTER OVERVIEW

In the previous chapter you learned how the application of diffuse, ambient, and specular colors, along with the addition of surface reflections, can add considerable realism to your models. In this chapter, we will take this focus on materials one-step further: the application of material textures.

Modeling geometry is created and stored as vectors, lines, and curves that define plane surfaces. Only when the geometry is rendered to the screen, as when you choose **Smooth + Highlights** for the viewport or actually render the scene, does 3D Studio actually fill in the surfaces with small elements, called pixels. This is where computer modeling and animation start to become more of an art than a science. But to effectively represent diverse materials, you are going to have to, at least temporarily, put on your artist's cap.

KEY COMMANDS AND TERMS

- **Alpha Source**—the RGB color that is treated as transparent when applying a decal in the diffuse color channel.
- **Ambient Channel**—a section of the **Material Editor** where a material map is loaded to represent the object in shade.
- **Bitmap**—a rectangular array of image data representing color or value; a picture or graphic comprised of individual bits.
- **Channels**—slots in the **Material Editor** into which raster material maps can be loaded.
- **Diffuse Channel**—a section of the **Material Editor** where a material map is loaded to represent the illuminated object.

- **Environment Map**—an option in the **Render** dialogue that allows bitmap images to be displayed in the open background of a rendering.
- **Generate Mapping Coords**—an option in the **Modify** roll-out that assigns default mapping coordinates to selected geometry.
- **Gray Scale**—a bitmap containing only value information, usually with a maximum of 8-bits or 256 shades of gray.
- **Mapping Coordinates**—the shape and orientation of the projector by which materials are mapped onto geometry.
- **Material Editor**—the interface that allows the creation of materials for assignment to geometric objects in the scene.
- **Material Library**—a collection of materials stored in a .mat file; the default 3D Studio Material Library.
- **Material Map**—bitmaps assembled into various channels in the Material Editor to represent a material.
- **Material Map Browser**—the interface that facilitates selecting material maps and their components.
- **Material Type**—provides choice of the structural design of the material map.
- **Multi/Sub–Object Material**—a material holding two or more submaterials that can be assigned to subselections of an object.
- **Opacity Channel**—a section of the **Material Edito**r where a material map is loaded to represent degrees of transparency.
- **Opacity Map**—a grayscale bitmap used to alter material maps during rendering; white bits have no impact, black bits are interpreted as transparent, and gray bits in between.
- **Raster**—an image being made of discrete bits of information.
- **Reflection Channel**—a section of the **Material Editor** where a material map is loaded to represent reflections.
- **Sample Temperature**—the assignment status of a material sample whether it is assigned and selected, assigned but not selected, or not assigned.
- **UVW Mapping Gizmo**—the representation of the shape by which the material is projected onto the selected object.

IMPORTANCE OF RASTER MATERIALS

A raster material is a two-dimensional matrix that contains color or *grayscale* information. Also called a *bitmap*, this is the kind of graphic produced by digital cameras, camcorders, and scanners. Bitmaps are edited in programs such as Photoshop and can represent any kind of material such as metal, wood, plastic, concrete, or glass. As an engineer or technologist you probably will not create your own material maps. Instead, you will make use of 3D Studio's extensive library of materials, pick from a universe of materials available on the Internet, or take pictures of textures in the world around you.

When you assign a material (galvanized metal, for example) to a part, you add a level of realism unavailable using color assignment alone. Consider the panel in Figure 10.1. In (a) the panel is assigned a medium blue color that represents galvanized metal. Unfortunately, it just as easily could be plastic or aluminum. Compare this to (b) where the raster material representation of galvanized metal has been applied to the surface. 3D Studio wraps the material (c) around the geometry to produce a realistic representation of galvanized metal using the instructions you give it.

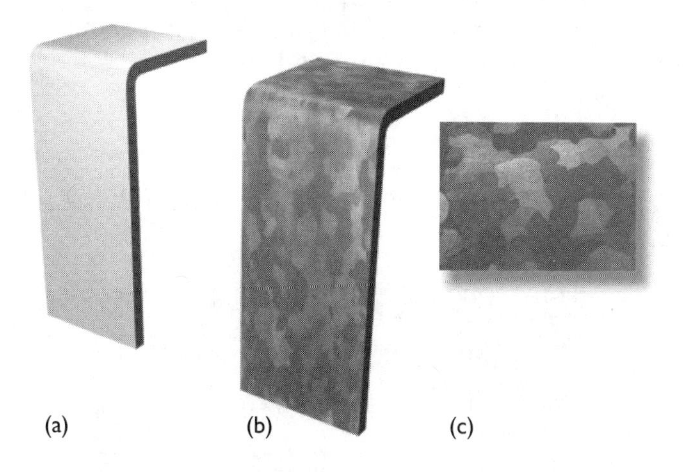

Figure 10.1 *Raster material map.*

There are other mapping methods available in 3D Studio to arrive at extremely sophisticated results. *Procedural maps* use mathematical instructions to create material representations. You can create *composite material maps* where the final effect is the result of several materials being overlayed, one on top of the other. You can also create *ray–traced materials* that faithfully reproduce reflections. This chapter will give you the tools to apply realistic materials to surfaces so that later, in the next two chapters, you can add the final detail that will make your models nearly photorealistic.

SOURCES OF RASTER MATERIALS

The first source of materials is the 3D Studio materials library that ships with your software. This library contains much of what you need from natural materials such as wood and concrete to metals, plastics, and environments. To peruse the material library:

STEP 1

Open the **Material Editor** and click on the **Type:Standard** button. The **Material Map Browser** opens (Figure 10.2). Click on the **Mtl Library** radio button. A list of the materials is shown.

STEP 2

Click on one of the materials and a preview appears in the window. Type a descriptor (like "metal") in the field above the sample and if a match from the list is found, it is displayed in the window.

STEP 3

Click **OK** and the material is assigned to the currently selected sample in the **Material Editor**.

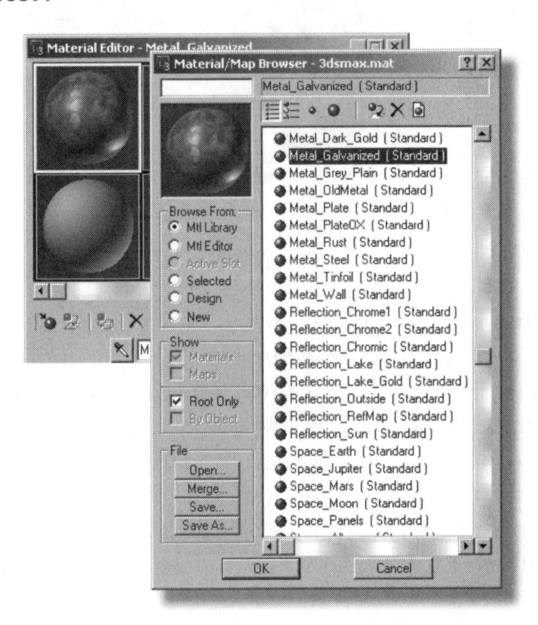

Figure 10.2 *Material library.*

Materials in the editor can be "hot," "assigned," or "cold." A hot material, (Figure 10.3(a)), has been assigned to an object in the scene and that object is currently selected. It has solid white triangles in the corners. Any changes in the material will be automatically updated in the object. An assigned material (b) is assigned to object(s) not selected in the scene but changes will be updated. This is identified by open triangles in the corners. A cold material (c) has yet to be assigned to any object in the scene. This is identified by no markings in the corners.

Tip: If you want to make changes to a hot or assigned material without changing objects in the scene, drag the material sample and drop it on an empty sample. This creates a cold copy.

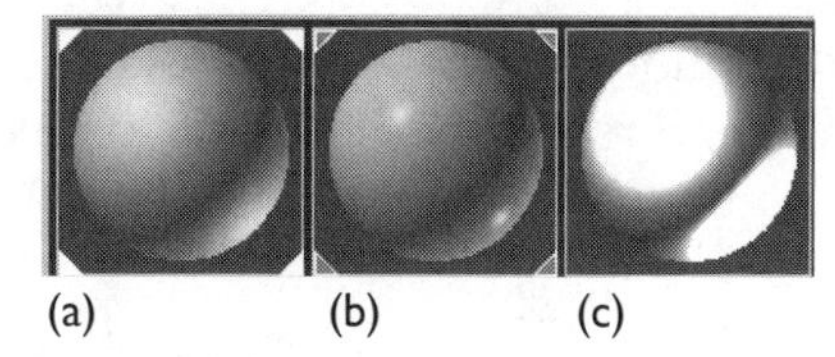

Figure 10.3 *Assignment status of material samples.*

Another source of raster materials is, as you might imagine, the Web. If you go to *http://www.grsites.com*, you'll find literally thousands of raster patterns and textures that can be used in your material maps (Figure 10.4). Some of these textures are tiles and as such can be repeated across a surface with no visible seams.

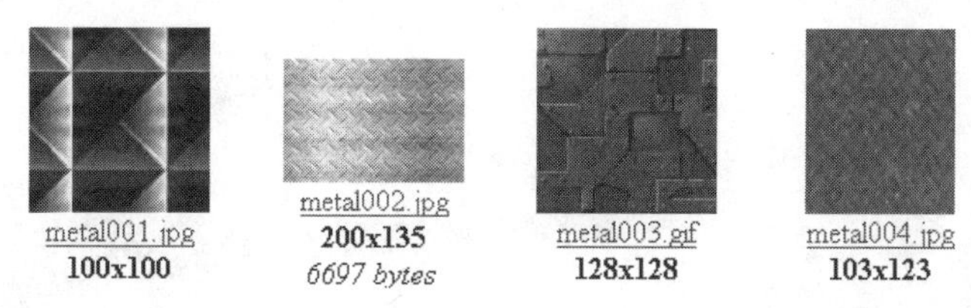

Figure 10.4 *Material source at http://www.grsites.com.*

Note: Be careful when appropriating textures from the Web. Make sure that the images are offered "copyright free" or are in the public domain.

Another source of raster textures is as near as your digital camera. Figure 10.5 shows a raster material as captured by a digital camera and its appearance when successfully mapped onto an object in 3D Studio. Materials that you make can be saved to the default 3D Studio library or into a custom library that holds textures by project or material type.

Note: 3D Studio doesn't actually store the textures in the .max file. Instead, it stores the path to the libraries that are used. For this reason, it is paramount that libraries are not relocated. Likewise, if you work in two locations, you'll have to update your libraries as you add to them.

Two factors are important when using pictures from a digital camera: camera-target angle and file format. Because material maps will usually be projected perpendicular to a surface, the digital camera shot should be as perpendicular as possible to the subject. In 3D Studio, you can use a variety of raster file formats as materials. JPG images can have a natural "noise" built into them so if irregularities (artifacts) are of no concern, JPG is fine. On decals or images that have large areas of solid color, use GIF or BMP formats.

In Figure 10.5, a digital photograph is taken of a socket (a). This photograph is edited into a "material chip" (b) and loaded into the 3D Studio Material Editor. When applied to a metallic handle (c), a realistic metal material is displayed.

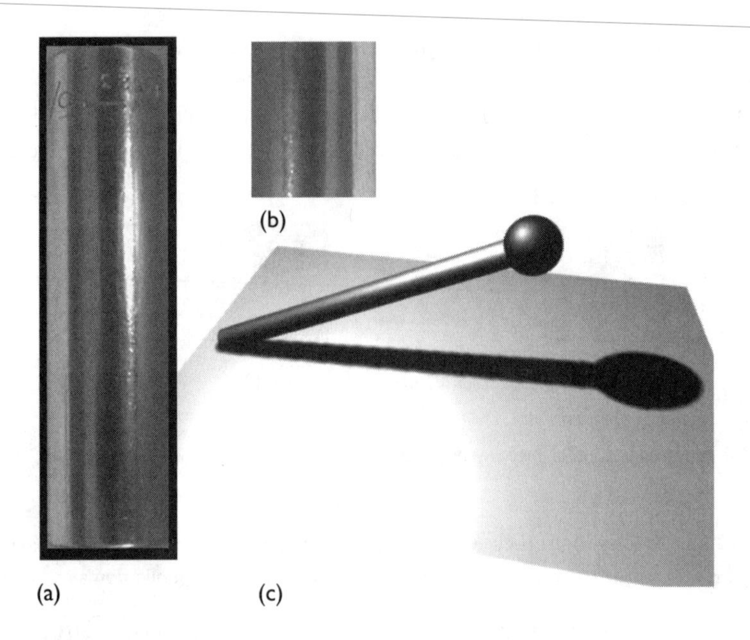

Figure 10.5 *Material captured with digital camera.*

MATERIAL MAP GUIDELINES

- The 3D Studio VIZ and MAX material libraries are compatible and contain a slightly different mix of materials and maps.
- Materials and lights lengthen rendering time considerably. It is counterproductive to apply materials to objects in a scene that are either in the periphery or at a distance. It is better to use a basic material color.
- The effectiveness of a material map is determined in large part by the manner in which it is illuminated.
- Don't use a TIF file for a material map. The TIF file contains more information (instructions for high–quality printing devices) than is needed for material mapping. Use JPG, BMP, and GIF.
- The physical size of the bitmap determines the resolution of the pattern as it is applied to geometry. Bitmaps are fixed resolution so if you enlarge the **UVW Gizmo**, you enlarge the individual picture elements.
- It's always better to match the dimensions of a bitmap material to the geometry. It's counterproductive to try to apply a long narrow material to a square object.
- If given the choice, apply tiling in the UVW map roll-out. This way, it's easier to use the same material for more than one tiling application in the same scene.
- Opacity maps should only be grayscale. A color bitmap will work, but it's difficult to predict results when 3D Studio reads the value component of HSV color.
- Most geometry respond well to the automatic generation of mapping coordinates in the **Modify** roll-out.
- If you create a material that's particularly effective, save it to a material library so it will be readily available.

MAPPING COORDINATES

When a raster material map is assigned to an object, 3D Studio needs to know how to project the bitmap onto the geometry. To do this, mapping coordinates are assigned to the geometry. These *mapping coordinates* are expressed in terms of local object UVW coordinates. The map itself is displayed as a gizmo and the position of the gizmo relative to the object controls the manner in which the material map is projected. If you do not assign mapping coordinates to an object carrying raster materials, you will receive a warning when you try to render the scene.

When you create an object, you are given the option of assigning mapping coordinates. Objects that are prisms, cylinders, spheres, cones, or other primitive shapes can be mapped quite effectively this way. If you do this, 3D Studio analyzes the object and assigns what it feels is an appropriate map (Figure 10.6). If the results are acceptable, you may never need to address the material's UVW map gizmo.

STEP 1

Create an object. In this case, a spline shape was lathed about the Y-axis.

STEP 2

Assign mapping coordinates. At the very bottom of the **Modify** roll-out, check **Generate Mapping Coords** (Figure 10.7). Mapping coordinates are assigned to the lathe. Raster materials can now be applied to the lathe.

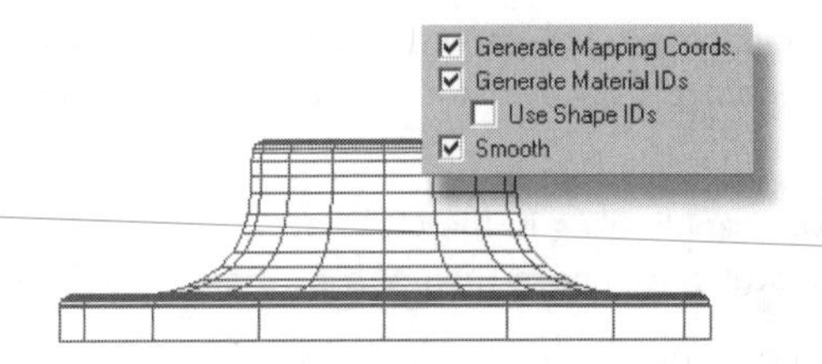

Figure 10.6 *Assign material map coordinates when an object is created.*

You can always explicitly assign mapping coordinates. Many times you will want to customize how materials are applied when you create objects highly irregular in shape. To explicitly assign mapping coordinates to geometry, follow these steps.

STEP 1

Select the object. Choose **Modify|UVW Map**. Choose **Sub Object|Gizmo** to see the mapping gizmo.

STEP 2

Choose the mapping type. Some mapping applications are obvious and some require a certain amount of experimentation. In this roll-out you can align, scale, rotate, and tile the material. The handle on the gizmo (Figure 10.7) represents the top of the raster map.

MAPS IN THE MATERIAL EDITOR

In the last chapter we spent our time in the top portion of the Material Editor. There, we adjusted the ambient, diffuse, and specular colors and reflective characteristics that cause the surface to be hard, soft, or transparent. With an understanding of where raster materials come from and how they are mapped onto 3D Studio geometry, we can now use various material channels to load and then apply materials to objects in our scene. Figure 10.8 shows the lower portion of the **Material Editor** with the material channel roll-out opened up. In this chapter we will concentrate on three of the material channels: ambient, diffuse, and opacity. Later, we will address the displacement and bump mapping channels. If you observe Figure 10.8, you can see that these channels can be turned on and off and the effect of the raster map adjusted by the numerical spinner fields.

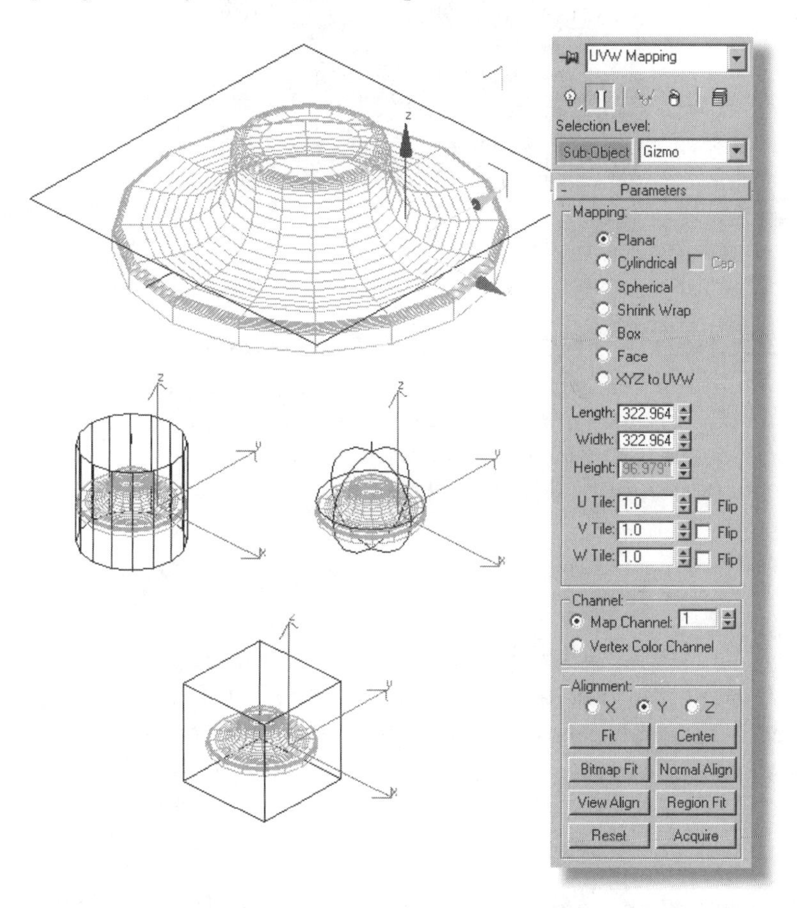

Figure 10.7 *Assign material map coordinates when an object is created.*

Tip: Only one set of mapping coordinates can be active for a given object. You can switch and render to see the effect, but only one can be active. For irregular geometry the most effective mapping type may not be evident. Remember that the material is "projected" from the gizmo to the geometry. This can lead to unusual results as planes on the object become parallel to the projection. For highly irregular geometry, try the "Shrink Wrap" or "Spherical" options.

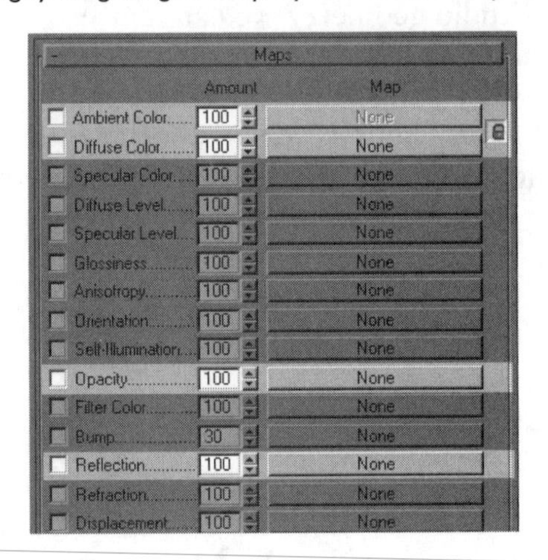

Figure 10.8 *Materials section of the Material Editor.*

USING SCANS OF ENGINEERING DRAWINGS

Before we build materials that represent metal, plastic, rubber, or glass, a simpler, but nonetheless important, use is to place a scanned engineering drawing on a surface so that it can be used to guide modeling. The resolution of the scan should be 150 to 300 dpi and the scan physically as large as possible.

STEP 1

Scan the drawing. The drawing should be scanned in grayscale with sharp contrast. Use the highest quality drawing available because the scan will be another generation removed from the original drawing. Make note of the physical dimensions of the scan. Save the scan in BMP format.

STEP 2

Make the scanned drawing a material. Open the **Material Editor** and **Maps|Diffuse Color|None**. In the **Material/Map Browser**, click

New|Bitmap and browse to the scan. In the **Coordinates** roll-out, deselect both **Tile** check boxes.

STEP 3

Create the drawing plane. In the Front view, create a **Plane** that matches the dimensions of the scan. This plane is created at Y=0 world, because you will be drawing on the Y=0 X-Z plane. Move this drawing plane rearward to Y=1.0 world. At the bottom of the **Modify** roll-out, check **Generate Mapping Coords**.

STEP 4

Apply the drawing material map to the plane. Select the drawing plane in the Front view. In the **Material Editor** toolbar, click **Assign Material to Selection** and then **Show Map in Viewport**. The scanned drawing is displayed on the drawing plane.

STEP 5

Use the scan. Set the grid to a division that corresponds to the measurements on the drawing (decimal inches, fractional inches, millimeters). Move the drawing plane until a major feature (like the center of a circle) is aligned to the grid. By turning **Snap to Grid** on and off, you can accurately overdraw the scan with 3D Studio geometry. Figure 10.9 shows a drawing in position and 3D Studio **Spline** objects in place.

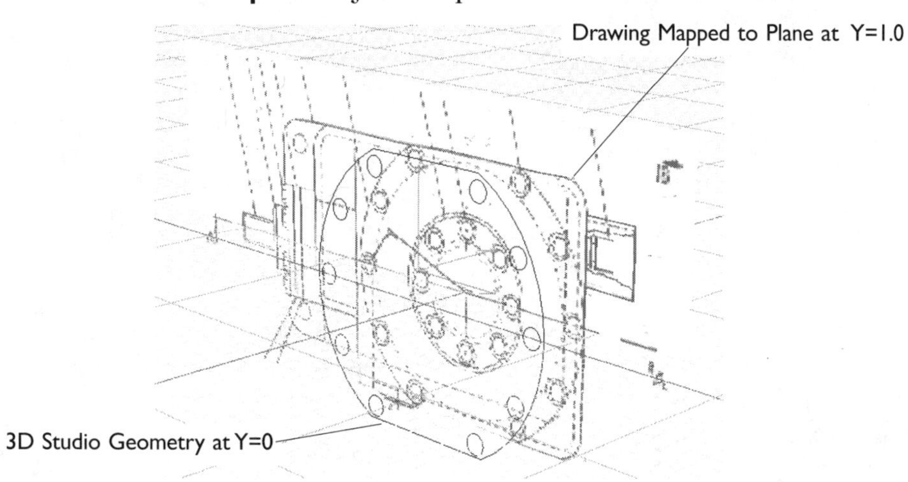

Figure 10.9 *Engineering drawing for overdrawing as material mapped to plane.*

MATERIAL IN THE DIFFUSE AND AMBIENT CHANNELS

In most cases you will want to lock the ambient and diffuse channels so that the same raster map is loaded into each. The **Lock Icon** straddles the **Ambient** and **Diffuse** channels in the **Maps** roll-out. By doing so, you intensify the diffuse map's impact under all lighting conditions. The ambient channel is applied in the shade, where light objects don't illuminate; the diffuse channel is applied to the illuminated portion. In both cases, any **Amount** less than 100 will mix the maps with their underlying ambient and diffuse colors so choosing these colors correctly is very important. Figures 10.10 through 10.13 outline how a diffuse material of pitted and rusted metal is applied to the inner race of a bearing.

STEP 1

Build the material. Choose the desired shading method. Assign **Ambient**, **Diffuse**, and **Specular** colors appropriate for the desired diffuse map as well as **Specular Highlight** settings (Figure 10.10). This is the basic object color that will mix with the diffuse map.

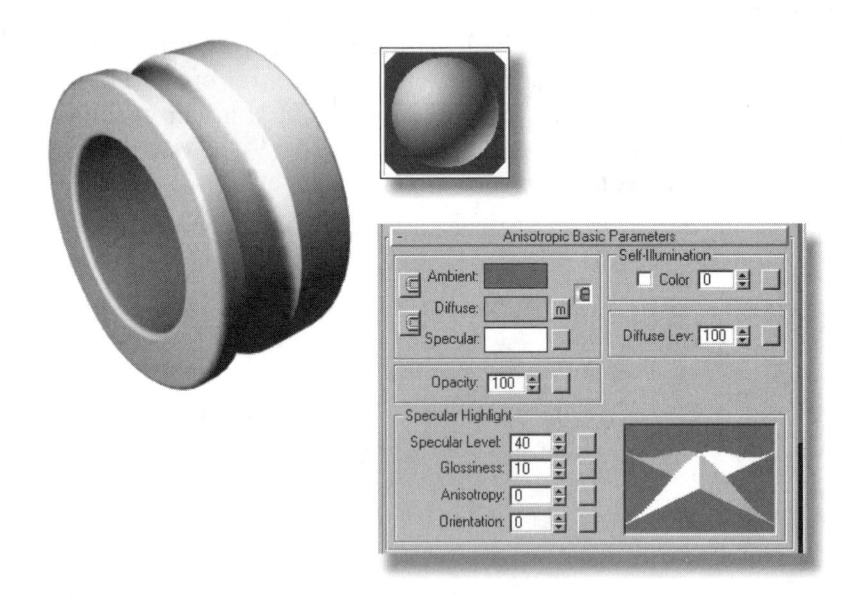

Figure 10.10 *Basic object color.*

STEP 2

Assign map to diffuse channel. Activate the sample slot that will hold the material. Open the **Maps** roll-out in the **Material Editor**. Click on the **Diffuse Color|None** button. Select the desired diffuse map. Click on **Assign Material to Selection** in the **Material Editor** toolbar and preview the material with a quick render.

STEP 3

Adjust the map. Choose **Modify|UVW Map|Sub Object|Gizmo** and select a map type. In this case a **Planar** map is aligned to the X-axis. The size of the diffuse map can be adjusted by scaling the gizmo (Figure 10.11).

This scales the bitmap pattern as it is projected onto the bearing. The number of times the map repeats across the object is controlled by **Tiling** settings in either the **UVW** or **Diffuse Map** roll-outs (Figure 10.12). The final mapped bearing race is shown in Figure 10-13.

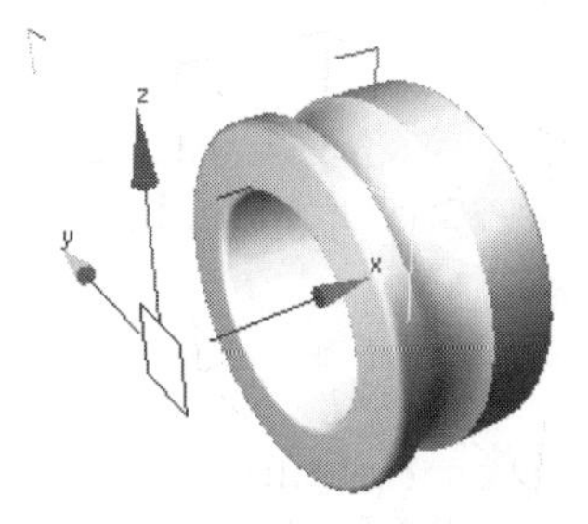

Figure 10.11 *Adjust the size of the material map.*

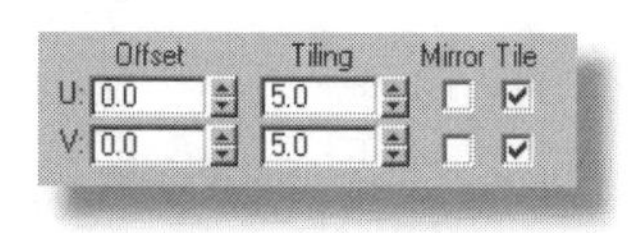

Figure 10.12 *Adjust repeating with **Tile**.*

Figure 10.13 *Bearing race with material applied.*

APPLYING A DECAL

The diffuse channel can also be used to apply a decal, a bitmap with a transparent background. Without involving composite maps (where you can overlay as many bitmaps as you wish), this is slightly problematic because once the diffuse channel is occupied by the decal, it can't be used for a general object material. Figure 10.14 shows a raster bitmap that will serve as a decal on a galvanized bucket. The portion of the image you want transparent must be black. Grays will be interpreted as varying levels of transparency. Colors will be decaled as colors.

Figure 10.14 *Decal bitmap.*

STEP 1

Set the basic color parameters. Choose the **Blinn** shader and set **Ambient** and **Diffuse** to **H=0**, **S=0**, **V=90**. Set **Specular Level=50** and **Glossiness=35**.

STEP 2

Load the decal. In the **Maps** roll-out, choose **Diffuse Color**. Click on **None** and open the **Material/Map Browser**. Choose **New** and browse to the decal bitmap.

STEP 3

Turn tiling off. In the **Coordinates** roll-out, uncheck both boxes for **Tile**. With tiling off, the decal will be applied once.

STEP 4

Specify the alpha source. In the **Bitmap Parameters** roll-out, choose **RGB Intensity** as the **Alpha Source**. By doing so, 3D Studio will look at pure black as transparent.

Apply the decal to the bucket. Depending on the physical size of the decal you may need to choose **Modify|UVW Map|Sub Object|Gizmo** and uniformly scale the decal. When properly scaled, the decal appears on the bucket in Figure 10.15.

Figure 10.15 *Decal applied and scaled.*

Because the diffuse channel is occupied holding the decal, we have to use another channel to hold the galvanized material. The ambient channel won't work because its material is illuminated only by ambient light. Our scene is well-illuminated by omni lights.

STEP 5

Load the galvanized material into the reflection channel. Turn down reflection strength to 50 so that the reflection doesn't wash out the decal. The decaled galvanized bucket is shown in Figure 10.16.

Figure 10.16 *Galvanized material and decal.*

OPENINGS WITH THE OPACITY CHANNEL

The opacity channel uses grayscale values to determine opacity: white is opaque, black is transparent, and shades of gray are interpreted as more or less transparent. It is important to note that geometry is not removed so this is not a substitute for modeling operations such as extrusions with nested shapes (openings) or Boolean subtraction. However, with opacity maps you can simplify the model for openings that are not critically inspected.

The opacity map blocks the rendering of material maps so that the affected area appears to be clear. Specular level and glossiness impact how reflections appear in this area.

Take the boiler panel in Figure 10.17 for example. The inspection window could be removed by Boolean subtraction. But this, of course, could increase the complexity of the part. The solution is to use an opacity map.

STEP 1

Create the opacity map. Figure 10.18 shows this map. (The black line around the outside represents the limits of the map and doesn't actually exist.) The black will be transparent and the white will be opaque. It is important to make the overall dimensions of the opacity map larger than the object. Any portion of the object outside the map will be treated as transparent.

STEP 2

Load the opacity map into the **Opacity Channel**. Turn off **Tile** options. Select the panel and turn on **Generate Mapping Coords** in the **Modify** rollout. Test render the scene.

STEP 3

Reposition if necessary. If the opening is not in the correct location, choose **Modify|UVW Map|Sub Object|Gizmo** and scale and move the mapping gizmo. It should be a **Planar** map aligned with the **X Axis**.

Figure 10.17 *Boiler panel with inspection window.*

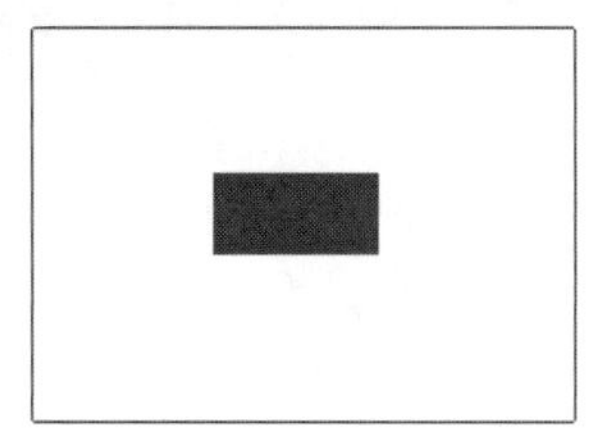

Figure 10.18 *Opacity map for inspection window.*

Note: For an opacity map to work, a material map must be applied in one of the map channels. With our boiler panel, a rusted material has been applied in the diffuse channel.

ENVIRONMENT WITH THE REFLECTION CHANNEL

True reflections are created by ray-traced materials. During raytracing, the impact of every light ray from every light source striking or passing through every surface is calculated. It produces photographic results but is computationally intensive and of less value in engineering and technology than in entertainment or industrial design. However, with a little planning, realistic reflections can be created using raster maps in the **Reflection Channel**.

Bitmaps loaded into this channel are overlayed on ambient and diffuse maps, giving the appearance of reflecting the surrounding environment. Take the pressure vessel in Figure 10.19. As one would expect, the stainless steel material would reflect the outdoor environment in which it is placed.

Figure 10.19 *Reflection map with environment.*

STEP 1

Create the reflection map (Figure 10.20). This map started life as *sky.jpg* from the 3D Studio **Materials Library** (a). This sky was edited in Photoshop to produce a sky, horizon, and gravel foreground (b). This becomes the reflection map.

STEP 2

Create the seam map. This map is simply a white background with a black grid representing the seams (c). Loaded into the **Materials Editor** the diffuse and reflection maps appear as in Figure 10.19.

STEP 3

Apply the material maps. Note in Figure 10.20 that two additional materials have been created from the 3D Studio **Materials Library** : **Foliage** and **Gravel**. These are applied to the plane objects defining the ground and mountains.

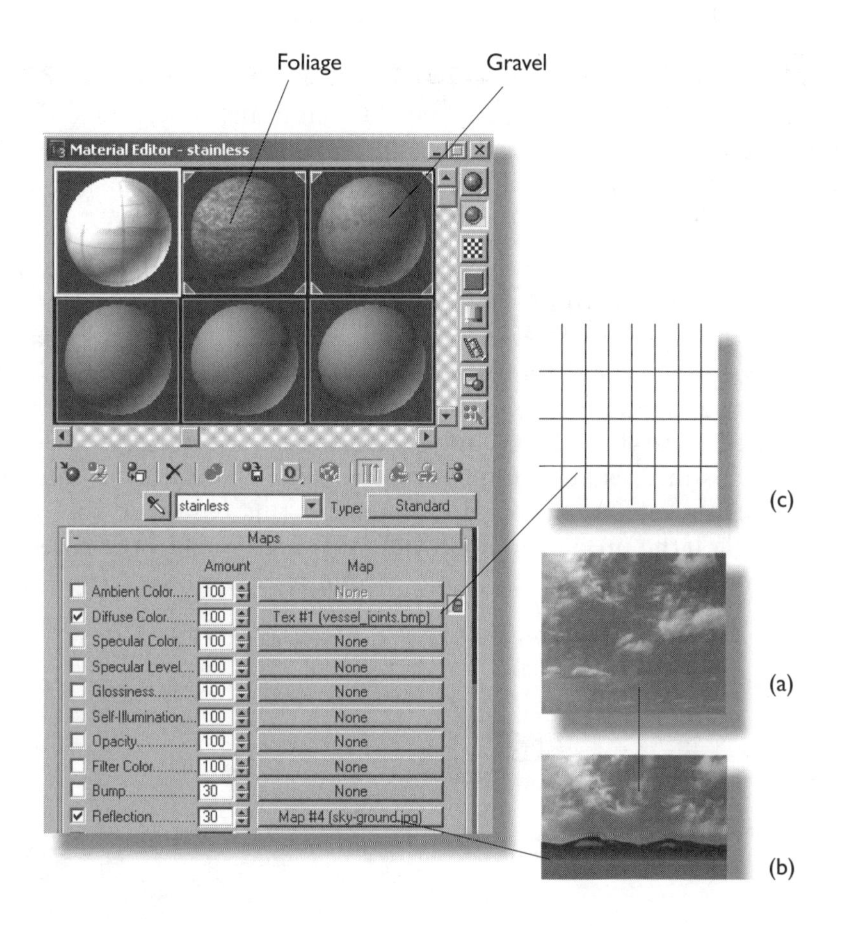

Figure 10.20 *Diffuse and reflection maps in editor.*

STEP 4

Use an environment map in rendering. One of the things that makes reflections look like reflections is if the surrounding environment mirrors the reflection. That's why an irregular horizon was built in the reflection along with the gravel under the pressure vessel. But if we use the same sky in the background of the rendering, we add even more realism. In VIZ, choose **Rendering|Environment|Environment Map** and load *sky.jpg* (**Render|Environment|Environment Map** in MAX). This will fill the background of the rendering above the horizon with the same sky used in the reflection map.

EXAMPLE: HAND TOOL—MULTI/SUB-OBJECT MATERIALS

An object in 3D Studio can have only one material. To overcome this limitation a special material type called *multi/sub-object material* can be created. Essentially, this is a single material that holds other materials so that they can be assigned to subselections of a single object.

There are two ways to create a multi/sub-object material: set the material up before assigning it to the object or create the multi/sub-object material on the fly. To better understand the procedure we will set the material up first. Consider the hand tool in Figure 10.21. It is comprised of a rubber grip, a plastic handle, and chrome tool. One way of modeling would be to create three separate objects and assign a separate material to each. Not only does this increase the number of objects in the scene, it unnecessarily complicates movement during animation. In this case, the tool is formed from a single lathe profile. To apply different materials to the three parts of the tool, a multi/sub-object material is required.

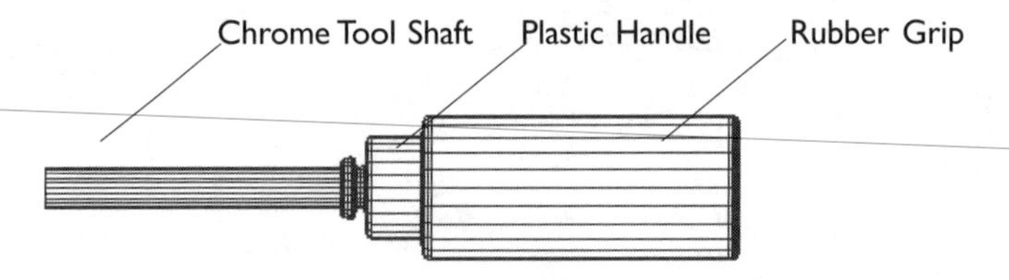

Figure 10.21 *A single object for multi/sub-object material.*

STEP 1

Create the multi/sub-object material. Activate a sample in the **Material Editor**. Click on the **Type** button (**Standard** by default) and choose **Multi/Sub Object**. Because you have yet to make the first material, choose **Discard Old Material**.

The standard **Material Editor** roll-outs are replaced by channels that will hold the various sub-object materials.

STEP 2

Set up the materials. Click **Set Number** and enter **3**. The list is reduced to the number of sub-object materials you will be using. In the open fields to the left, enter appropriate names for the sub-object materials (Figure 10.22). Click on the color chip beside each material and assign an appropriate diffuse color. Check boxes to the right turn the material on and off after assignment. The sample is subdivided to reflect the color assignment of the three materials.

STEP 3

Set up the individual materials. Each sub-object material button contains the typical material roll-outs. For the grip a mottled gray/black diffuse map is added. For the plastic handle, opacity is reduced to a basic yellow plastic material color. No diffuse map is used for the yellow plastic. The tool shaft carries a chrome reflection map (*chromesky.jpg* from the 3D Studio Materials library).

Note: You can always add materials by increasing **Set Number**. But if you reduce the number below the materials you are using, you lose the material and 3D Studio assigns black to that sub-object.

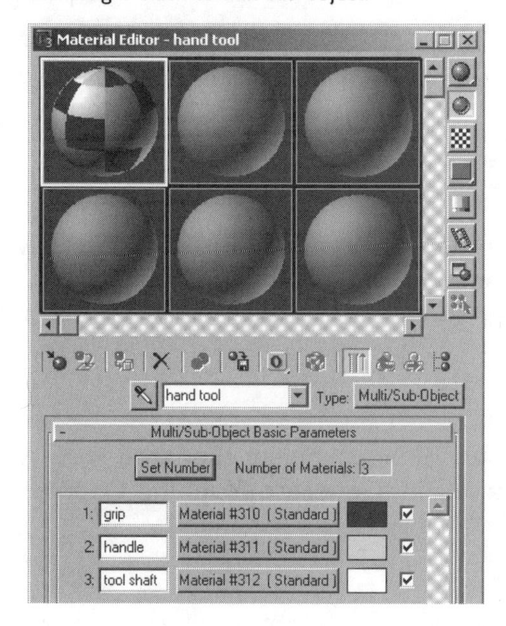

Figure 10.22 *The multi/sub-object interface.*

STEP 4

Create selection sets. In the scene, select the Hand Tool and choose **Modify|Edit Mesh|Sub Object|Face** and create three selection sets: grip, handle, and tool shaft (see pages 175-176 in Chapter 9). In Figure 10.23 the grip has been selected. At the bottom of the roll-out is the **Material** section. Assign the grip the same material ID number as was established in Figure 10.22. Continue this process for the handle and the tool shaft.

Note: You don't have to create selection sets to us the multi/sub-object materials but if you ever have to reselect the sub-objects, having the selection sets can really speed things up.

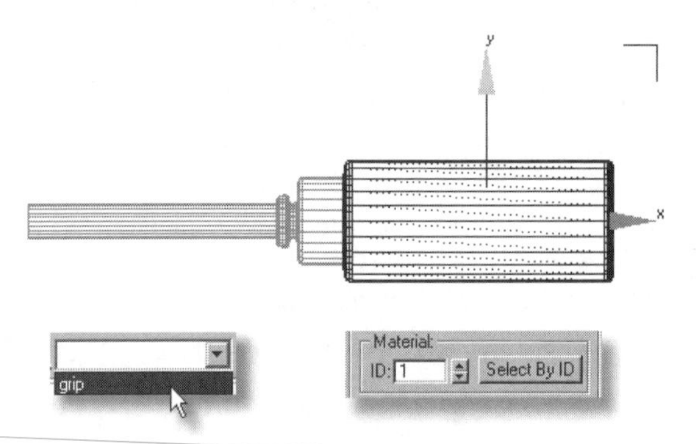

Figure 10.23 *Selection set and material ID.*

Tip: To select sub-object faces you will often have to switch between rectangular, circular, and irregular selection shapes. Also, hold down the **CTRL** key to add to a selection; hold down the **ALT** key to subtract from a selection.

STEP 5

Generate mapping coordinates. Select the Hand Tool. At the **Lathe** level of the **Modifier Stack**, check **Generate Mapping Coords**.

STEP 6

Assign the multi/sub-object material to the Hand Tool. With the Hand Tool selected, click on the **Assign Material to Selection** icon in the **Material Editor** toolbar; 3D Studio assigns the sub-object materials to the selection sets by material ID number (Figure 10.24).

Tip: If the geometry is an editable mesh, you can create a multi/sub-object material "on the fly." When you drop a material on a selected group of faces, that material becomes the first entry (ID=1) in a **Multi/Sub–Object Material**. Subsequent materials can be dropped on the material channels in the **Multi/Sub–Object Material** roll-out and then assigned by ID number to face selections. If you create selection sets, this process is facilitated.

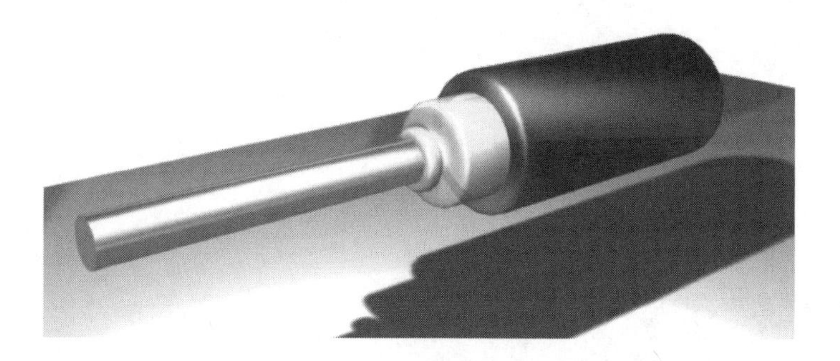

Figure 10.24 ***Multi/Sub–Object Material*** *assigned to selection sets of the Hand Tool.*

PROBLEMS

Problems appropriate for material mapping are found on the following pages and of course, you can use any of your previously completed assignments as subjects for material mapping. Create materials as assigned by your instructor. In each case, be prepared to document your material development as was done in Figure 10.20. Assign appropriate names to all materials, objects, and selection sets.

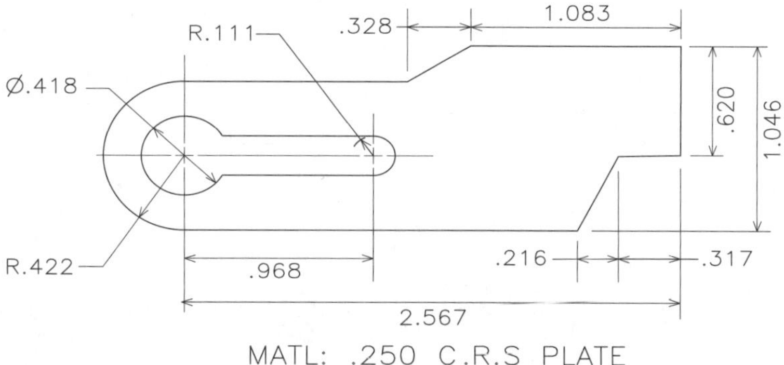

Problem 10.1 *Support Arm.*

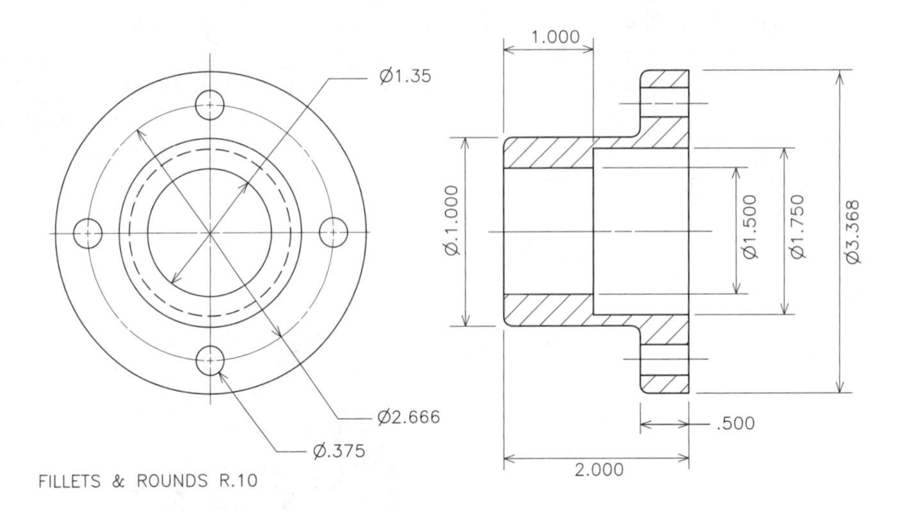

Problem 10.2 *Outer Cap End.*

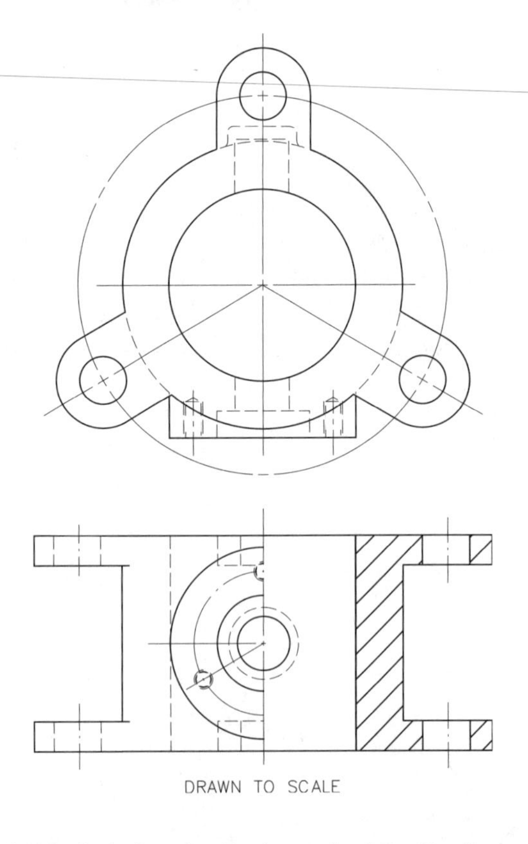

Problem 10.3 *Butterfly Valve Body (Bertoline: Fundamentals of Graphics Communications).*

Modeling with Displacement Maps

CHAPTER OVERVIEW

In Chapter 10 you were introduced to the Material Editor and how raster material maps are used to increase the realism of products you model in 3D Studio. In this chapter you will learn that raster maps can also be used to alter geometry.

A displacement map interprets bitmap grayscale information as changes in elevation perpendicular to map coordinates. The displacement map displaces the underlying mesh based on the gray value corresponding to each node of the mesh. This is similar to an opacity map where grayscale is interpreted as varying levels of transparency. When the modifier stack is collapsed, the impact of the displacement map is frozen into an editable mesh and subsequent modifiers can be applied.

A displacement map depends on the underlying geometry for its effectiveness. For that reason, displacement maps are best applied as large, gently changing surface modifications. For smaller and sharper details, it is better to use bump maps (covered in the next chapter, Chapter 12, Modeling with Bump Maps). Bump maps do not alter object geometry. Instead, their grayscale information is interpreted at rendering time and affects the position of individual pixels.

KEY COMMANDS AND TERMS

- **Blur**—an option within the **Displace** roll-out that allows the impact of the displacement to be softened.
- **Displace**—a modifier that interprets a raster map to move or displace mesh nodes a distance corresponding to the brightness value of a pixel in alignment with the node.

- **Displacement Map**—a bitmap used to displace an editable mesh.
- **Displacement Strength**—the displacement distance, expressed in current units, between black and white pixels in the displacement map.
- **Optimize**—a modifier that provides options to reduce the complexity of an editable mesh.
- **Patch Grid**—a geometry object often used as the base plane for a displacement map.
- **Rasterize**—to convert into a bitmap; to open a vector file in a raster editor.
- **Smoothing**—a modifier that provides options to increase the complexity of an editable mesh.

IMPORTANCE OF DISPLACEMENT MAPS

It is difficult to edit mesh geometry with consistent and gently gradated surface changes by using **Sub–Object** modifiers and moving selections by hand. Take Figure 11.1 for example where a lofted NURBS surface (a) has been converted to an editable mesh. Its surface has been altered by a displacement map (b) into the new surface (c). Cylindrical mapping coordinates have been rotated around the world Y-axis until the bulge is in the desired position. You can see from the wireframe (d) that the geometry has indeed been changed by the displacement map.

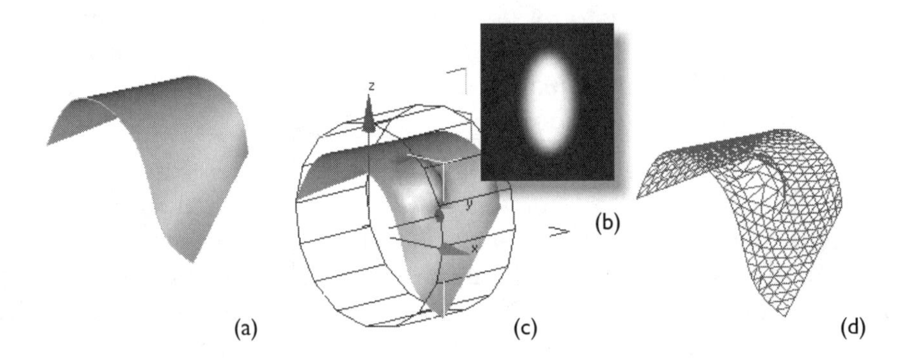

Figure 11.1 *The impact of a displacement map on mesh geometry.*

You can also see from the wireframe in Figure 11.1(d) that the displacement map is sensitive to the resolution of the underlying mesh. Mesh nodes on the original surface and their spacing have been displaced to the point that too few may be available to accurately define the change. Some of this coarseness may be taken care of with **Smoothing**, but irregularities may still persist unless the underlying mesh has both sufficient subdivisions and a mesh configuration that matches the displacement map.

Some elevation information is naturally in grayscale or can easily be converted to grayscale. Shadows from satellite photographs can be interpreted into various grays that represent elevation change. Terrain features are more commonly encountered in architecture than in manufacturing. Still, civil or construction engineers have to place their projects on a site, usually with some terrain elevation change. Terrains can be interpreted from aerial photographs into a displacement map that accurately models the surface. Or intervals from contour maps can be assigned values representing relative heights. When mapped to a *patch grid*, the displacement map forms the terrain.

Figure 11.2 shows a contour map and its grayscale raster interpretation. In the previous chapter you were presented with the topic of material mapping scanned engineering drawings as the basis for overdrawing. A scanned terrain map (a) provides the contour intervals for gray assignment (b). The blurring (c) in a raster editor such as Photoshop provides a smooth transition between elevations.

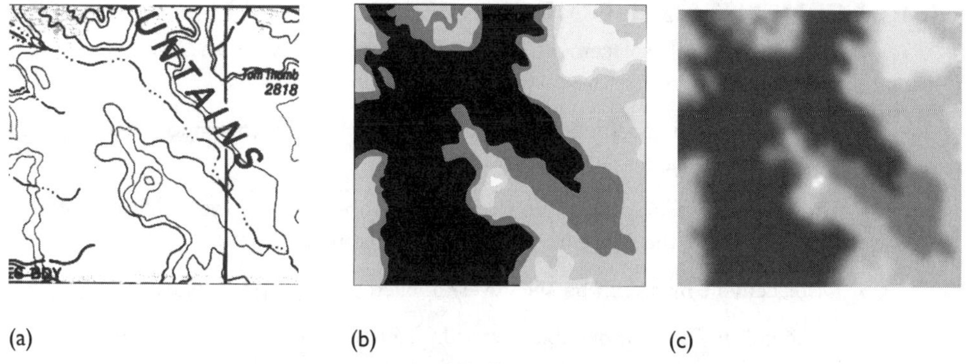

Figure 11.2 *Contour intervals and corresponding grayscale displacement map.*

Tip: You can create a "hot link" between the displacement map and the Materials Editor. Load the displacement map in **Modify|More|Displace|Image|Map** and drag the button with the map's name and drop it on an empty sample in the Materials Editor. The material type becomes **Bitmap** and you can use the map controls to crop the image without going back to the raster editor,

When this displacement map is applied to a patch grid (Figure 11.3) the underlying mesh is displaced (a) relative to the grayscale values in the map, producing the desired terrain (b).

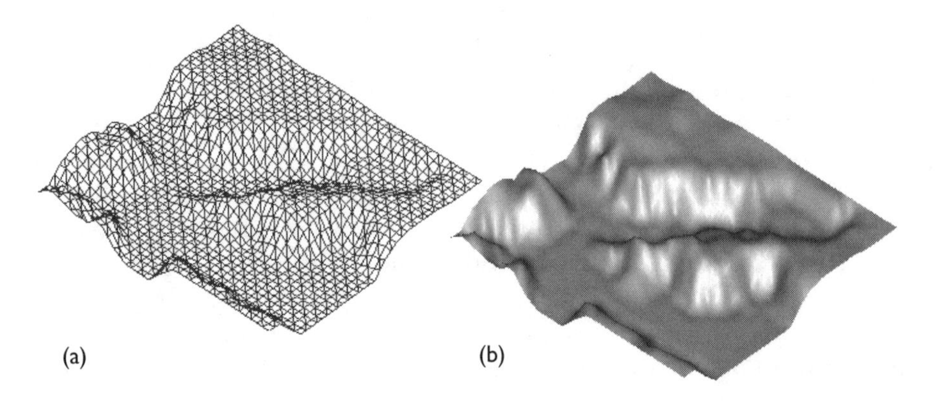

Figure 11.3 *Effect of displacement map.*

Tip: The displacement map needs to either be the exact dimensions of the surface it is applied to or sufficiently larger so that its edges fall outside the limits of the geometry.

DISPLACEMENT MAP GUIDELINES

- The greater the displacement, the less effective the results because the underlyinmg mesh has to cover more area.
- The finer the displacement, the finer the mesh must be to support the displacement map.
- The finer the resolution of the bitmap, the more accurate the displacement; the finer the mesh, the more accurate the displacement.
- Displacement maps can be selectively applied to sub-object selections.
- Once collapsed to an editable mesh, displacement mapping coordinates are discarded and material mapping coordinates can be applied.
- Make sure you are operating in units that make sense for the displacement.
- Make the white areas of a displacement map a little larger than anticipated. When the map is blurred, the blurring will "borrow" some of the white material, narrowing the raised portion.

INTERPRETATION OF RASTER MAP

A raster displacement map should be grayscale. If the map is color, 3D Studio will read the value component of HSV colors and affect the displacement. However, as has been mentioned before, it is difficult for humans to visually extract value from color; color maps should be reduced to grayscale before application.

Grayscale is independent of file format. You can have a grayscale TIF, BMP, JPG, or GIF file. There is no reason to use a TIF file. When reduced to grayscale, a BMP file will contain 8 bits of value information per pixel. The GIF format is structurally limited to 8 bits of data per pixel.

An 8-bit grayscale image contains a maximum of 256 separate gray values, more than the human eye can register, and more than even the finest mesh can probably reflect. Two variables work together:

- The grayscale value. White is interpreted as extending the surface; black is interpreted as no displacement of the surface; a middle gray (H=0, S=0, V=127) is interpreted as extending the surface 50% of the displacement distance.
- The **Displacement|Strength.** The number you enter in this field is expressed in currently active units and refers to the displacement of mesh nodes that match spatially with white pixels in the displacement map.

Study the displacement map in Figure 11.4. It contains three values: white, middle gray, and black. We would expect this map to displace a plane to two elevations: the white portion to the full distance and the gray portion to half this distance. The black portion would not effect the plane at all. Figure 11.5 shows a thin surface at Z=0 world with the displacement map applied and a value of **10'-0"** entered in the **Displacement|Strength** field. You can see from the Front view that the white pixels (left side) have displaced the grid +10'-0" the middle gray +5'-0" and the black pixels 0'-0". Figure 11.6 shows the original Y=0 position of the plane and its displaced position. Figure 11.7 shows the rendered surface.

Figure 11.4 *Three-ramp displacement map.*

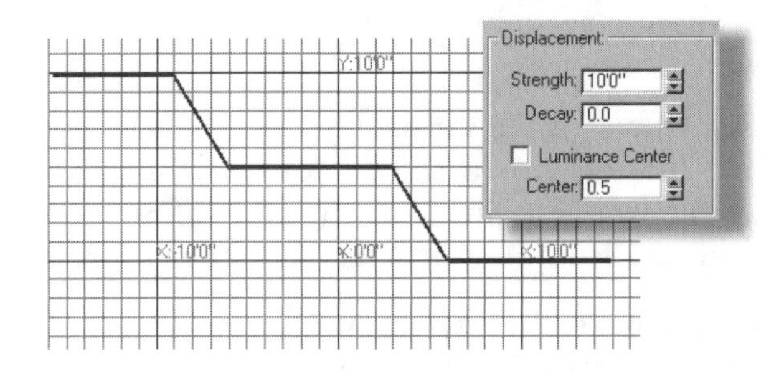

Figure 11.5 ***Displacement Strength*** *acts on the displacement map.*

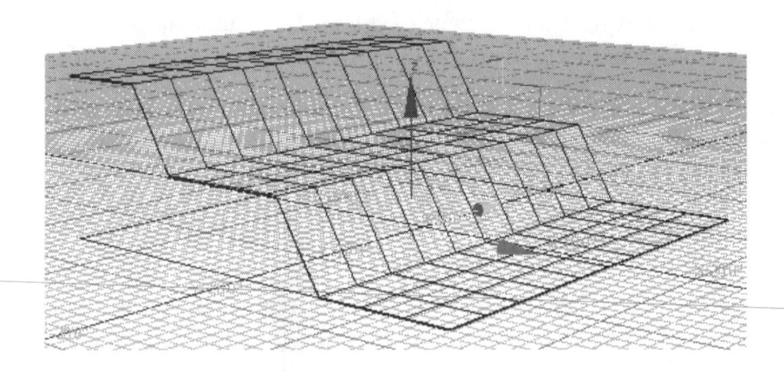

Figure 11.6 *Plane is displaced from Y=0 position.*

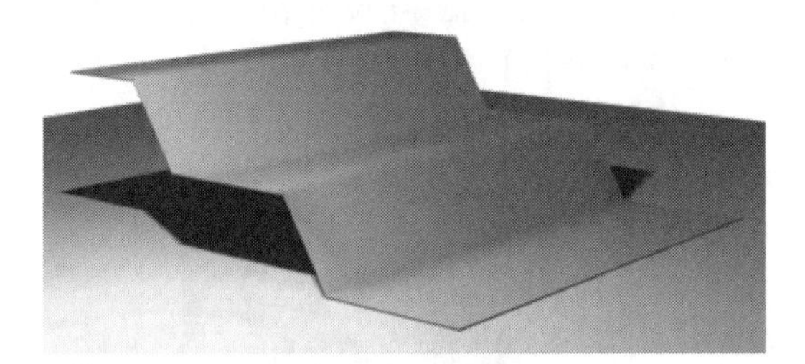

Figure 11.7 *Displaced plane rendered.*

Tip: Even though the displacement map will be white lines and shapes on a black background it is usually easier to work black on white (like you are used to drawing) and then invert the image when done,

GEOMETRIC ANALYSIS

Displacement maps provide an alternative, although nongeometric, method for modeling. True, many displacement map applications could be modeled by a combination of other methods (NURBS or Booleans, for example). But viewing geometry in terms of changes in elevation is often overlooked, and displacement maps provide an elegant method of affecting geometry, if you can see the relationship between shape and grayscale.

Figure 11.8 shows a Pipe Plug with a square end for gripping with a wrench. Because the plug is cylindrical and the end a square prism, a single lathe won't produce this part. The part could be modeled by:

- A combination of cylinder and chamfer box primitives.
- A lathed plug shape with the box pulled as sub-object faces from the top surface of the cylinder.
- A NURBS loft with square and circular cross–sectional curves.

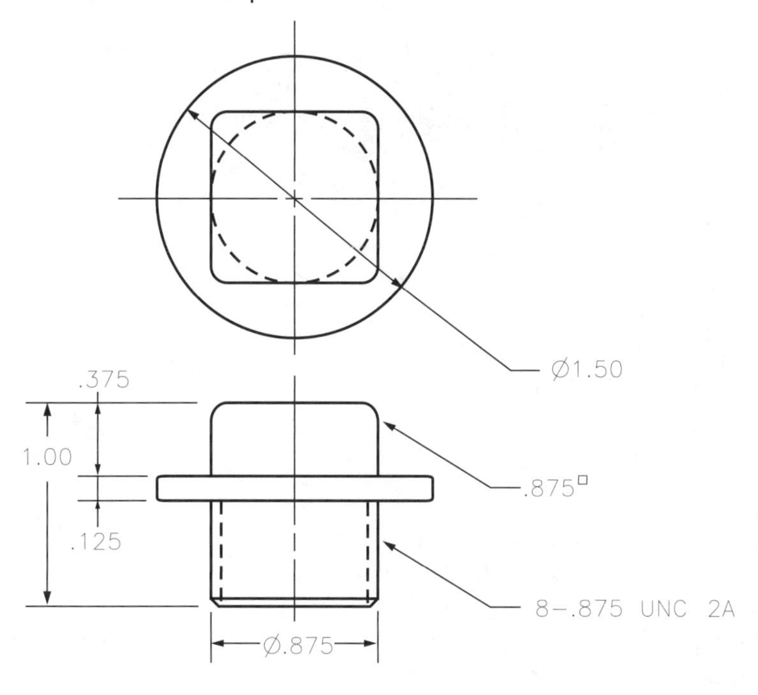

Figure 11.8 *Pipe Plug.*

But because modeling by displacement map may not be evident at first, let's see how this might be done.

STEP 1

Model the Pipe Plug. Create a profile that defines the cylindrical part of the plug. **Lathe** this profile about the Y-axis with **48 Steps** and relocate its **Pivot** to X=0 world.

After the lathe, the top surface has 48 pie-shaped polygons. This is insufficient for a displacement map so the mesh of top surface must be increased. Collapse the stack so that the plug becomes an editable mesh. Choose **Modify|Sub Object|Faces** and in the Front view, select the top surface. Rotate the view slightly to assure that all faces have been selected. Choose **Modify|More|Tessellate** and specify **Tension=0** and **Iterations=3**. The top surface is subdivided into a fine mesh (Figure 11.9). Collapse the stack again.

Note: Depending on how the mesh aligns with the displacement map you may be able to set iterations to 1 or 2.

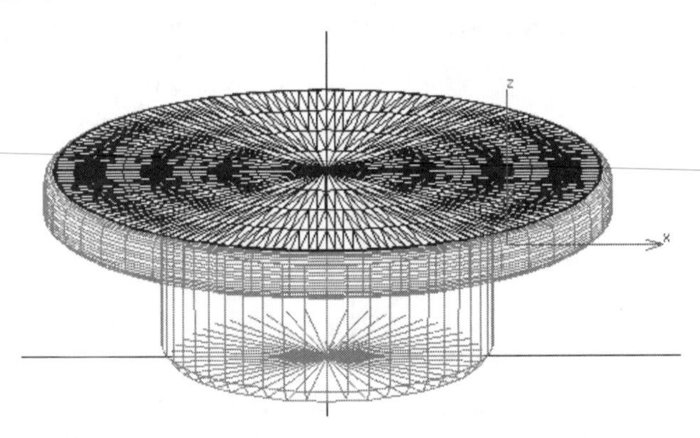

Figure 11.9 *Top surface prepared for displacement map.*

STEP 2

Size the displcement map. The top of the Pipe Plug is 1.50" diameter and the wrench box is .875" square. The resolution of the displacement map should be 300 dpi (so a correct grayscale pixel will have the best chance to fall on a mesh node). The physical dimensions of the raster displacement map just need to be square, so we will use 1.50" x 2=3.0".

STEP 3

Create the displacement map. In a raster editor such as Photoshop, create a canvas 3.00 x 3.00 at 300 dpi. Use a white background to start out (Figure 11.10). Using rulers and guides, create the basic Top view geometry of the plug and wrench box. You can try this freehand but you'll find that if you use the guides to create a *path* (Figure 11.11), you will be more accurate. Using the path as a selection, you can fill the wrench box with black (Figure 11.12).

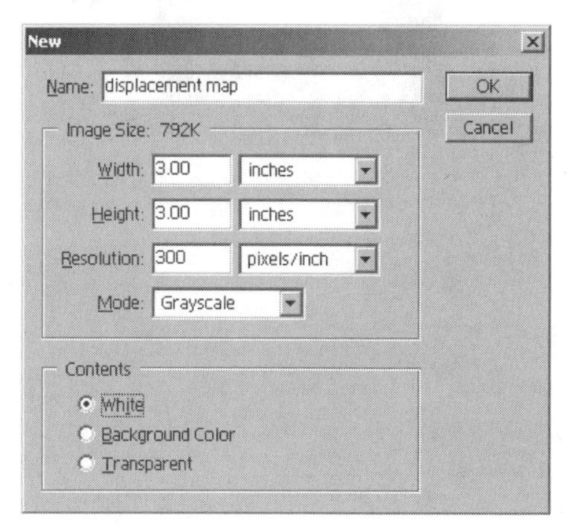

Figure 11.10 *Dimensions of displacement map.*

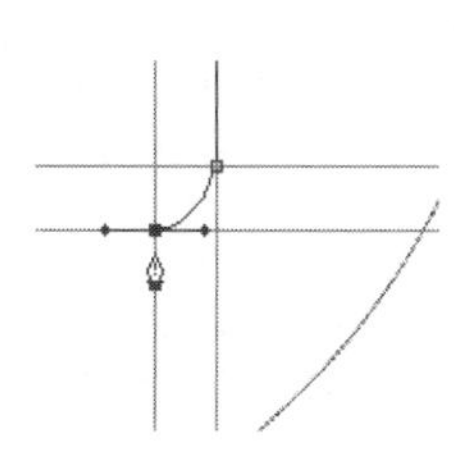

Figure 11.11 *Guides are used to control path.*

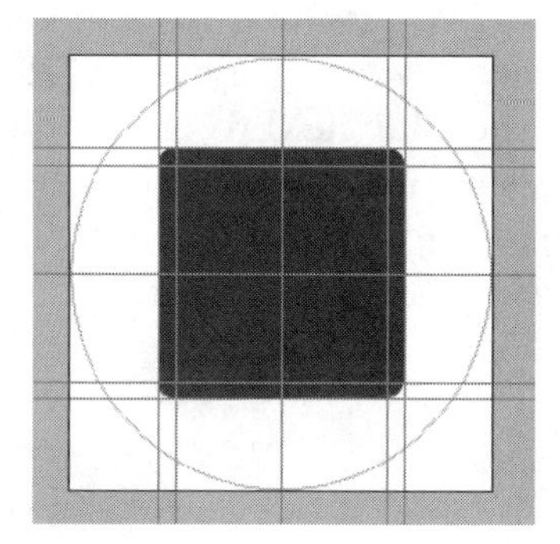

Figure 11.12 *Wrench box filled.*

We know that black pixels in the displacement map will not affect the mesh, while white pixels will move mesh nodes a specified distance. Invert the bitmap (making white pixels black and black pixels white. Remove the plug circle (blacken in) and you have a displacement map (Figure 11.13).

Figure 11.13 *Final displacement map.*

STEP 4

Apply the displacement map. In the Front view, choose **Modify|Edit Mesh|Sub Object|Face** and select the top of the Pipe Plug. Press the **Space Bar** to **Lock** this selection. Switch to the Top view. Choose **Modify|More|Displace** and specify **Planar** mapping and **Alignment:Y|Fit**. The mapping **Gizmo** is the same size as the Pipe Plug. Rotate the view slightly to see the effect of the displacement map.

In the **Modify|Displace** roll-out click on **Image|Map|None** and browse to the displacement map. Enter **–.375** in the **Displacement|Strength** field. Enter **.5** in the **Blur** field to create fillets and rounds. Based on the complexity of the mesh, a more or less smooth wrench box is formed. Figure 11.14 shows the planar mapping coordinates in place and the result of the displacement mapping.

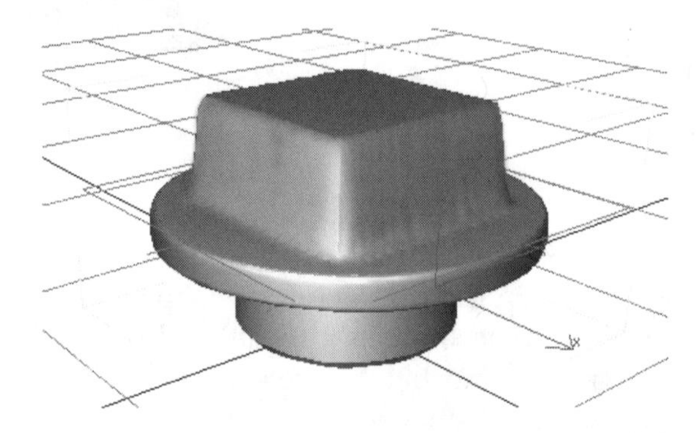

Figure 11.14 *Result of displacement.*

Tip: Once the mesh has been altered by the displacement map you may want to use two other modifiers—**Mesh Smooth**, or **Optimize**—to reduce imperfections in the displaced portion. Be aware, though, that smoothing may take a while and you can optimize the shape away.

EXAMPLE: COVER PLATE

Displacement maps must be created in a raster editor. However, for technical purposes, raster editors do not generally have the tools necessary for technical constructions. For that reason, you may choose to lay out the displacement map in a CAD or illustration program. You can then rasterize the vectors in a paint program as the basis of your displacement map. Consider the views of the Cover Plate in Figure 11.15. The strengthening ribs and ring are candidates for a displacement map. The ring will carry the full displacement distance. The ribs will begin white and blend to black, forming the angular displacement.

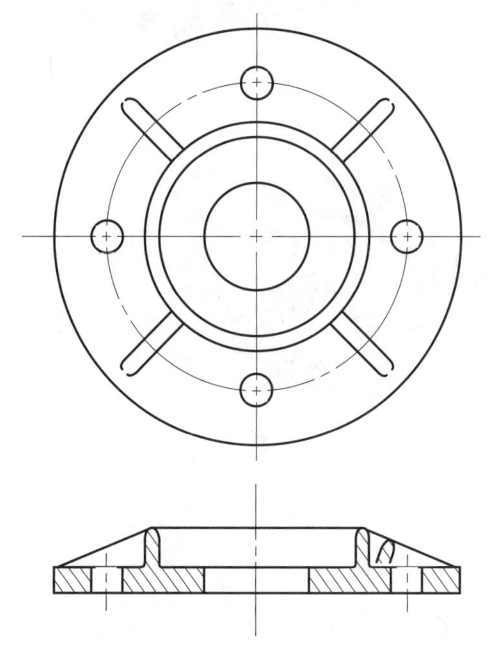

Figure 11.15 *Cover Plate.*

STEP 1

Create the plate. The base plate (without ring or ribs) can be modeled by extrusion. However, extrusion provides no option for specifying capping segments and in this case, a sufficient mesh is required to support the displacement map.

So here, cylinder primitives (hole tools) are subtracted from a cylinder having **48 Sides** and **48 Cap Segments**. **Height Segments** are unimportant (Figure 11.16).

Tip: It is convenient to have a segment align with a feature (like a rib) you are forming with the displacement map. This assures that the maximum number of mesh nodes will be centered on the feature. If you are displacing four ribs, choose a number of segments evenly divisible by 4.

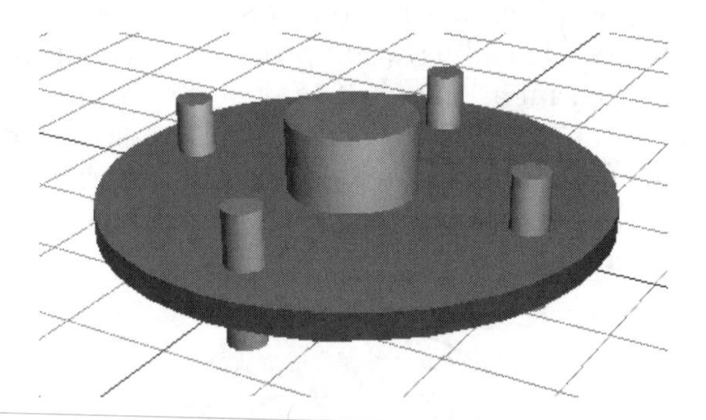

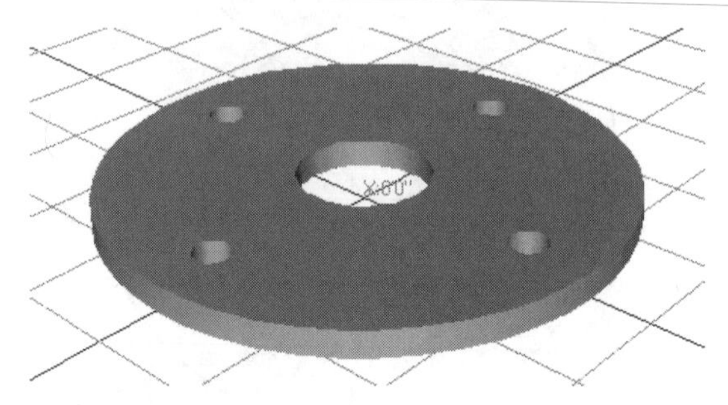

Figure 11.16 *Base Cover Plate by Boolean subtraction.*

Tip: If you have visibility problems during multiple Boolean subtractions (holes disappear, subtractions are not complete), decrease the geometric complexity of the cylinder tools to 1 cap and 1 height segments.

STEP 2

Create the displacement map. The base elevation will be black and the ring white. The ribs will have a gradient fill, white to black, from the ring to the base elevation. It may be easier to rasterize the vector top view and use this as a template in Photoshop.

Figure 11.17 shows the displacement map. A Gaussian blur has been applied to the entire image to provide fillets and rounds in the ring and rib intersections.

Figure 11.17 *Cover Plate displacement map.*

> **Tip:** To move geometry from 3D Studio to a raster editor you can simply do a screen grab (**CTRL-SHIFT-Print Screen**). The screen image is placed on the system clipboard. A new file's dimensions in a raster editor will automatically be the dimensions of the screen grab. Paste the screen grab and crop to the desired geometry. Or, if you need greater than screen resolution, choose **Export Selected** and specify **Adobe Illustrator (.AI)** format. When opened in a raster editor such as Photoshop, the vectors can be rasterized at a user-specified resolution.

STEP 3

Apply the displacement map. Select the plate and collapse the stack to an editable mesh. Choose **Modify|More|Displace** and choose **Planar|Fit** mapping coordinates aligned with the face of the plate. Choose **Image|Map|None** and browse to the displacement map. In the **Displacement|Strength** field, enter a value equal to the height of the ring above the top surface of the plate.

Figure 11.18 shows the result of the displacement map being applied to the editable mesh. Compare the result to the views in Figure 11.15. The ring has been raised an amount equal to its height above the Cover Plate and in doing so, the ribs have been displaced an amount proportional to the descending gray values. This creates ribs that angle from the top of the ring to the top surface of the base.

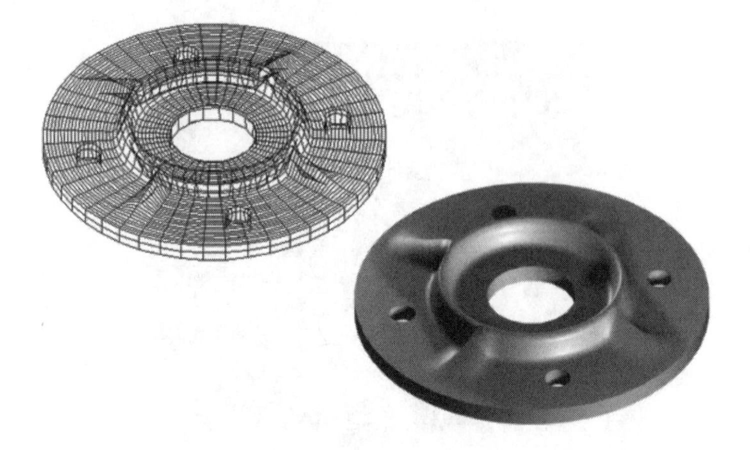

Figure 11.18 *Cover Plate wireframe and rendering.*

PROBLEMS

Problems appropriate for displacement mapping are found on the following pages. Although each could be modeled by a variety of methods, analyze each in light of displacing a **Editable Mesh** to arrive at an acceptable result. In each case, be prepared to document your displacement map development. Assign appropriate names to all materials, objects, and selection sets.

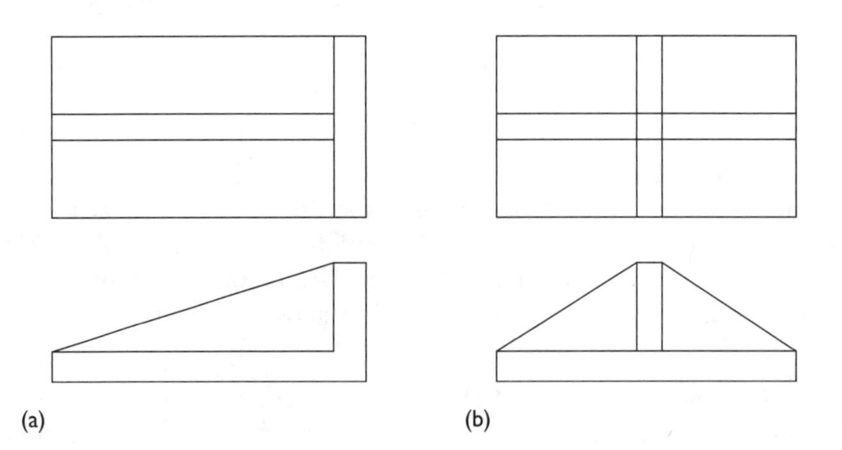

Problem 11.1 *Objects for displacement mapping.*

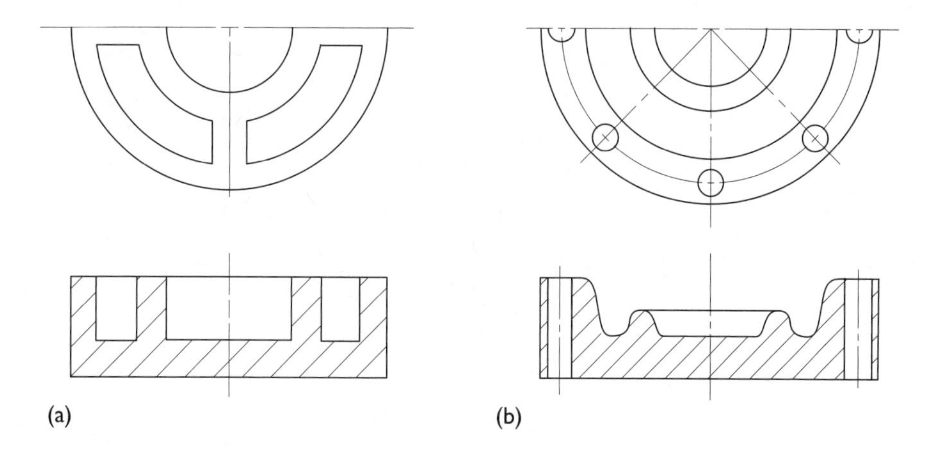

Problem 11.2 *Objects for displacement mapping.*

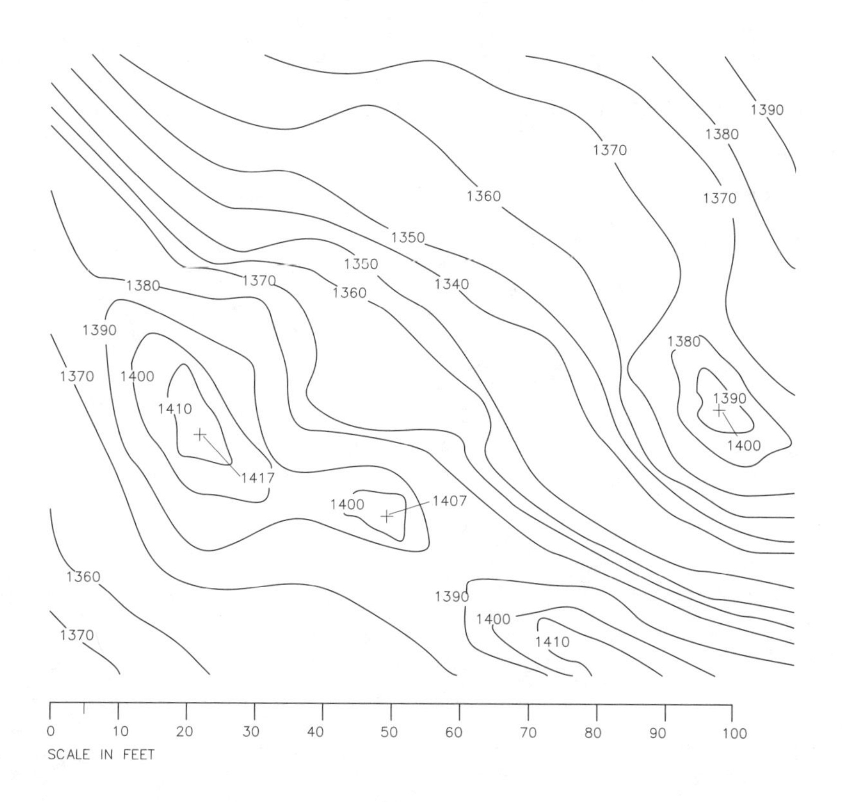

Problem 11.3 *Terrain Map. Develop a grayscale displacement map based on the contour intervals shown. The EPS vector file can be found on the CD-ROM in chapters\chapter_11.*

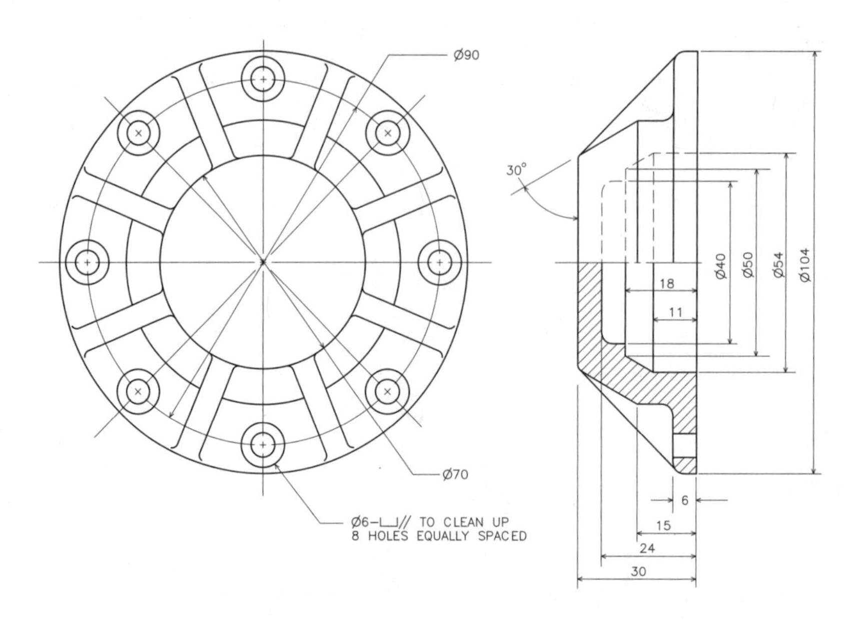

Problem 11.4 *Cutting Head Cover (Luzadder and Duff:* Fundamentals of Engineering Graphics*).*

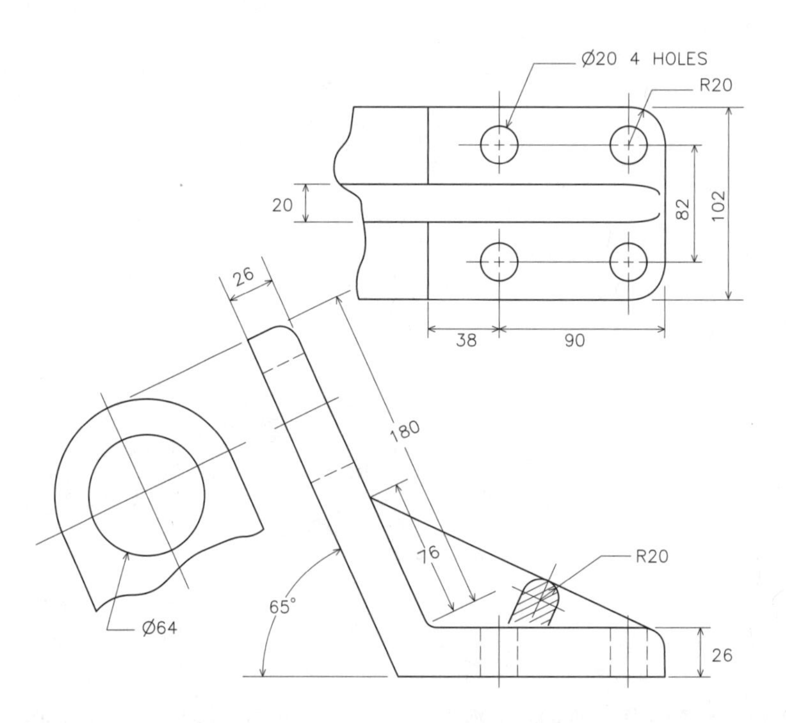

Problem 11.5 *Cable Anchor (Earl:* Graphics for Engineers*).*

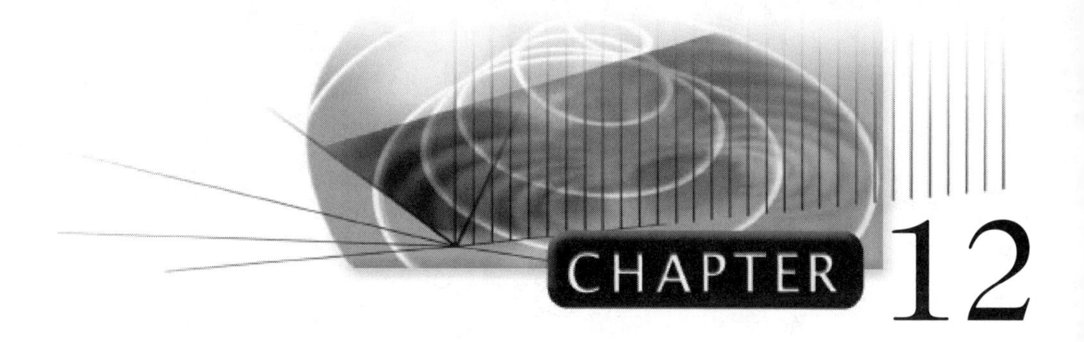

Modeling with Bump Maps

CHAPTER OVERVIEW

The first eight chapters of this text were devoted to technical modeling because every effective demonstration, presentation, or simulation is first based on accurate geometry. You then saw how the addition of object color and raster materials added to surface realism. But even after these efforts you may find that your models lack realism. This is because objects in the real world are not computer-perfect. They have imperfections and are stained, tarnished, dented, and scratched. Castings have sprue and riser marks, flashing and parting lines, and surfaces that take the impression of the mold material. Machined or turned surfaces often bare the marks of the tools that created them. Fasteners have threads and splines, and handles can have knurls or abraded areas for grip. Taken together, these surface characteristics can make the difference between realistic computer models and more crude representations.

Tip: Not every computer model needs photorealistic attention. Apply the "TGE" rule. When you can say "that's good enough" based on how the model will be used, any further texture mapping, ray tracing, or bump mapping just adds to the cost of the project.

In the previous chapter you were presented with a method where a raster map was used to displace an editable mesh. This was fine for large deformations but unsuitable for detail as fine as the threads on a fastener. For these fine details we need to model at the pixel level, much finer than the underlying geometry. Because computer models don't have pixels assigned to surfaces until rendering time, fine detail is not modeled, rather rendered. The technique that accomplishes this, like displacement maps, uses a grayscale bitmap to displace pixels and is called a *bump map*. Because the resolution of a rendered scene can be controlled, even the finest detail can be represented.

KEY COMMANDS AND TERMS

- **Anti-alias**—the process by which rough raster outlines are smoothed by the blending of adjacent pixels.
- **Bump Map**—a grayscale raster graphic used to affect rendered pixel position perpendicular to the map. The underlying model geometry is unaffected and surface deflections occur only at rendering time.
- **BMP Format**—a raster format native to the Windows operating system appropriate for bump maps. Though capable of lossless run length encoding (RLE compression) 3D Studio does not recognize BMP files that have been compressed.
- **Bump Map Amount**—a relative number that corresponds to the bump of white pixels in map. Middle gray is bumped half this distance. Black pixels are not bumped and remain at the level of the plane.
- **Bump Map Channel**—the area of the **Maps** roll-out where a bump map is activated.
- **Raster Editor**—a program that creates and edits raster graphics.
- **Reload**—the area of the **Bump** roll-out that causes the previously loaded bump map to be reloaded.
- **Render Time**—when surfaces, their textures, and effects of lighting are calculated and the scene represented by a fixed resolution raster file.
- **Show Material in Window**—an icon in the **Material Editor** toolbar that when at the material map level allows map to be visible in the scene window where it can be adjusted.

IMPORTANCE OF BUMP MAPS

Bump maps, because they are affected at rendering time, do not add to the geometric complexity of your models. In fact, because bump maps work at the pixel level, they are not, like displacement maps, dependent on the density of the underlying geometric mesh. This means that the same (and very simple) geometry can have different bump maps for different effects. The same part geometry can be cast, machined, or threaded. This dramatically simplifies modeling and the number of parts you have to keep track of.

Because bump maps are applied in a material map channel, their effect can easily be controlled. The effect of the bump can be turned up or down, or turned off completely. If the bump were modeled, you would be stuck with complexity that you couldn't mitigate or remove. However, because the bump map is a material channel, its precise placement can be problematic. You may want to consider using a multi/sub-object material to place a bump map, using its own sub selection mapping coordinates.

INTERPRETATION OF BUMP MAP

It may be helpful to see the impact of white, gray, and black pixels on a bump map. In the previous chapter we used a three-ramp map to determine the effect of a displacement map. Figure 12.1 shows the effect of this bitmap as a bump map applied to a plane. The mapping gizmo has been scaled and positioned so the bump map is centered on the plane. A value of **500** has been entered in the **Amount** field.

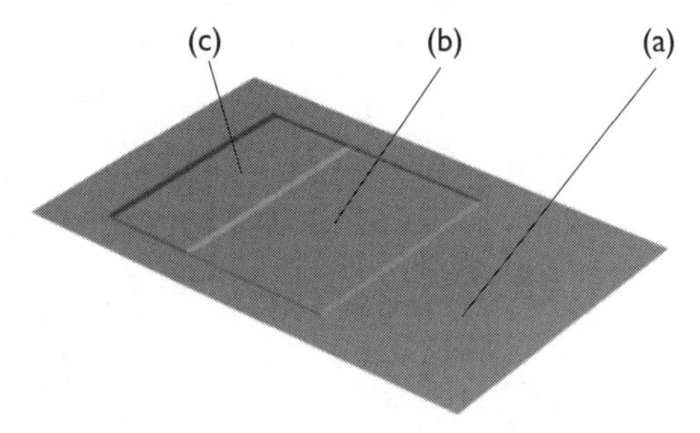

Figure 12.1 *Three-value bump map.*

Black pixels (a) are interpreted as having no effect on the rendered plane. Surface pixels corresponding to white values (c) in the bump map are raised a distance determined by the value in the **Amount** field. Middle gray pixels (b) impact the surface at 50 percent of white distance.

You can also see the difference between a displacement map and a bump map in this example. The edges of the bump map do not alter the underlying geometry as does a displacement map. For this reason, the bump map does not have to be carefully sized in order to avoid the edges of the map being seen in the bump.

BASIC BUMP MAP PROCEDURE

It is common that identifying serial numbers or a company name be cast into a part. Figure 12.2 shows a casting with identifying company name and part number. How can this manufacturing detail be added to the model? Because of the small scale of the lettering, a displacement map is inappropriate. As an alternative, individual letters could be extruded from **Spline|Text** objects, but this would add considerable geometric complexity to an otherwise simplistic model. The answer is a bump map.

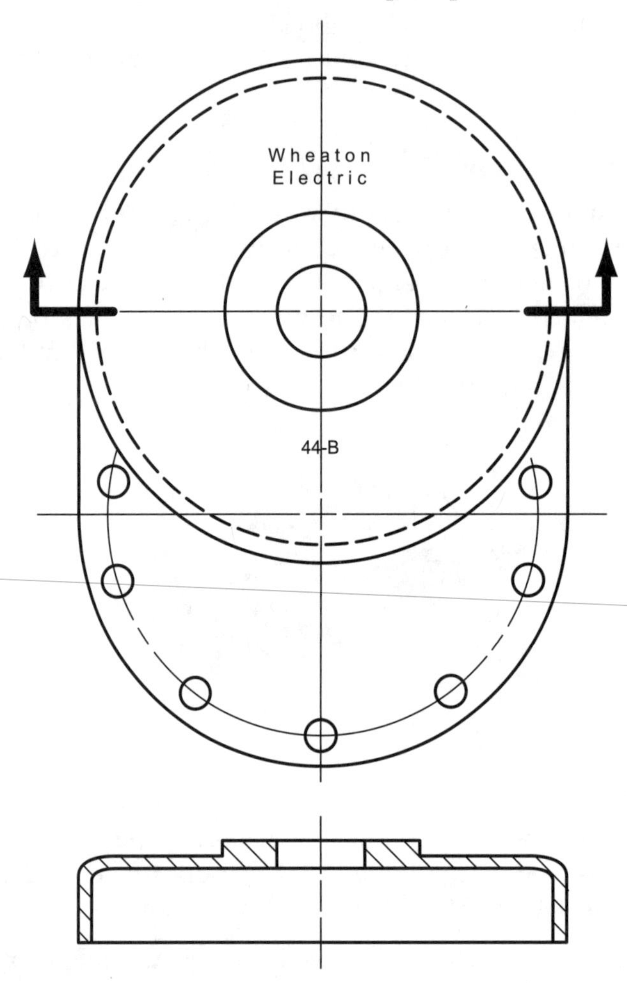

Figure 12.2 *Casting with identification.*

STEP 1

Model the part. Model the part in the Y-X world plane with the object's height represented by Z world. Figure 12.3 shows the part with basic material color and a single omni light. Unlike with displacement maps, you don't have to create complex surface tessellation to bump the detail.

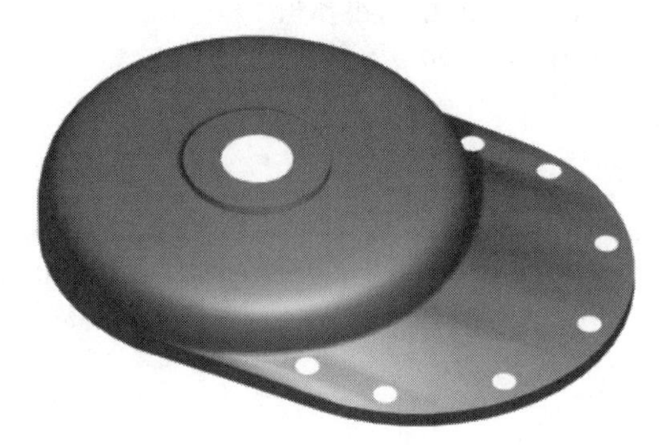

Figure 12.3 *Part ready to be bump mapped.*

STEP 2

Create the bump map. Note that the part of the casting carrying the bump map has equal height and width. The canvas (Figure 12.4) that holds the name and part number is then also unitary. Work at first white on black and then invert the pixels. Save this file in BMP format.

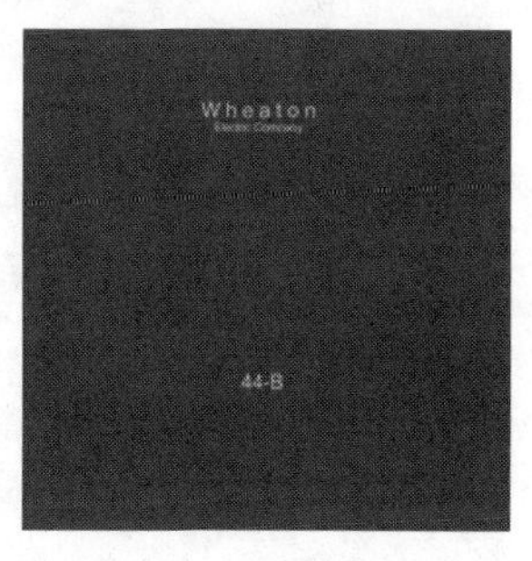

Figure 12.4 *Bump Map.*

Note: Allow the raster editor to anti-alias the text when it renders the letterforms. This provides a small blend between white and black creating filleted and rounded text, characteristic of castings.

STEP 3

Create the bump map material. In the **Material Editor** open the MAPS roll-out, click on **Bump|None**, and browse to the bump map. Enter **100** in the **Amount** field to begin. (This is a relative number and unlike a displacement map, does not correspond to the current units). Turn off tiling so the bump map will be applied only once.

STEP 4

Apply material mapping. Because this material contains only the bump, choose **Modify|UVW Map|Planar|Alignment Y** and uniformly rescale the **Gizmo** and position it over the portion to be bump mapped.

STEP 5

Assign the bump map to the casting. In the **Material Editor** click on the **Assign Material to Selection** icon or drag the material and drop it on the casting in the scene window. The pixels on the surface of the casting corresponding to the white pixels in the bump map are raised (Figure 12.5). The inset detail shows the impact of **Specular Level** and **Glossiness** on the raised pixels, as well as the resolution of the bump map under enlargement.

Figure 12.5 *Part after application of bump map.*

Note: Click on the bump map name in **Maps|Bump** and activate the **Show Material in Window** icon in the **Material Editor** toolbar. This displays the map in the scene, allowing you to interactively move the mapping gizmo for precise positioning. This way, you don't have to render-move gizmo-render-move the mapping gizmo to get the lettering where you want it.

COMBINING DIFFUSE AND BUMP MAPS

Though we can see from Figure 12.5 that bump maps do not require a diffuse map, by combining a diffuse map and a bump map (Figure 12.6) a even more realistic part is obtained. Here, a "dirt map" has been applied to simulate weathering. In fact, all mapping channels are applied before the bump, save the opacity channel. The surface responds as if the geometry was, in fact, altered.

Figure 12.6 *Diffuse and bump map both applied.*

BUMP MAP GUIDELINES

- The coarser the resolution of the bump map, the coarser the result. An ideal bump map contains one pixel for each pixel on the rendering being bumped.
- Bump maps are heavily influenced by lighting and basic reflectivity parameters.
- Combining a diffuse map based on the same image can increase the effect of a bump map. Depressions can be made to appear darker and raised portions lighter.
- Work interactively between your raster editor and 3D Studio's **Material Editor**. Once you load the bump map, you can adjust and **Reload** and then re-render to see the results.
- To get areas higher and lower than the surface, the surface itself must be bumped with gray.
- Use the **Show Material in Window** sparingly. Dark bump or opacity maps obscure the shape and make other operations difficult.

BUMP MAPPING SURFACE FINISH

Cast Material. The application of a bump map to a cast surface is instrumental in accurate representation. Figure 12.7 shows the effect of applying a bump map to represent a cast surface. The Manifold (a) has received basic color material. The Manifold (b) has had a bump map applied to the surface, representing cast material. The procedure is straightforward, using a standard material from the **Material Library**:

STEP 1

Create the material. Select an empty sample in the **Material Editor**. Select **Blinn** as the shader type. Choose **Maps|Bump|None** and in the **Material/ Map Browser** choose **Material Library** and find *Bump: Tex #9(OLDSKILT.JPG)*. Increase the **Amount** to **70**.

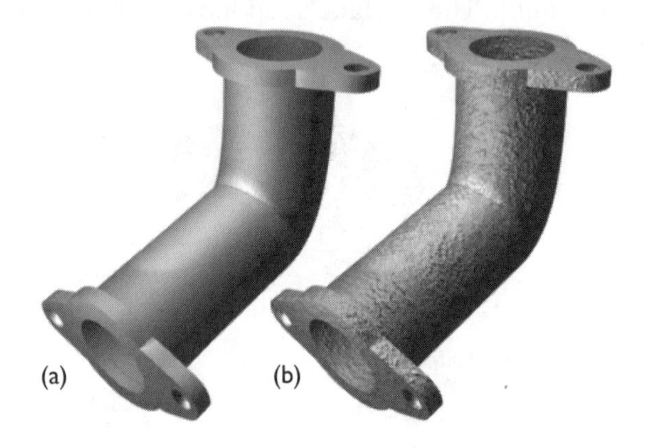

Figure 12.7 *Before and after cast surface bump map.*

The mounting surfaces of the flanges will be machined, necessitating a multi/sub-object material. This will be done later in the chapter.

Engraved Pebble Grained Plastic. Figure 12.8 shows a plastic control panel with engraved text and recesses for buttons. This requires a custom bump map, created in a raster editor.

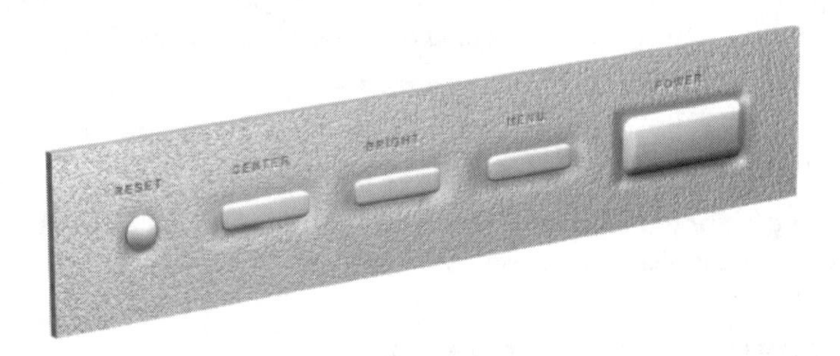

Figure 12.8 *Pebble grained with recesses and engraved text.*

STEP 1

Create the pebble grain. Open a white canvas in grayscale mode that is the same proportion as the panel. Our panel is 25" x 5" so that's the size of our canvas. Add 70 percent uniform noise to the canvas. Adjust the brightness +35. Because you want the panel to be softly grained, you need gray, not black pebbles (Figure 12.9). The white background is bumped out. The gray pebble is slightly indented from the white.

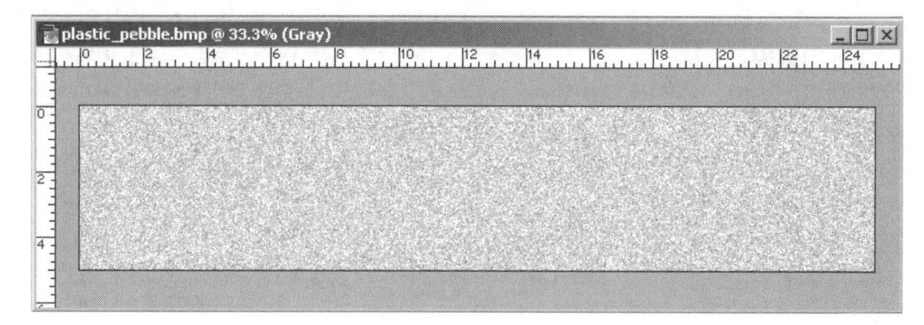

Figure 12.9 *Light pebble grain.*

Note: It's a good idea to work back and forth between your raster editor and 3D Studio. Once you load the bump map in, you can render, adjust in the raster editor, and render again.

STEP 2

Create depressions for buttons. Using guides, create shapes to hold buttons on a separate layer. These will be black, and depressed from the pebble grained surface. Feather the selections before filling with black so the depressions are blended in (Figure 12.10).

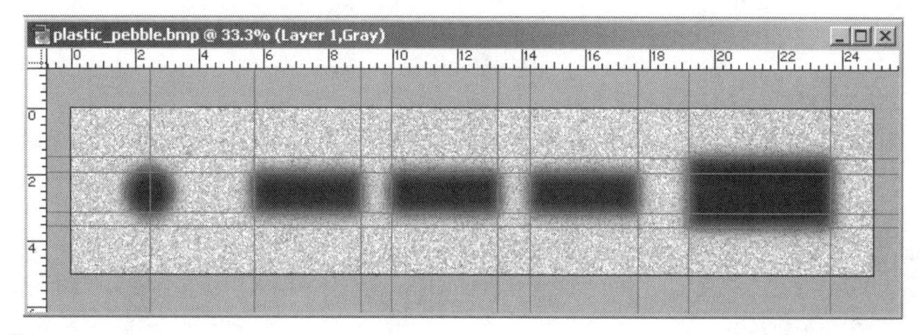

Figure 12.10 *Depressions for buttons.*

STEP 3

Create engraved text. For the text to be engraved, it needs to be black. Set the text with crisp anti-alias. When the text is complete, flatten the layers and save in BMP format.

STEP 4

Create the material. In the **Material Editor**, create appropriate plastic colors and highlights. (There is no diffuse map so the color of the plastic comes solely from ambient, diffuse, and specular colors.) **Choose Maps|Bump|None** and select **New** in the **Material/Map Browser**. Browse to the panel bump map. Start with a bump **Amount** of **50** and adjust to your taste. The panel without buttons is shown in Figure 12.11. Compare this to the finished panel with buttons in Figure 12.8. Unless it's absolutely necessary, you don't even have to punch holes in the panel.

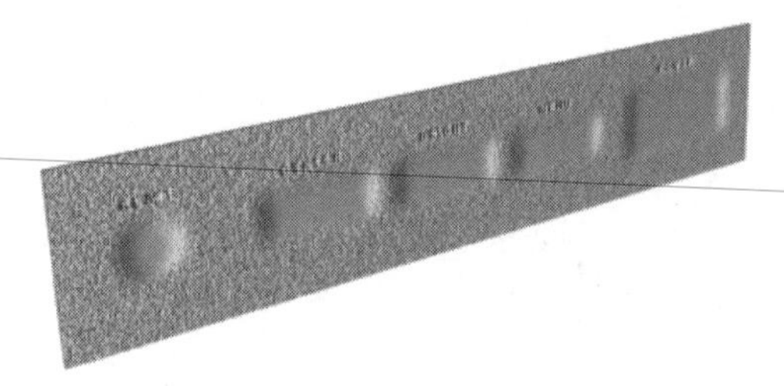

Figure 12.11 *Finished panel ready for buttons.*

Tip: Always save a version of your material maps in the native raster editor format with layers intact. That way you can go back and adjust components of the bump map separately, saving the bump map with flattened layers in BMP or JPG format.

INDUSTRIAL SURFACES

Several surfaces show up in many engineering scenes and the accurate depiction of these surfaces can mean the difference between successful and unsuccessful presentations. The bump maps used in these examples can be found on the accompanying CD-ROM in *chapters\chapter_12\maps*. The examples that follow should give you an idea of the limitless possibilities that bump maps afford.

Flooring. Many industrial floors are covered with a thick, durable, nonskid surface. The key to an effective bump map is the map tile. That is, a small bump map can be repeated over a surface, with no discernable edges (unless you want them, and then you build them into the bump map).

Figure 12.12 shows the bump map for an industrial floor comprised of raised circular disks. Note that the space at the outside of the bump map is exactly one-half the total space between the raised disks. This way when the bump map is tiled (Figure 12.13), the flooring appears contiguous, and without seams.

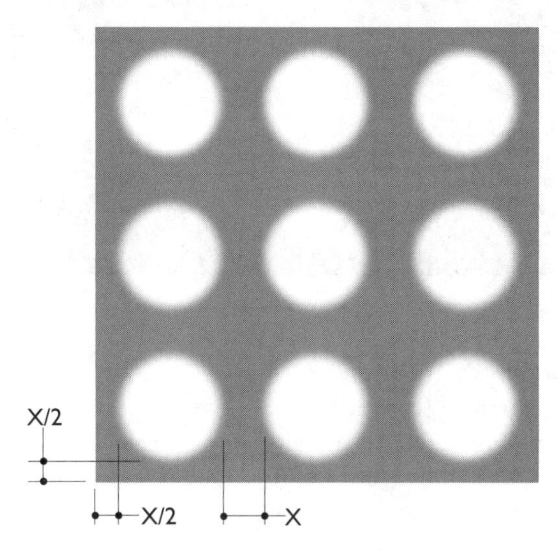

Figure 12.12 *Floor bump.*

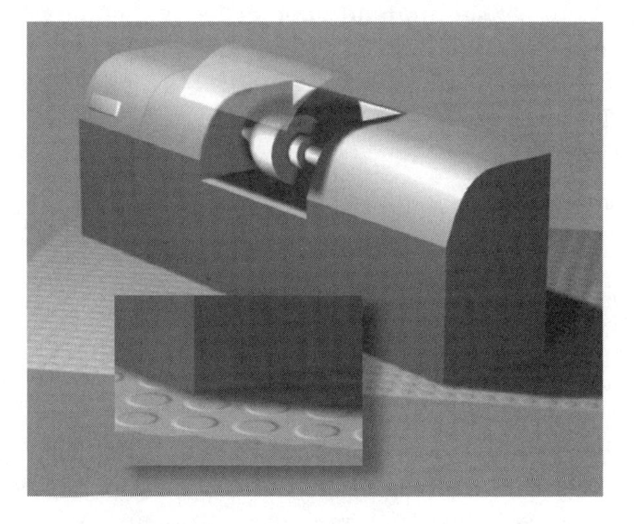

Figure 12.13 *Seamless industrial floor.*

Diamond Surface. A common industrial surface is covered with small raised diamonds (Figure 12.14). These diamonds create both a nonskid surface and add rigidity with a small weight penalty. Because of the multiple orientation of the diamonds, the surface is rigid in all directions, something that a simple ribbed surface is not.

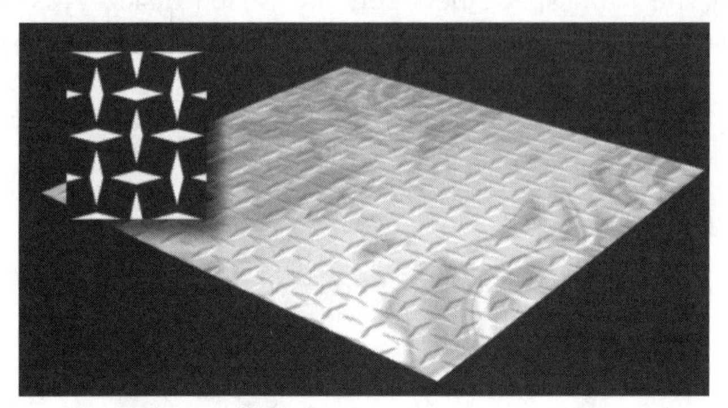

Figure 12.14 *Diamond surface and bump map.*

Expanded Metal Flooring. Walkways and stairs are often made from sheet steel that has been perforated, and the edges of those perforations raised out. This creates a surface that is nonskid but with the added benefit that liquid (like rain or sea water) falls through. You'll find this kind of surface in petrochemical plants, oil rigs, and ships.

This material requires both a bump map, to raise the edges, and an opacity map, to punch the holes. To do this as geometry (with Booleans, for instance) would be both time-consuming and geometrically complex. By using bump and opacity maps, the effect is nearly the same without the penalty of complex geometry. Figure 12.15 shows a cross section of the expanded metal decking with the bump–mapping strategy.

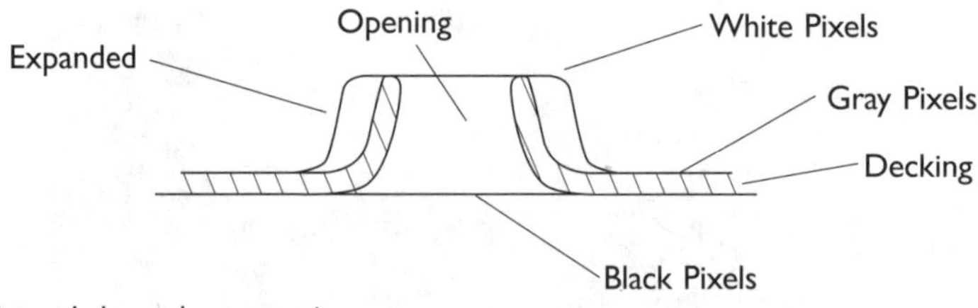

Figure 12.15 *Expanded metal cross section.*

Figure 12.16 shows the two maps that are required. The first map (b) bumps the surface up. It uses three brightness values: black for the depressed opening, white for the raised lip, and gray for the top surface. The opacity map (a) is made from the bump map by adjusting contrast to result in black and white. The black portions of the opacity map form the openings, on the lowest portion of the bump. The final surface is shown in Figure 12.17. You can see the raised portion formed by the white area of the bump map. When the opacity map is applied to the lowest area, the bump forms the thickness.

The trade-off with opacity maps is that lights still affect shadows. If you look below the walkway in Figure 12.17 you'll see a solid shadow, formed by the solid plane. There are several masking techniques that can take care of this but oftentimes, such shadows are either out of the field of view or during an animation, visible for only brief periods of time. The modeling benefit usually outweighs this limitation.

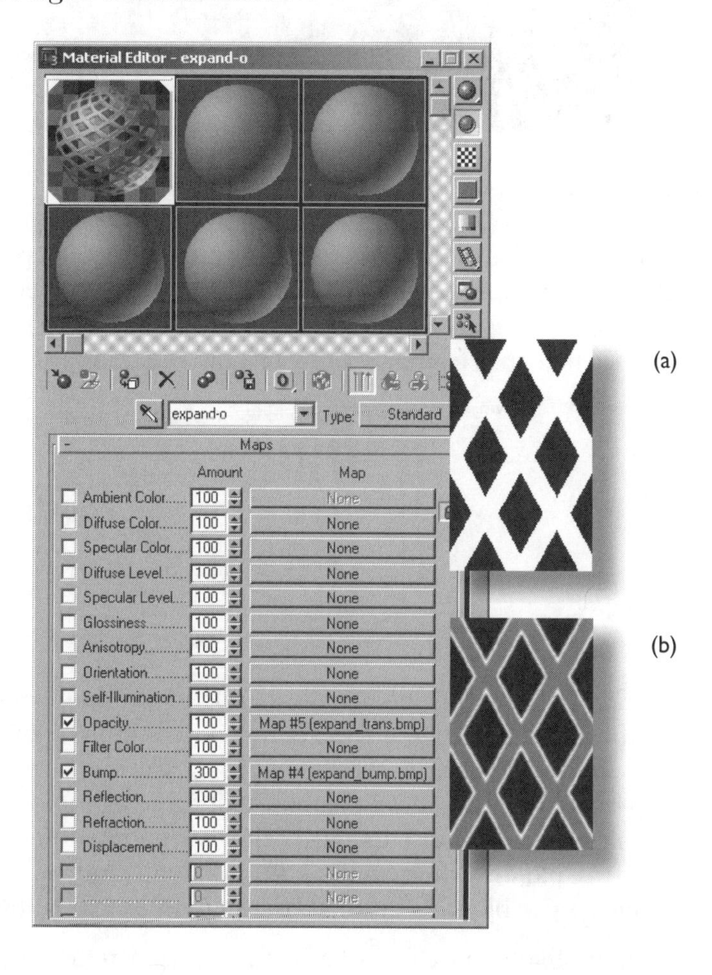

Figure 12.16 *Expanded metal material maps.*

Note: An opacity map doesn't show any thickness as it creates its transparency. Revisit Figure 10.17 and propose a way of creating thickness in this example.

Figure 12.17 *Expanded metal walkway surface.*

Riveted Panels. Many industrial products are covered by riveted panels. These panels are thin and flexible, and show the effect of being attached to a framework by slightly deforming at each rivet. Figure 12.18 shows a cross section of two panels attached to a framework. Note how the panel bends in at the rivet.

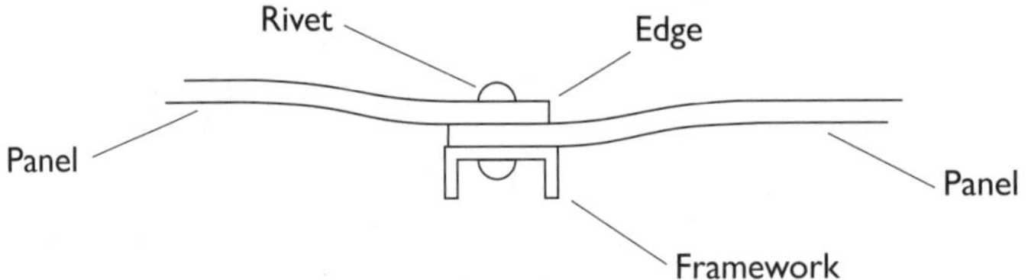

Figure 12.18 *Bump mapping strategy for riveted panels.*

We could, of course, create multiple panels and little semihemispherical rivet heads at the expense of increasing the geometric complexity of our model. Instead, we will create the panels, the depressions, and the rivets as a bump map. The bump map in Figure 12.19 combines all these elements. Each panel blends from light gray to white. This forms the white-gray difference and an overlap of the two panels where one is bumped up and one is bumped halfway up. Because the blend is gradual, the panels appear to overlap.

The irregular dark shapes on each side of the overlap represent the slight deflection of each panel at the rivets. The rivets themselves are bumped out from these depressions by white pixels. The result (Figure 12.20) includes a diffuse map for the panel material.

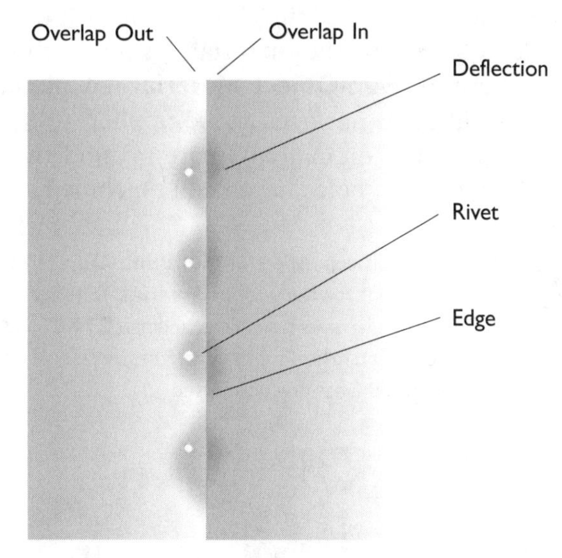

Figure 12.19 *Bump map for riveted panel.*

Figure 12.20 *Final riveted panel.*

BUMP MAPS AND MULTI/SUB-OBJECT MATERIALS

The cast manifold in Figure 12.7 represented the part as it was cast without machined surfaces. But of course, the mounting flanges would be machined (so gaskets would seal), mounting holes cleaned up (so fasteners would fit), and the inside polished (so the material flowing in the manifold would be unobstructed). The **Multi/Sub-Object Material** technique presented in Chapter 10 allows these different surfaces to be applied to sub selections of a part. Figure 12.21 shows the original cast part (a) and the finished part (b) after the mounting flanges and holes have been machined and the interior polished.

Note: The difficult aspect of **Multi/Sub-Object Material**s is to select only the faces you want for the various materials. You may have to use rectangular, circular, and irregular selection areas, adding (**CTRL**) and subtracting (**ALT**) to get what you want. When subtracting faces, unselected faces are not toggled on. Only selected faces are deselected.

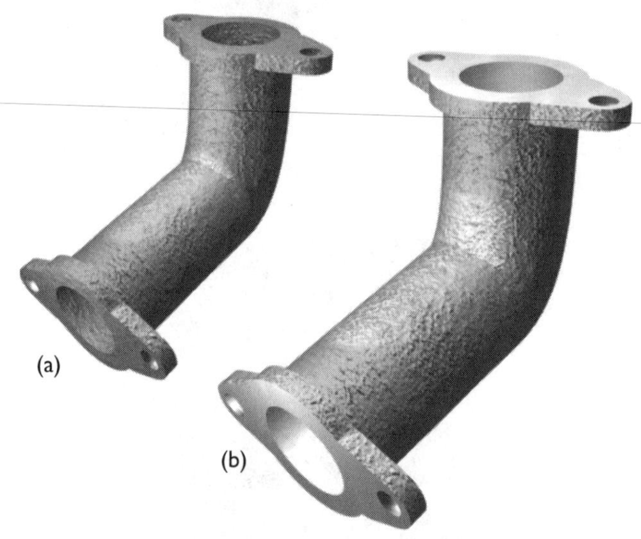

Figure 12.21 ***Multi/Sub-Object Materials.***

THREADS WITH BUMP MAPS

Threaded parts and fasteners are commonly encountered in industrial products. Unless threads are the subject of your presentation, they should never be modeled. Threads formed by lofting a profile along a helical path, included in a profile for lathing, or subtracted with a Boolean tool, are simply not worth the added geometric complexity. The solution is to bump map the threads.

Two types of threads are encountered: acme (square) threads and the more common sharp V-groove threads. In either case, bumped mapped threads should be designed as parallel rings, without the lead of real screw threads. Because the lead is quite small, and you will hardly ever have a truly orthographic view of them, making parallel vertical threads is visually sufficient. If accuracy is important you can make the threads of the bump map correspond to the number of threads per inch or millimeter. Both thread types can be effectively bump mapped. Observe the Adjusting Screw in Figure 12.22. First, an Acme, or square thread, will be applied as a bump map.

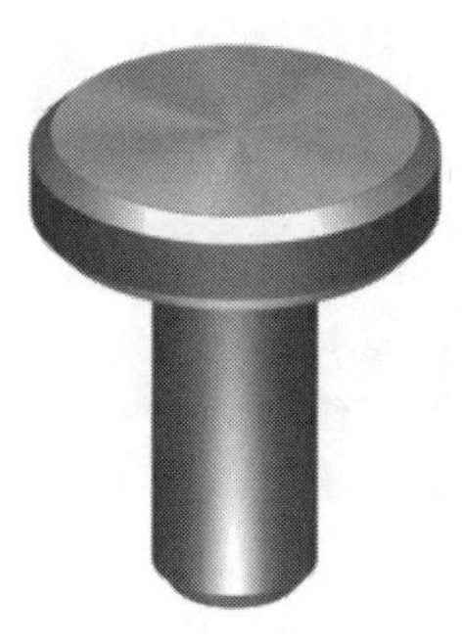

Figure 12.22 *Adjusting Screw.*

The Acme thread, because it is square, is simply a pattern of black and white. When applied as a multi/sub-object material, the screw shaft becomes threaded. An inspection of Figure 12.23(a) reveals that as suspected, the bump map does not alter geometry.

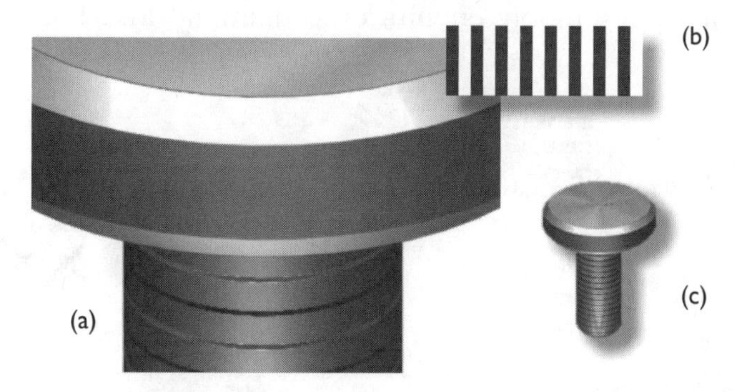

Figure 12.23 *Bumped square thread.*

The outside of the threaded shaft is not deflected as you would normally expect on threads of this scale. With **Bump|Amount** set to **300**, the square threads are formed. A sharp V-thread requires a smooth transition between the in (black) and the out (white) as shown in Figure 12.24(b). This creates the illusion of angular threads. Under close inspection (a), the threads appear simply as a pattern of light and dark. But when viewed in its entirety (c) the bump map patterns of light and dark appear as realistic-looking threads.

Tip: Always leave a little extra white to pull the threads out. Also, make the black of the groove a little wider to push this in.

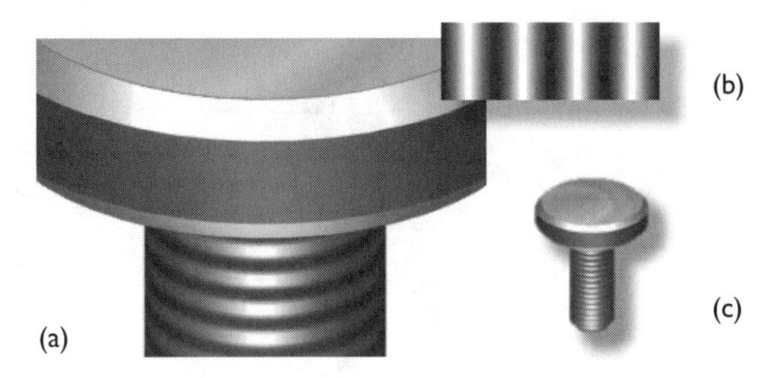

Figure 12.24 *Bumped V-threads.*

KNURLS

Our final bump mapping is applying knurls to a surface. Knurls provide an antislip surface and are cut into the surface by a tool. Knurls can be diamond or straight. When they are straight (Figure 12.25) they are often called splines. The more commonly encountered diamond knurls are shown in Figure 12.26.

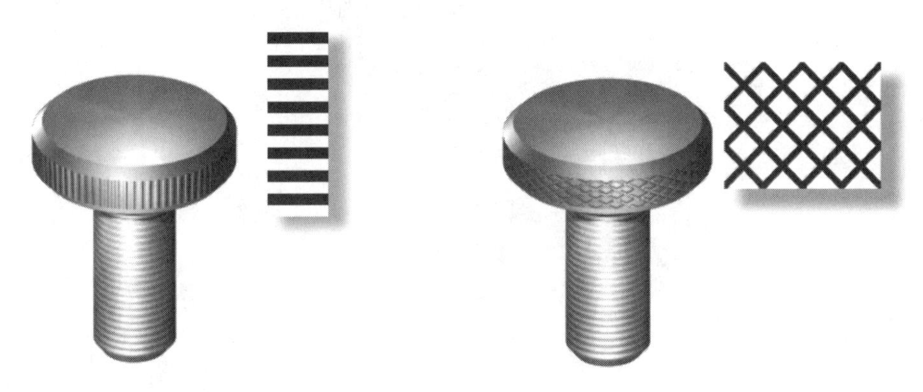

Figure 12.25 *Straight knurl (spline) and bump map.*

Figure 12.26 *Diamond knurl and bump map.*

Tip: Thread bump maps can be considered general purpose maps. Many surface details (ribs, grooves, joints, etc.) can be created by either editing existing thread bump maps or by simply scaling, cropping, or rotating the map right in the Material Editor. Because the bump map is grayscale, it can be applied to any surface without impacting other material map colors.

INCREASING BUMP MAP EFFECT

If the same map is used in both the bump and diffuse channels, the effect of the bump can be heightened. Figure 12.27 shows the effect of loading the same diamond knurl map in bump and diffuse channels (and in the ambient channel also because it is locked to the diffuse channel). The raised surfaces of the diamond knurl are mapped with white while the cuts are mapped with black. The result is brighter, more effective bumps.

Note: When diffuse and bump maps are the same they must be tiled, cropped, and rotated the same. Likewise, their UVW mapping coordinates must be identical.

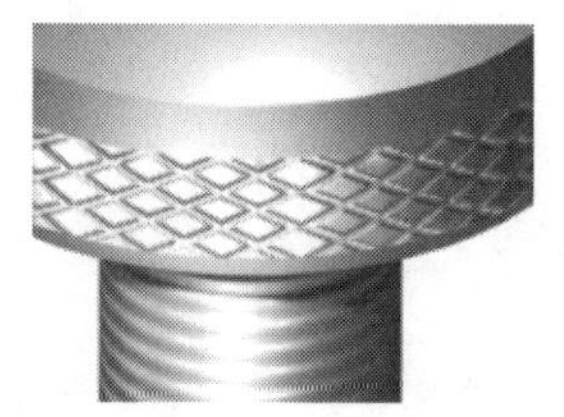

Figure 12.27 *Diamond knurl loaded into diffuse and bump channels.*

PROBLEMS

Problems appropriate for bump mapping are found on the following pages. You may use the bump maps found in *chapters\chapter_12\maps* or create your own. In either case, be prepared to document your bump map development as was done in Figure 12.16. Assign appropriate names to all materials, objects, and selection sets.

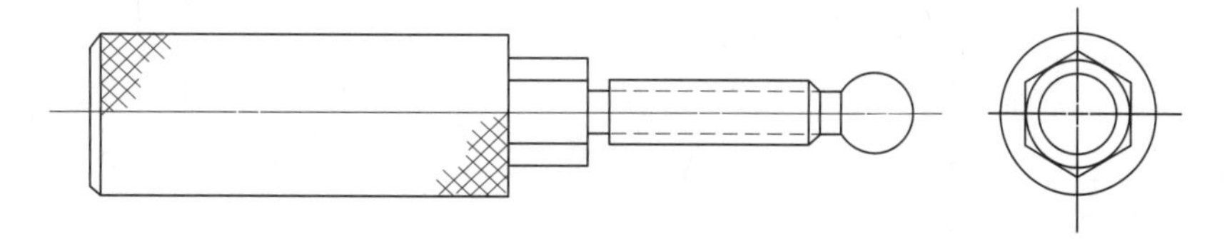

Problem 12.1 *Handle. Create knurl and thread bump maps in a multi/sub–object material. The base material is steel and the ball is highly polished (refer to Figure 1.9).*

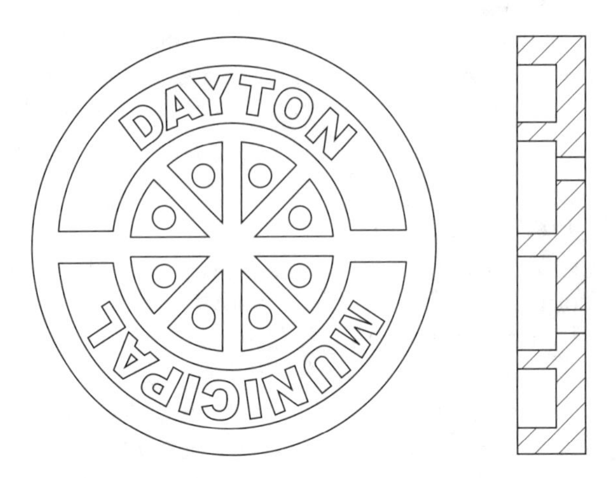

Problem 12.2 *Manhole Cover. Create a simple cylinder primitive and apply a bump map to create the cast detail (along with distress). Complete with cover with a diffuse cast metal dirt map. The EPS file can be found on the CD-ROM in problems\chapter_12.*

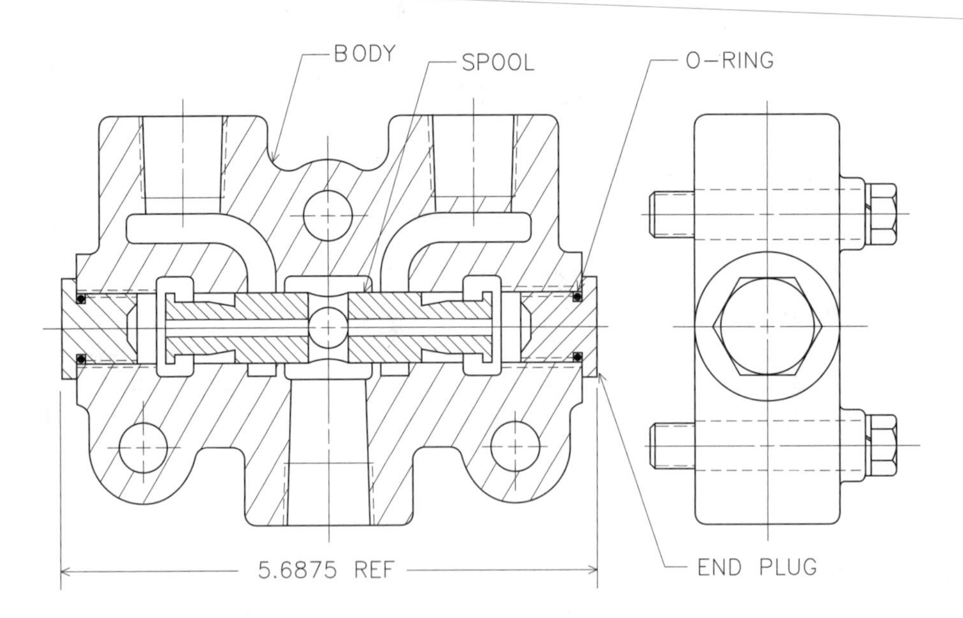

Problem 12.3 *Flow Valve (Duff:* Technical Illustration with Computer Applications*). The exterior of the valve body is unfinished cast metal. All interior passageways are smooth cast. Surfaces for end plugs and fasteners have been machined. O-rings are black rubber and the spool has been machined and finished.*

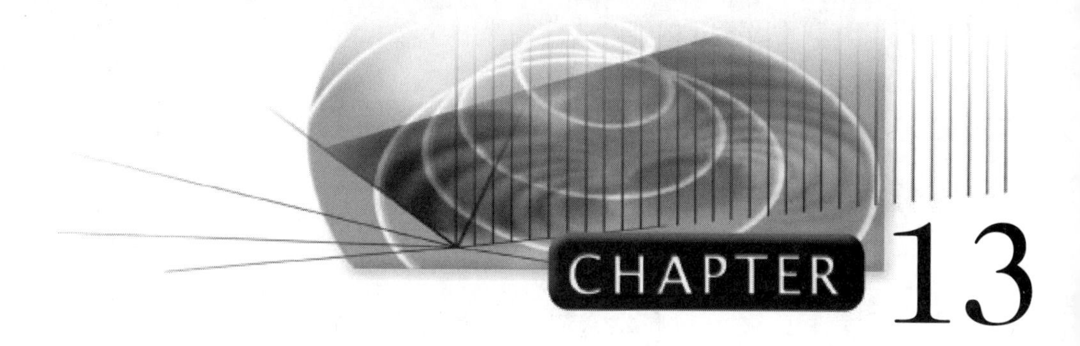

Putting Modeling and Mapping Together

CHAPTER OVERVIEW

You now have experience with the majority of modeling techniques available in 3D Studio as well as insight into how detail and realism can be added with displacement, material, and bump maps. Now it's time to put modeling and mapping together.

Subjects for modeling and mapping should be chosen so that several materials are present. This chapter will use an automotive wheel and tire as the subject because together they have multiple materials, multiple surface finishes, and the type of contour detail appropriate for displacement or bump mapping. Other manufactured products could just as easily be used such as scientific equipment or consumer products. The problems at the end of this chapter are alternative possibilities. Larger scale raster files for tracing are on the accompanying CD-ROM in *chapters\chapter_13*. Additionally, you'll find the files you need to follow along with the example in this chapter in the same directory.

To shorten the automotive wheel and tire example we will be using the wheel developed in Chapter 7, Modeling with Modifiers. Figure 13.1 shows the views of the Wheel and Tire. The Wheel is comprised of two objects: a Rim and a Hub.

Tip: Material mapping, especially bump mapping, adds a considerable overhead to rendering scenes in 3D Studio. You should carefully analyze the benefits of adding detail, especially if such detail is not critical in understanding the model. Mapping channels can be selectively turned on and off as required. You may want to consider creating two sets of materials: one comprised of basic colors and surface characteristics, and a copy of these with maps.

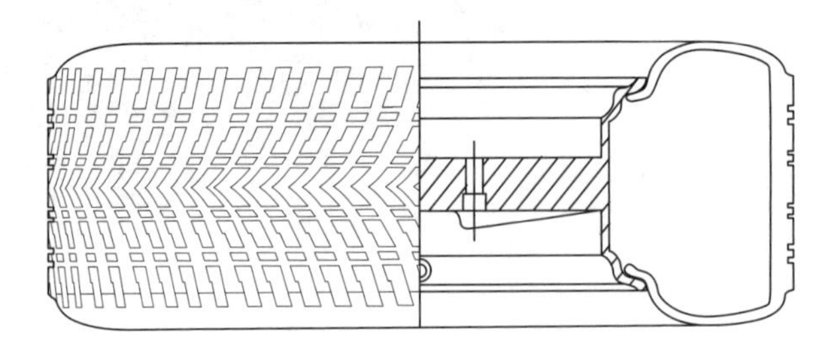

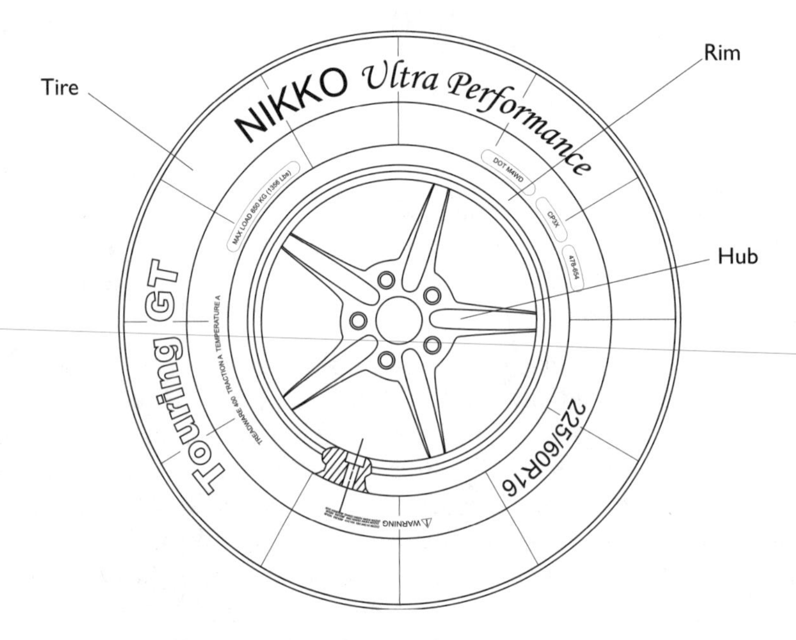

Figure 13.1 *Wheel and tire for modeling and mapping.*

TIRE MODELING STRATEGY

An analysis of the tire reveals it is theoretically comprised of three separate parts: the Inner and Outer Sidewalls, and the Tread. Although the tire is molded in a single piece by splitting the tire into inner and outer sidewalls and tread, it becomes simpler to create maps and assign UVW mapping coordinates. For example, the tread should have cylindrical mapping coordinates because it is, essentially, a cylinder. The sidewalls, though curved in a partial torical shape, would best be mapped with a planar UVW map. It would be unnecessarily difficult to create a single raster material map that when wrapped around the tire with a single UVW mapping coordinate would produce the desired materials and bumps on sidewalls and tread. The tire could be modeled as a single unit with *multi/sub-object materials* applied to subselections. You have experience doing this in Chapter 10, Raster Material Maps. As an alternative, separate tread and sidewall objects could be created and mapped individually. Because we have already implemented the multi/sub-object material, we will model the sidewalls and tread separately.

MODELING THE TIRE COMPONENTS

The Tread and Sidewalls are surfaces of revolution. Figure 13.2 shows the spline shapes for the sidewalls and tread superimposed over the half section Top view. The Tread portion provides a shoulder over which the tread shape will wrap and has been slightly curved to represent the contour in an inflated state. Because the Sidewalls are separate objects, we don't have to worry about keeping the tread off the sidewall. Once lathed (Figure 13.3) the three shapes form the carcass of the tire.

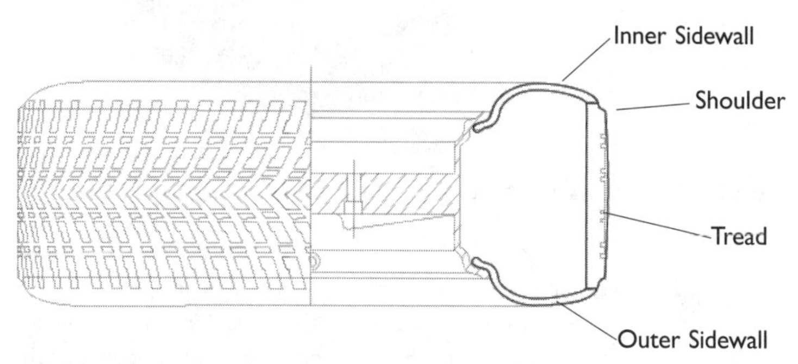

Figure 13.2 *Spline curves for surfaces of revolution.*

Figure 13.3 *Lathed shapes comprise the carcass of the tire.*

Because consistent measurements were used from Figure 13.1, when the Wheel is merged into the tire scene, it matches perfectly with the Tire (Figure 13.4). The unmapped geometry creates an accurate representation of the wheel and tire. Were the wheel and tire applied to a vehicle that was seen at a distance, or that appeared only briefly in an animation, material mapping would not be necessary. You could get by with simple color assignment and basic surface finish.

Figure 13.4 *Wheel and tire in position.*

Tip: This is a good time to practice file management. Once you assure that the wheel and tire are in correct spatial position and at the correct scale, save a wheel file (without the tire) and a tire file (without the wheel). This way, you can work on the tire without the overhead (even if hidden) of the wheel geometry being in the same file. The tire can be merged with the wheel after both are mapped.

THE TREAD MAP

The Tread will be mapped with a raster diffuse and bump map. The diffuse map will be a low intensity application of the bump map so the tread blocks will be lighter than the grooves. This intensifies the effect of the bump map.

The bump map is made from the Top view of the treads in Figure 13.1. Because the tread becomes foreshortened as it bends around the tire, we will use the first full tread pattern (Figure 13.5) and repeat this to make multiple copies. Because the tread pattern will tile around the tread shape, only a small number of treads are needed to create a tiling shape.

Note: The easiest way to make sure that the tread pattern will tile is to start and stop the pattern at exactly the same spot at the right and left edges.

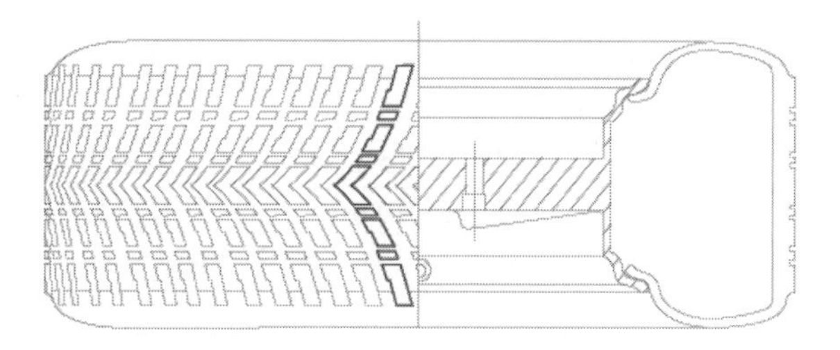

Figure 13.5 *The tread pattern used for the tread bump map.*

You want the tread blocks out (white) and the tread grooves in (black) for the bump map (Figure 13.6). Note that the edges of the tread blocks blend to black so that as they wrap around the tread, the blocks will taper. The tread map is high resolution, 1,200 dpi.

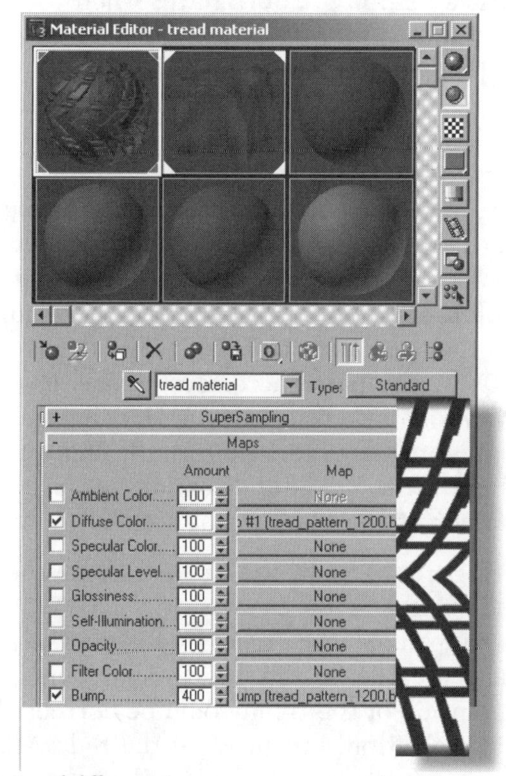

Figure 13.6 *Tread bump and diffuse map.*

STEP 1

Hide the Sidewalls. This allows you to work on the Tread and its UVW mapping without the distraction of other geometry.

STEP 2

Create the tread color. Choose the **Blinn** shader. Create a **Diffuse**, **Ambient**, and **Specular** color scheme appropriate for the tire (gray with a touch of brown). Keep the diffuse and ambient colors locked and the specular color only slightly lighter; set **Specular Level** at **60** and **Glossiness** at **20**.

STEP 3

Create the tread material. Load the tread map into the **Diffuse** mapping channel and set **Amount** to **10**. This allows the Tread to be mostly basic color with just a hint of lightness in the tread blocks. Load the tread bump map into the **Bump** channel and set the **Amount** to **300**.

If you did not **Generate Mapping Coordinates** when you lathed the Tread, select the Tread and choose **Modify|UVW Map** and choose **Mapping:Cylindrical|Alignment:X**.

STEP 4

Apply the tread map. With the Tread selected, click **Apply Material To Selection** and for each of the maps click **Show Map In Viewport**. Because you have created the tread map intentionally to tile, the pattern is repeating around the circumference, only not the correct number of times. Figure 13.7(a) shows the tread with **U:Tiling** at the default value of **1**. When **U:Tiling** is set to **10** (b) the pattern correctly maps around the tread. In the **Modify|UVW Map** roll-out, enter **5** in the **U:Tiling** field. Make sure that the same tiling parameters are entered into both **Diffuse** and **Bump** channels. There is no tiling in the **V:Tiling** direction because the tread pattern is repeated only once across the width of the tread.

A quick rendering (Figure 13.8) of the Tread reveals the blending of the tread blocks into the shoulder. You can adjust the UVW mapping gizmo to fine-tune where this happens on the Tread. By scaling the gizmo's width you reposition where the map falls on the tread's shoulders. The fact that the Tread appears on the inside of the tread could be corrected by applying the map to an outside subselection, but because the Sidewalls and Wheel will obscure this, it is of no concern.

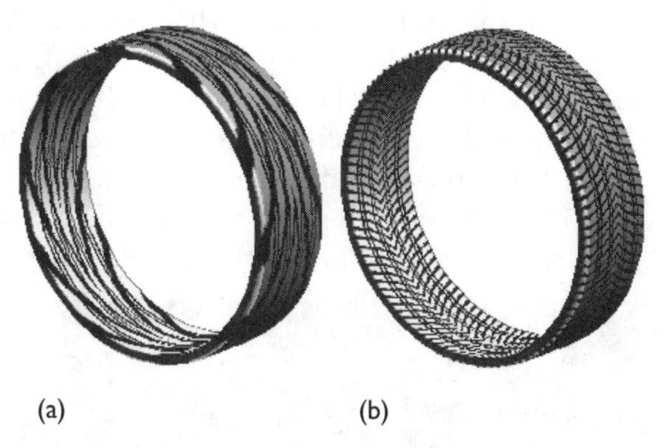

Figure 13.7 *Default (a) and reset tiling (b).*

Figure 13.8 *Pattern successfully bumped on tread.*

THE SIDEWALL MAP

Hide the Tread and display the Outer Sidewall. Unless you will be viewing the Tire and Wheel from all sides you won't need to map the Inner Sidewall. The Sidewall will receive a diffuse map to dirty up the rubber, and a bump map to model the sidewall detail as shown in Figure 13.1.

Note: Although the tire is comprised of the same material throughout, the tread will be shown lighter. Start with the same **Diffuse**, **Ambient**, and **Specular** colors as were used for the tread and adjust from there.

The sidewall bump map (Figure 13.9) is the inverse of the engineering drawing in Figure 13.1 and has been blurred slightly to encourage the bump to have rounded edges. To insure maximum clarity of the fine sidewall detail this map was created at 1,200 dpi. The material itself (Figure 13.10) combines the diffuse *dirt map* and the bump map to produce an effective sidewall. The **Diffuse** setting of 80 allows the basic color to mix with the dirt map. The **Bump** setting of 100 is a starting point for the sidewall detail. You want more subtle detail on the Sidewall than on the Tread.

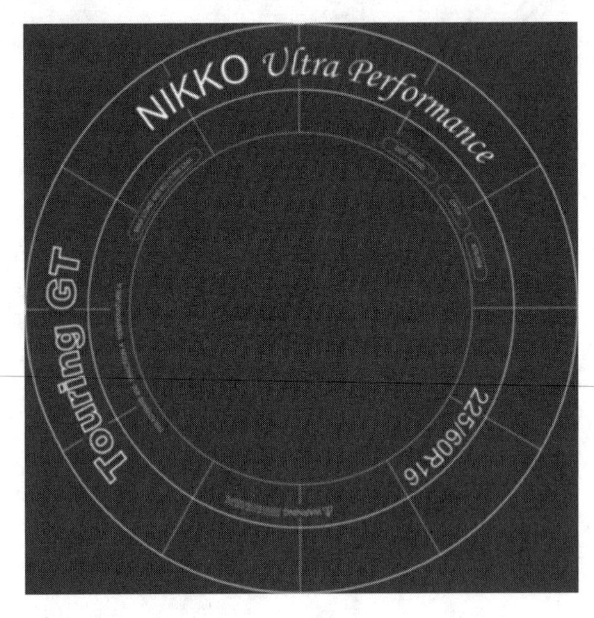

Figure 13.9 *The sidewall bump map.*

STEP 1

Create the dirt map. This is intended only to provide a degree of randomness to the sidewall color. Were this a 4 x 4 tire covered with mud, the diffuse map (and the bump map for the grime) would take on greater importance.

STEP 2

Create the sidewall bump map. The sidewall detail of Figure 13.1 is rasterized at 1,200 dpi, inverted in a raster editor, and slightly blurred.

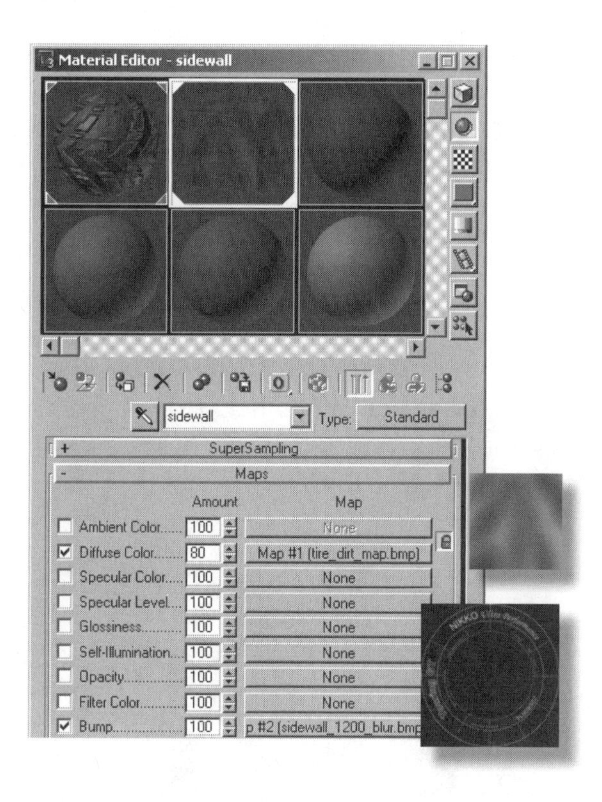

Figure 13.10 *The sidewall material.*

STEP 3

Create the material (Figure 13.10). In the **Material Editor** load the dirt map into the **Diffuse** channel and set **Amount** to **80**. This lets a little of the underlying color through. Turn off tiling if the dirt map is large enough to cover the sidewall. If not, you may have unfortunate tiling edges.

Note: The planar UVW map must be correctly scaled for both the diffuse and sidewall bump maps. Enlarging the coordinates to correct tiling problems will cause the bump to be misaligned.

In the **Bump** channel load the sidewall bump map (having identical width and height to match the shape of the sidewall) and set **Amount** to **100**. This is a good staring point for the sidewall detail. You want enough bump to form the lines and lettering, but not so much to make the bump rough. Turn off tiling because you will apply the bump map once.

STEP 4

Map the material. Select the sidewall and choose **Modify|UVW Map|Mapping:Planar|Alignment:Y** and click on **Fit**. The map is unitary as arethe UVW mapping coordinates. If you want to make the name of the tire more readable, you can rotate the UVW mapping gizmo about Y world to get the lettering in an advantageous position.

STEP 5

Apply the sidewall map. With the sidewall selected click on **Assign Material to Selection** in the **Material Editor** toolbar. The mapped sidewall is shown in Figure 13.11. Perform a quick rendering to see the effect of the bump. Adjust **Map|Bump|Amount** as required.

Figure 13.11 *The sidewall after diffuse and bump mapping.*

Tip: Bump maps depend on **Specular Highlights** and correctly positioned lights for their effect. Position light sources so that they illuminate along the surface, rather than in front of the surface. Shallow lighting angles bring out the highlights and shadows of the bump.

The Inner Sidewall has not been mapped. It carries only the basic **Ambient**, **Diffuse**, and **Specular** colors of the sidewall material. When the Sidewalls and the Tread are combined, an effective Tire is presented (Figure 13.12). If the pattern needs to come down the shoulder more, select the Tread and choose **Modify|UVW Map|Gizmo** and scale along Y world.

Figure 13.12 *The finished tire.*

ALTERNATE METHOD WITH DISPLACEMENT MAP

An alternate method for creating the treads is by using a displacement map. However, after an inspection of the results, we find that Chapter 11, Modeling with Displacement Maps was correct: displacement maps are best reserved for large displacements, larger than the tire treads in Figure 13.1.

Figure 13.13 displays the tread lathe, increased to 128 segments, with a 1,200 dpi tread displacement map applied. (The same map as was used in the bump.) We know that displacement mapping is dependent on the resolution of the grayscale map as well as the density of the underlying mesh. To increase the density of the mesh the tread was selected and **Mesh Smooth** with **Iterations:2** was applied. The same tiling options were used as with the bump map. Observe the complexity of the geometry necessary to produce the displacement shown in Figure 13.14. Even with this increase in mesh complexity, the final result (Figure 13.15) is more appropriate for tractor tire treads than passenger car tire treads.

Note: By using a displacement map, the benefits of pixel-level modeling are lost, namely reduced geometric complexity and the ability to address each pixel of a rendering.

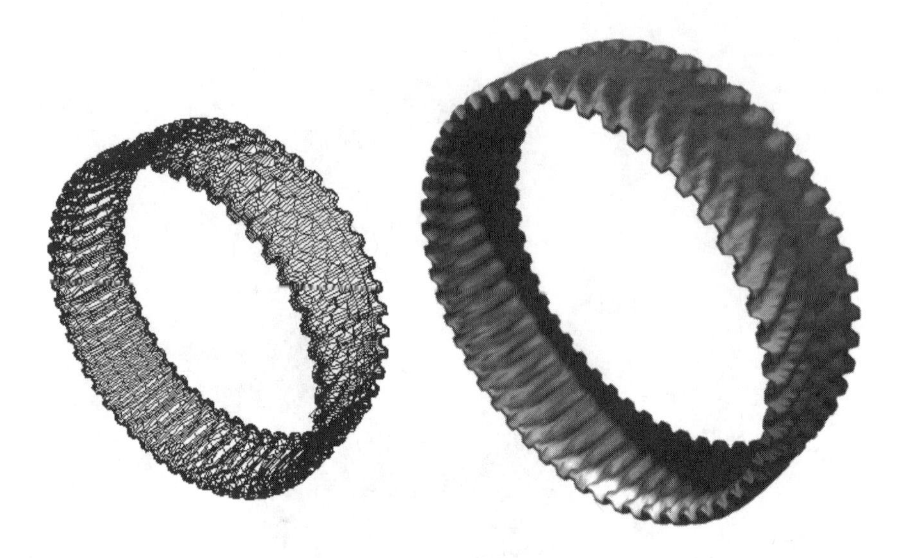

Figure 13.13 *Lathed tread displaced.* **Figure 13.14** *Treads formed by displacement map.*

Figure 13.15 *Displacement mapped tread and bump mapped sidewall.*

MAPPING THE WHEEL

Save the Tire and reload the Wheel. The Wheel can be finished in a number of ways:

- The entire Wheel can be painted with a high-gloss silver paint (Figure 13.16).
- The entire Wheel can be a sand-cast material (Figure 13.17).

- The Hub of the Wheel can be sand-cast and the Rim polished or chromed (Figure 13.18).

Figure 13.16. *Painted wheel.*

Figure 13.17. *Cast wheel.*

Figure 13.18. *Cast wheel with chrome rim.*

Because we want the most expressive materials, we will choose the cast and chromed option.

The cast part of the Wheel uses a material you have not used before: a **Noise** material. This material isn't based on a raster material map. Instead, the map is created procedurally.

STEP 1

Create a **Noise** material. Begin with a basic color appropriate for cast alloy and the **Blinn** shader. Enter **30** in the **Specular** field and **10** in **Glossiness**. Select an empty sample and in the **Bump** channel click **None**. Select **Matl Library** and **Bump: Tex#3 (Noise)**. Enter **8** in all three tiling fields and **20** in **Noise Parameters|Size**.

STEP 2

Map the Hub material. Select the Hub and choose **Modify|UVW ap|Mapping:Planar|Alignment:Z**. This will map the **Noise** material to the face of the Hub.

STEP 3

Assign the material to the Hub. With the Hub selected, click on the **Assign Material to Selection** button in the **Material Editor** toolbar. Figure 13.19 shows the effect of **Specular Highlights** and **Noise** on the Hub.

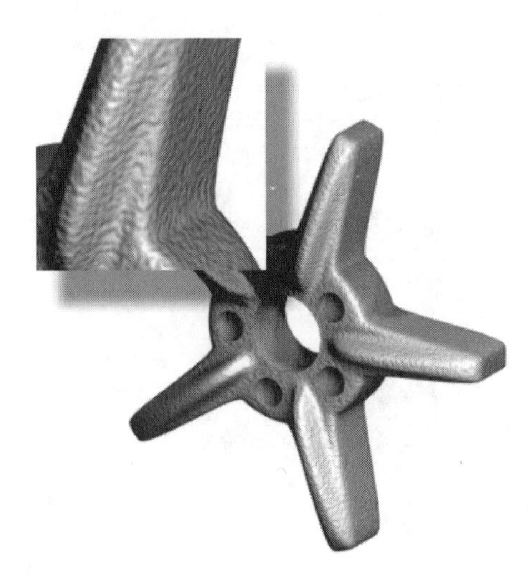

Figure 13.19 *Procedurally created bump map.*

STEP 4

Create the chrome material. Select an empty sample. Choose the **Metal** shader and set **Specular Level** and **Glossiness** each to **10**. Load *METAL7.JPG* into the **Diffuse** channel and set **Amount** to **30**.

In the **Reflection** channel load *CHROMBLU.JPG* and set the **Amount** to **20**.

STEP 5

Map the Rim material. Select the Rim and choose **Modify|UVW ap|Mapping:lPanar|Alignment:Z**. This maps the chrome material to the face of the Rim and streaks it along its depth.

The Rim requires a Valve Stem to complete the tire assembly. Because of scale, little Valve Stem detail is necessary. The valve stem and rubber grommet are lathed from a profile and a multi/sub-object material containing rubber (from the Tire) and chrome (from the Rim) complete the mapping. Figure 13.20 displays the rim and a detail of the Valve Stem.

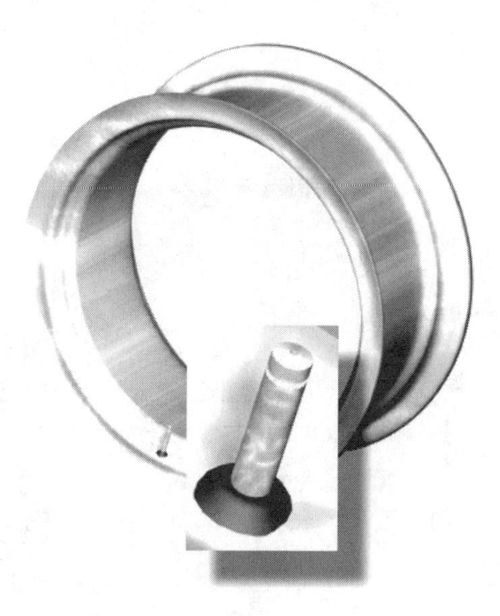

Figure 13.20 *Valve stem has multi/sub-object material.*

COMBINE THE WHEEL AND TIRE

Choose **Edit|Merge** and browse to the tire file. Select the Tread, Outer Sidewall, and Inner Sidewall and merge these objects into the wheel scene. If you have been careful, the Tire components are merged in correct spatial location and scale. Additionally, the materials associated with the tire have also been merged.

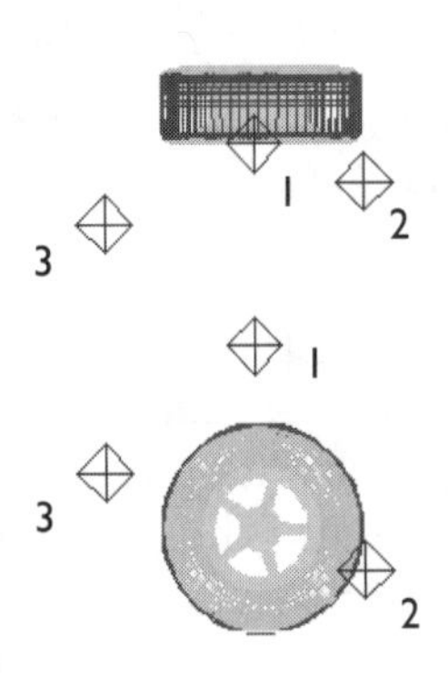

Lighting the subject is very important. Figure 13.21 shows the Top and Front views and the placement of three omni lights. Light #1 is above and almost in the plane of the Tire. This shallow angle illuminates the bump mapping of the Outer Sidewall. Omni light #2 illuminates the lower part of the Tread so its detail is observable in the shadow. Omni light #3 illuminates the front of the Outer Sidewall and the Rim. Notice that all three lights are tight to the tire. This intensifies the effect of bump mapping.

Figure 13.21 *Lights in Front and Top views.*

Figure 13.22 *Completed wheel and tire.*

GALLERY

Displayed on the following pages is a gallery of solutions by students in their second semester with 3D Studio. These wheel and tire combinations were from vehicles in the ASU East Technology parking lot and modeled from digital photographs, sketches, and measurements. You will find information on the CD-ROM for modeling several tires just like the ones you find here.

(a) Andrew Felecitti

(b) John Bartosch

(c) Michael Kelly

(d) Travis Zipper

Figure 13.23 *Gallery of wheel and tire modeling and mapping.*

PROBLEMS

The photographs of scientific equipment and consumer products on the following pages should challenge your modeling and mapping abilities. Large scale raster files for Problem 13.1 and Problem 13.2, as well as other assignments, can be found on the CD-ROM in *chapters\chapter_13*. You are encouraged to sketch your modeling strategies and document all materials through screen grabs.

Problem 13.1 *Automotive wheel and tire.*

Problem 13.2 *Coffee Maker.*

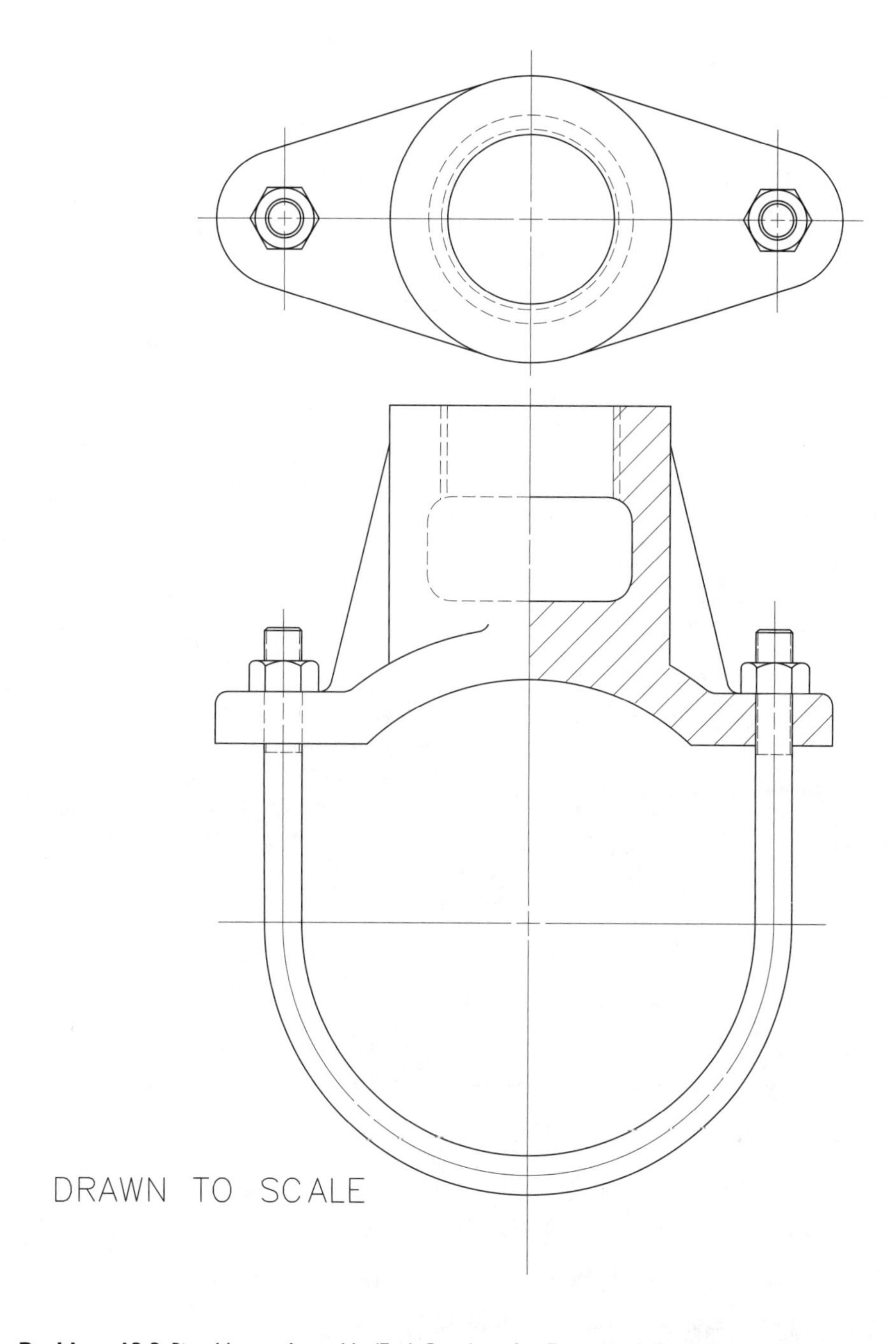

Problem 13.3 *Pipe Hanger Assembly (Earl:* Graphics for Engineers*). The body of the hanger is a rough cast metal to which a high-gloss epoxy coating has been applied. Use this hanger in a larger assembly by providing threaded mounting studs on the bottom of an I-beam, supporting a stainless steel pipe.*

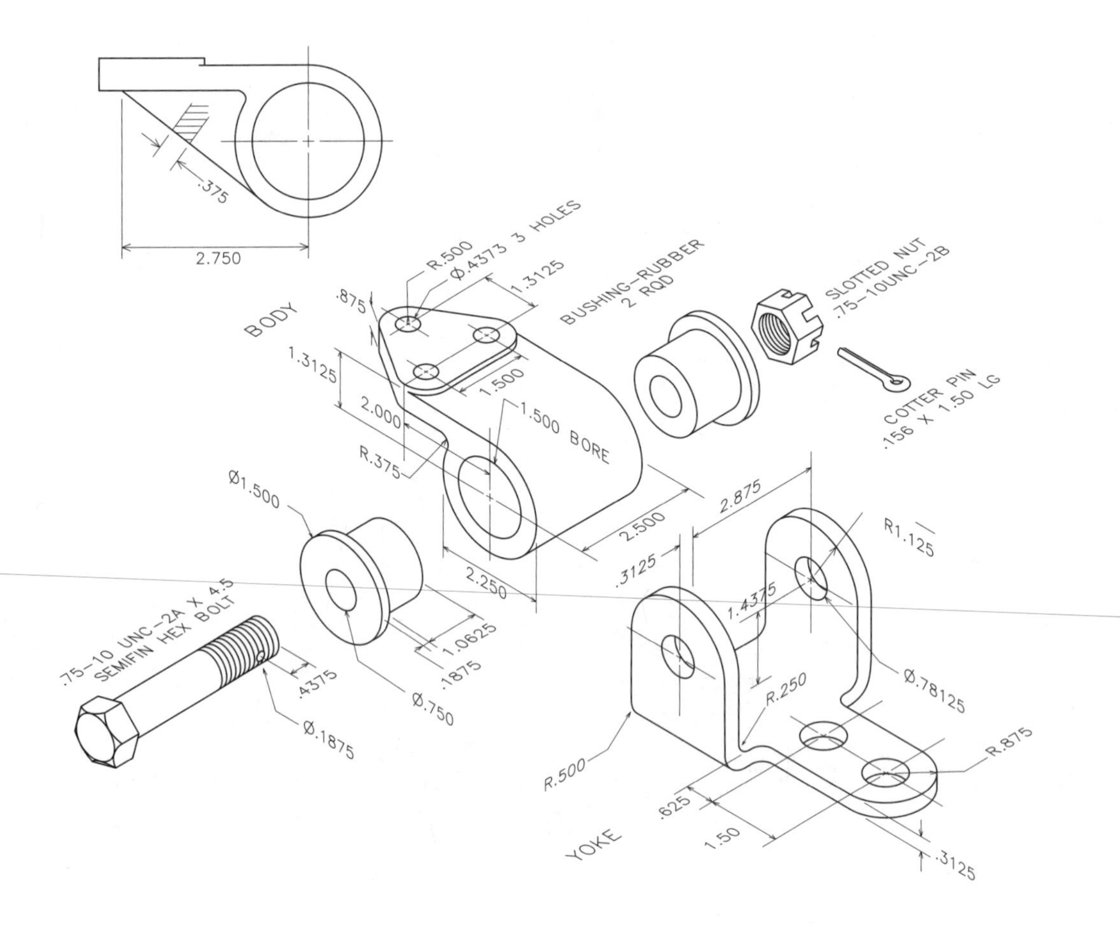

Problem 13.4 *Vibration Mount (French, Vierck, & Foster:* Engineering Drawing and Graphic Technology*).*
Model the Vibration Mount assigning the following materials to the parts.

Yoke—smooth metal plate with hole surfaces polished from drilling operations.

Body—cast metal with machined bores, holes, and mounting surfaces.

Bushings—black rubber casting (two piece mold).

Hex Head Bolt—nickel plated with threads, drilled hole for pin.

Slotted Nut—nickel plated with threads.

Cotter Pin—nickel plated.

Camera and Light Basics

CHAPTER OVERVIEW

The placement and specification of cameras and lights is instrumental in effective computer modeling presentations and animations. Both 3D Studio VIZ and MAX give you complete flexibility in view creation and with the lights that illuminate those views. Cameras in 3D Studio respond as they do in the real world. Lights can simulate sunlight or lighting fixtures. However, these objects are not intelligent. That is, they will respond to values that are outside the range of lights and cameras in nature. To keep your computer scenes realistic you must, to some degree, understand how physical lights and cameras work.

Of course, one of the strengths of computer graphics is to create situations that in real life would be impossible geometrically, spatially, or in terms of observer safety. It is your intelligence that keeps lights and cameras within the range of believability.

KEY COMMANDS AND TERMS

- **Ambient Illumination**—the level of general illumination in a scene without the influence of lights.
- **Axonometric**—a view where the line of sight is not parallel to any of the world axes. A User view.
- **Azimuth**—the angle of a camera in relation to the world Z-Y plane.
- **C Shortcut**—a single key entry that switches from an orthographic or User view to a Camera view.
- **Camera Icon**—the symbol that marks the position of a camera object.
- **Camera Object**—an object that creates a unique view of world space.

- **Camera Target**—the point at which the camera is focused. The distance between the camera and the view plane.
- **Camera View**—any view created by a camera object.
- **Cast Shadows**—an option in the **Modify|General Parameters** roll-out that determines whether a light casts shadows.
- **Central Line of Vision**—the line in the center of the field of vision connecting the camera and its target.
- **Convergence**—the characteristic of perspective that causes parallel features to appear to become closer in the distance.
- **Direct Light**—a light object that creates parallel light rays from a source in a single direction.
- **Dolly**—an option in the **Camera View** that moves the camera, its target, or the camera and target together along the central vision line.
- **Exclude**—an option in the **Modify|General Parameters** roll-out that allows geometric objects to be omitted from a light's effect.
- **Falloff**—the area of illumination cast by a light.
- **Field of View**—see **FOV**.
- **Focal Length**—the distance from the lens of a camera to its film plane measured in millimeters. A number that determines the degree of perspective convergence.
- **FOV**—the angle (horizontal, vertical, or diagonal) within which objects are included in a camera's view.
- **Free Camera**—a camera object that does not require a target.
- **Global Lighting Level**—the level of ambient illumination as set in the **Render|Environment** dialogue box.
- **Hotspot**—the part of a camera's illumination exhibiting the full strength of the multiplier.
- **Inclination**—the angle that a camera points upward or downward in relation to the world Y-X plane.
- **Lens**—the setting in the **Modify|General Parameters** roll-out that establishes the focal length of the camera.
- **Long Lens**—a camera whose lens value is greater that 55 mm.
- **Multiplier**—a number in the **Modify|General Parameters** roll-out that establishes the intensity of a light.
- **Normal Lens**—a lens of 55 mm in 35 mm film format.

- **Omni Light**—a light object that broadcasts light rays in all directions.
- **Orbit Camera**—an option in the **Camera View** that rotates the camera about its target.
- **Orthogonalized**—to make a perspective view orthogonal. To eliminate perspective convergence.
- **Orthographic Projection**—the toggle in the **Modify|General Parameters** roll-out that orthogonalized a perspective **Camera View**.
- **Pan Camera**—an option in the **Camera View** that keeps the camera in the same position but moves the target.
- **Perspective** (camera control)—an option in the **Camera View** that increases or decreases the camera's lens length.
- **Roll**—an option in the **Camera View** that rotates the camera about its local Z-axis.
- **Short Lens**—a camera whose **Lens** value is less that 55 mm.
- **Spot Light**—a light whose light rays emanate from a single point.
- **Target Camera**—a camera object comprised of separate camera and target objects.
- **Truck Camera**—an option in the **Camera View** that moves the camera, its target, and view plane as a single unit.
- **View Plane**—the plane on which the camera view is projected.

THE CAMERA VIEW

A camera object creates a Camera view (viewport) through which world space is seen. This is, by default, a perspective view (Figure 14.1) and unlike orthogonal views, most dimensions cannot be directly compared. The amount of perspective convergence depends on the length of the camera's lens. The camera has two parameters upon creation: *lens*, the focal length, and *FOV*, the field of view. These two parameters are inversely related. By making the lens longer, you narrow the FOV. By making the lens shorter, you broaden the FOV. The characteristic of perspective is convergence, a condition where parallel features (edges, limits, and intersections) become visually closer together as they recede into the distance. Perspective is appropriate for large objects and inappropriate for small objects. The perspective view can be orthogonalized by choosing **Orthographic Projection** in the camera's **Modify|Parameters** roll–out (Figure 14.2). This drives the camera to infinity along the camera's local Z-axis. This toggle stays on for subsequent cameras you create.

Tip: You can make objects look larger by increasing the perspective convergence (decreasing the focal length). Likewise, you can make a product appear smaller by reducing convergence (increasing focal length).

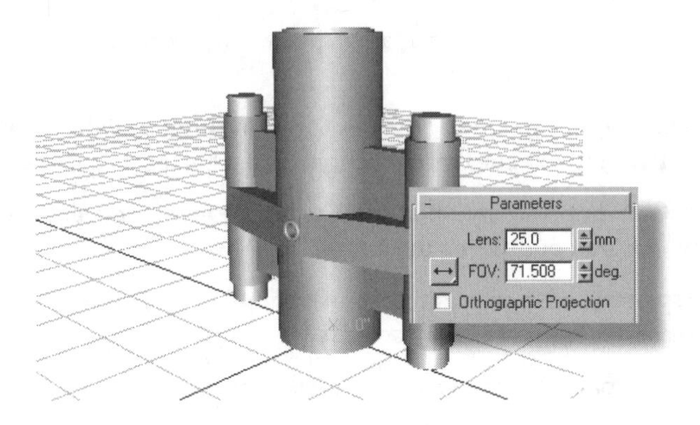

Figure 14.1 *The perspective of a Camera view.*

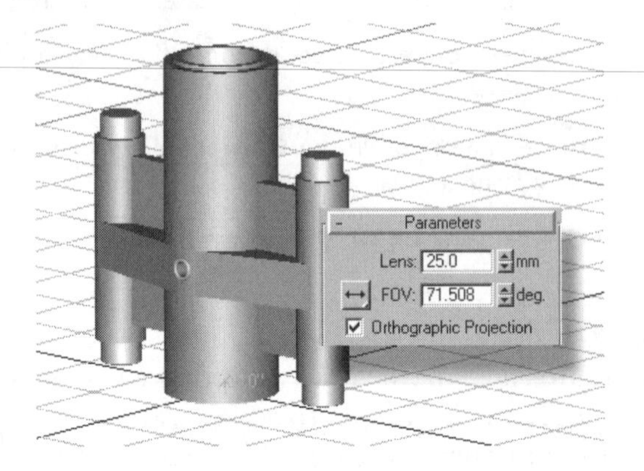

Figure 14.2 *The perspective view can be orthogonalized.*

Cameras are named "Camera01, Camera02, etc." as they are created but should be renamed with titles that make more sense. This is important because when switching to a camera viewport in a scene that has multiple cameras, a dialogue box (Figure 14.3) is presented for you to select the camera. Having descriptive names is a real help. You can switch from an orthogonal viewport to a camera viewport by using the **C** shortcut key. With a camera viewport active, you must right-click on the camera viewport label and choose **Views** and then the camera view you want.

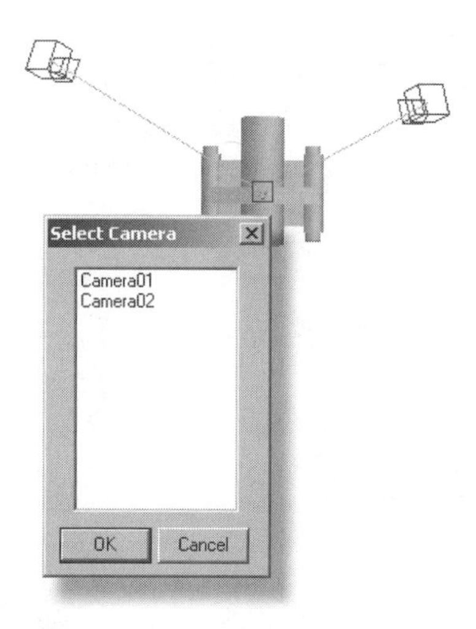

Figure 14.3 *Camera choice dialogue box.*

THE CAMERA AND ITS PARAMETERS

A camera icon represents the position of a camera. This position is recorded in absolute world space coordinates and can be transformed by rotating and moving. The camera's target is represented by a small box and is connected to the camera by a *central line of vision* that is concurrent with the camera's local Z-axis. The spatial location of the target can also be transformed, though rotation has no effect on the view. The camera's central line of vision is perpendicular to the *view plane* (Figure 14.4). The dimensions of the view plane are set in the **Rendering|Render** dialogue box.

By selecting the camera and right-clicking the **Select and Move** icon, two additional camera parameters are available: **Dolly** and **Roll**. These are explained later in this chapter in detail.

A camera is best created in an orthogonal viewport where its *azimuth* (angle around Z world) and its *inclination* (angle to X world) can easily be determined. In most designs you would use the Top view to set the azimuth and the Front or Side views to set the inclination. Cameras can be created in User, Perspective, or Camera viewports, but predicting ahead of time what view you'll get is problematic.

Tip: To see the effect of camera changes make one of your viewports a Camera view immediately after creating a camera. This way, any change of the camera position in an orthographic view is simultaneously reflected in the Camera view,

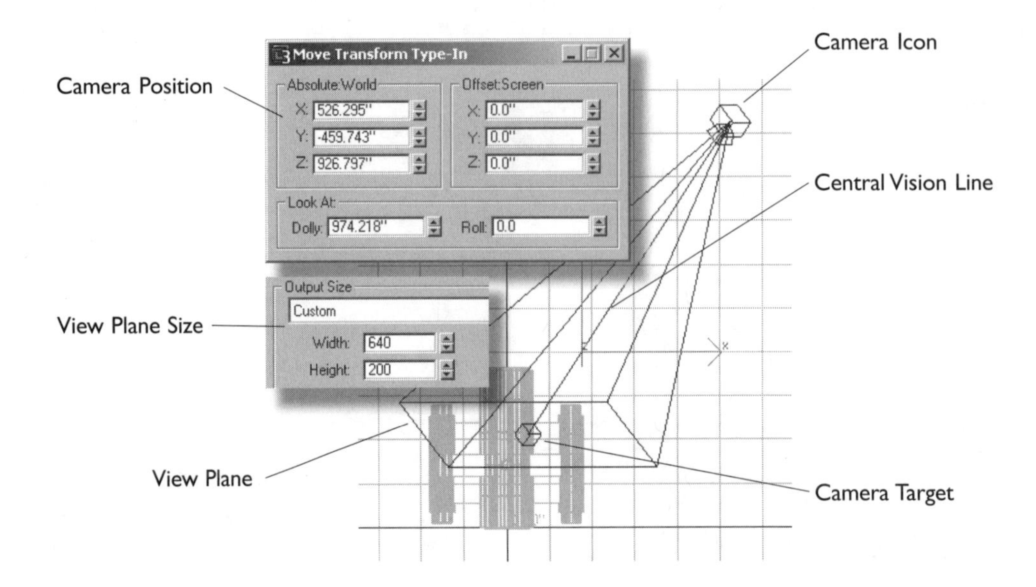

Figure 14.4 *Camera characteristics.*

- A **Target Camera** is aligned along the central line of vision. This line of vision connects the camera and the Camera Target and is created parallel to the active view (Figure 14.5). Use a **Target Camera** for most applications.
- A **Free Camera** is created with its central line of vision perpendicular to the active view (Figure 14.6). This type of camera has less use in technical applications because the orientation of its viewport, central line of vision, or target are not visible or predictable. However, a **Free Camera** is easier to animate along a path.

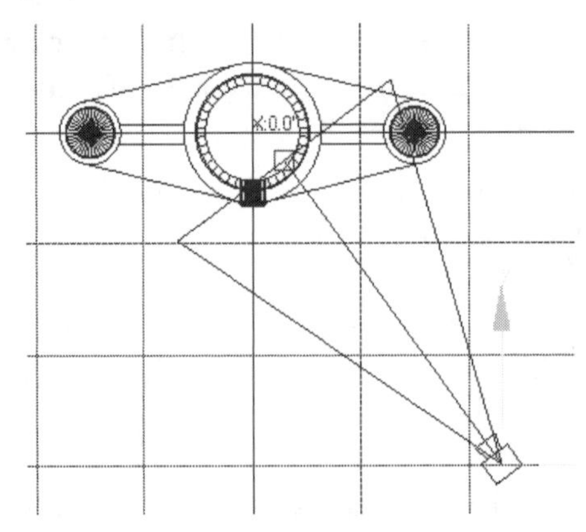

Figure 14.5 *The target camera.*

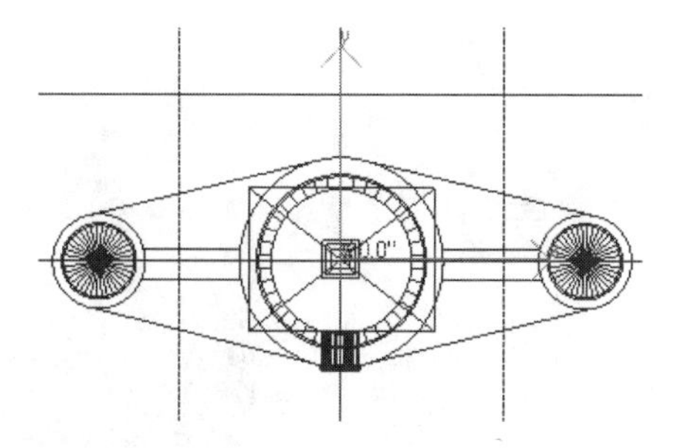

Figure 14.6 *The free camera.*

LENS AND FOCAL LENGTH

The *length* of the camera's lens determines the amount of perspective convergence in a scene. In 3D Studio, **Lens** corresponds to the focal length of 35 mm cameras. For example, a *normal lens* correlates to the perspective seen by the eye. In 35 mm format, this is approximately 55 mm. Values smaller than this are referred to as *short* or *wide-angle* while values greater than 55 mm are referred to as *long* or *telephoto*.

Tip: Avoid lens lengths below 25 mm as these cameras introduce unacceptable distortion into the view. Below 18 mm views take on a "fish eye" perspective and straight edges may actually appear curved.

Figure 14.7 displays a normal view. The object is contained within the view plane and the target is at the object's center. This places the object within the field of vision. Figure 14.8 displays a telephoto view. You can see that the camera must be further removed for the object to be contained within the view plane. Almost all perspective convergence is removed and an orthographic view is approached. Finally, Figure 14.9 shows the effect of severely reducing the length of the lens. The **FOV** is widened dramatically and an unnatural perspective convergence is introduced.

Note: Getting up close to a subject necessarily means that you have to decrease focal length (**Lens**) and widen the field of view (**FOV**) to see the same scene. For manufactured goods that you can hold in your hands, a focal length less than 35 mm is inappropriate. You may have to accept a smaller image in order to use a longer lens, generating less distortion.

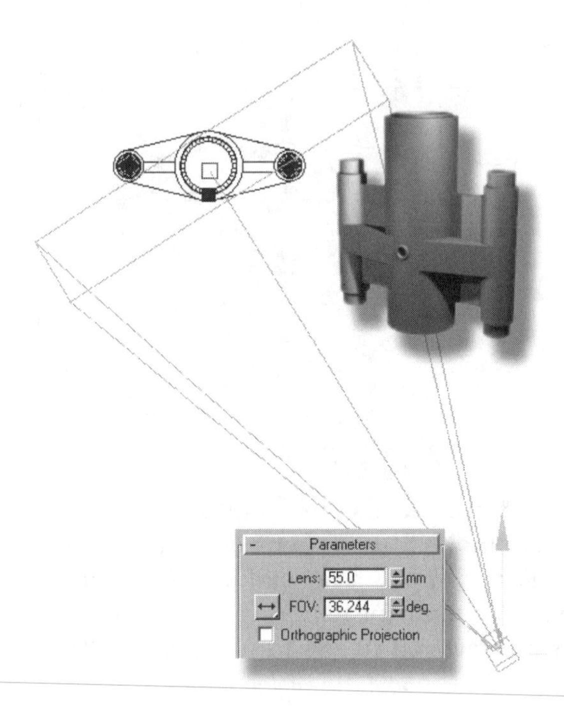

Figure 14.7 *Normal lens.*

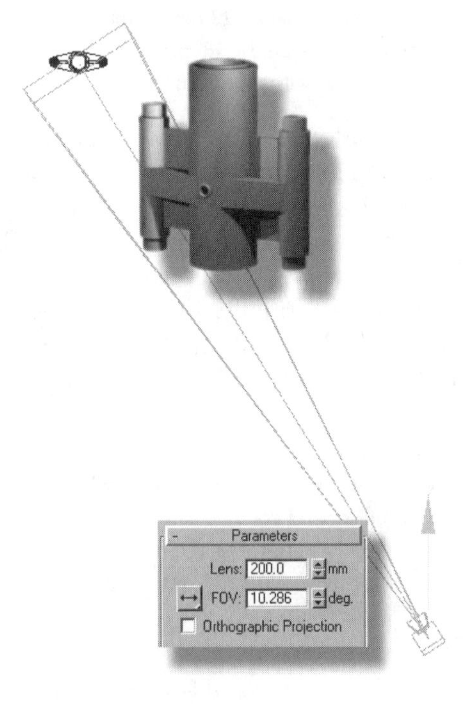

Figure 14.8 *Long or telephoto lens.*

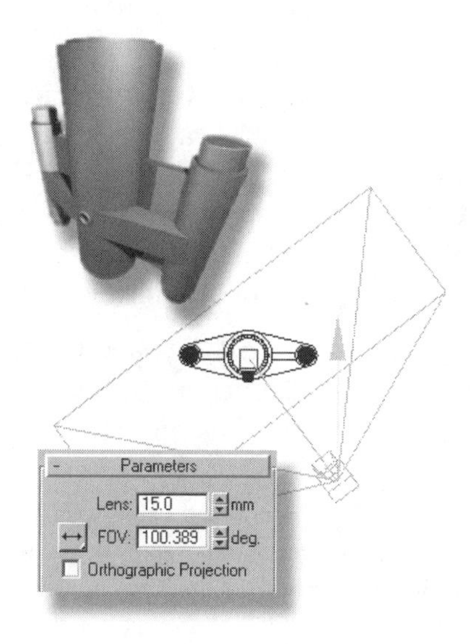

Figure 14.9 *Short or wide-angle lens.*

THE BASIC CAMERA METHOD

Technical subjects usually have a set orientation to the world Y-X plane (sitting on the floor, table, ground, or in the air a certain distance above the ground). Camera position is usually determined in the Top view first, where the angle (azimuth) of the view is apparent, and then in the Front view, where inclination is observable. To establish a basic camera, follow these steps.

STEP 1

Display multiple viewports. For simplicity, a two-viewport display is chosen. Choose **View|Configure|Layout** in VIZ and select two viewports. In MAX, choose **Customize|Viewport Configuration|Layout**.

STEP 2

Switch the left viewport to the Top view. Activate the left viewport by right–clicking in it. Type the **T** short cut to switch to the Top view. You can't switch the Right Viewport to the Camera Viewport because no camera exists in the scene.

STEP 3

Create a **Target Camera**. Choose **Create|Cameras|Target** and in the Top view click and drag to create the camera and its target. The click position establishes the position of the camera and as you drag, the central vision line is established by the target at your cursor position. Deposit the target at the center of the object as shown in Figure 14.10(a). You can use the Select and Move tool to reposition the camera or its target.

It may be difficult to select a camera's target when it is coincidental with geometric objects. You can use **Selection Filter|Cameras** from the main toolbar to restrict selections to cameras and their targets.

Tip: You can select a camera's target by first selecting the camera (this is easier to select because it is usually removed from geometric objects) and right-clicking. Choose Select **Target** from the pop-up menu. Hold down the **CTRL** key and deselect the camera. Only the target is selected.

STEP 4

Adjust inclination. The camera was created on the world Z-X plane as shown in Figure 14.10(b). Change to the Front view **(F)** and move the target so that it's in the center of the object. Move the camera to a position above the object.

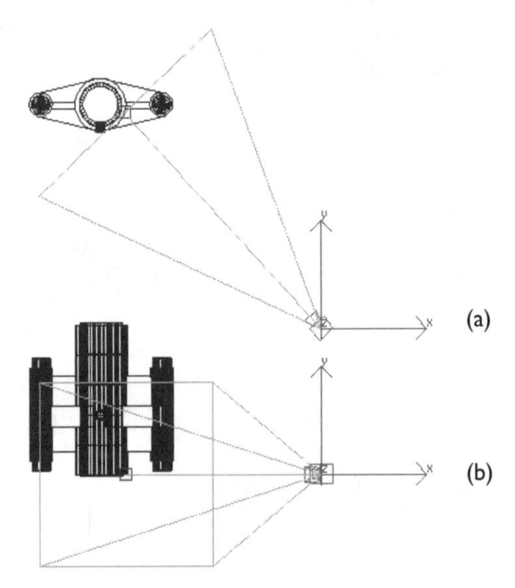

Figure 14.10 *Camera created first in Top view.*

Now you can switch the right viewport to the Camera Viewport. Right-click in the right viewport and type **C,** the shortcut for a Camera view. The view created by the camera is displayed.

STEP 5

Adjust inclination. With the Camera view displayed it is often easier to directly adjust the camera there than by switching back and forth between Top and Front viewports, adjusting the camera in directions parallel to each view. Camera controls (Figure 14.11) replace orthogonal viewport controls at the lower right of the interface.

- **FOV** changes the field of view and in doing so the length of the lens. This is like zooming in and out. The spatial position of the camera does not change.
- **Dolly Camera** moves the camera in and out along its central vision line. **Dolly Target** moves the target (and view plane) in and out along the central vision line. **Dolly Camera and Target** moves both together, again, along the central vision line.
- **Perspective** changes the length of the lens. Click-dragging the cursor downward increases lens length and reduces perspective. Click-dragging the cursor upward decreases lens length, increasing perspective convergence.
- **Roll Camera** rotates the camera about its local Z-axis (the central vision line).
- **Orbit Camera** moves the camera around the object, changing both azimuth and inclination, without moving the target.
- **Pan Camera** keeps the camera in the same position but moves the target, much like standing in the same place and looking around.
- **Truck Camera** moves the camera and its target together as a unit. The **FOV**, **Lens**, central vision line, and view plane are locked together as a unit as they are moved in space.

Tip: The benefit of using a Camera view that has been orthogonalized over a User view is that view manipulations are in relation to the camera. This keeps the subject in the field of view. If you **Orbit View** in a User view, the orbit happens about the world origin. If your subject is removed from the origin, it quickly is lost from the field of view.

As you adjust the camera in the Camera view, the position, **FOV**, and **Lens** are immediately adjusted and the effect is displayed in any orthogonal viewport.

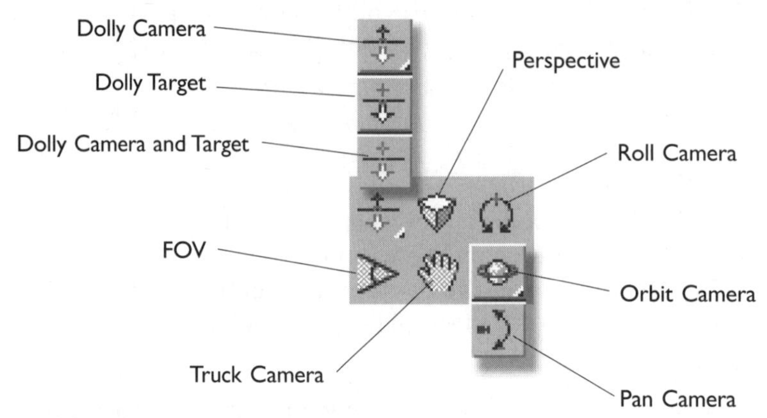

Figure 14.11 *Camera transformation controls.*

When you have arrived at the view you want, the camera's parameters can be checked in the **Modify|Parameters** roll-out.

IMPORTANCE OF LIGHTS

You'll quickly find that the default ambient illumination in 3D Studio may be sufficient for modeling, but not for rendering or animation. Lights are an important way to create contrast, between an object and its background, and on the object itself between features. Lights are also instrumental in allowing materials to be observed and without them certain kinds of mapping, bump mapping for example, are not effective.

Arriving at the correct amount of illumination is a difficult task. Too bright, and detail is washed out. Too dark, and detail is lost in shadow. One aspect of computer lighting that is problematic is arriving at a high level of general illumination. In the physical world, light bounces off various surfaces to illuminate areas that are away from the light source. Only in space (where there is nothing around to reflect light) is there a sharp line between illuminated and not illuminated. With computer lighting, these secondary light sources must often be explicitly placed in the scene.

Note: More realistic lighting and reflections can be achieved through a rendering technique called *ray tracing*. With ray tracing, light rays are traced to their source, bouncing off surfaces, picking up colors, being absorbed, and so on. Because this text concerns itself with engineering and technology applications of 3D Studio, ray tracing, though important, is left for other resources to explore.

BASIC LIGHTING METHOD

The level of ambient light, the light that's in the scene before adding any light objects, impacts general illumination. Ambient light casts no shadows, has no direction, and illuminates all objects evenly. Ambient light is adjusted in the **Rendering|Environment** dialogue box. By default, 3D Studio illuminates the scene with the lowest value of ambient light, black, and a level of 1. This produces the level of lighting in Figure 14.12, sufficient to evaluate your designs.

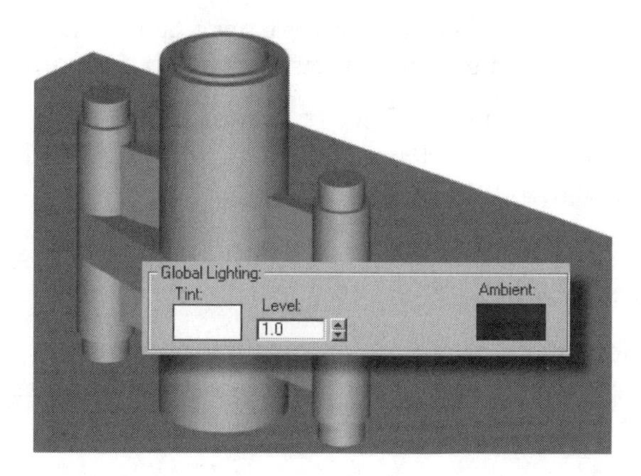

Figure 14.12 *Default ambient illumination.*

By increasing the level of ambient light, the sharp demarcation between light and shade can be softened. Start with a middle gray value of 128 and **Global Lighting|Level:1** (Figure 14.13). Notice that the ground plane (actually white) is now illuminated sufficiently.

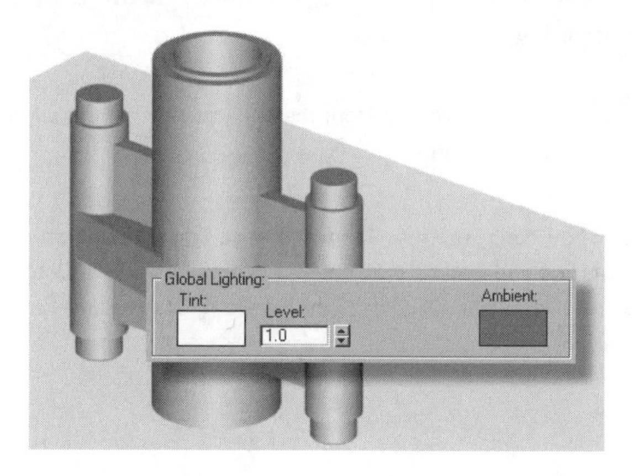

Figure 14.13 *Adjusted ambient illumination.*

Three types of standard light objects can be added to this general ambient illumination: **Direct** lights, **Omni** lights, and **Spot** lights.

- A **Direct** light casts parallel light rays and when its **Hotspot** and **Falloff** parameters are sufficiently large to take in the entire scene, the effect corresponds to that of the sun.
- An **Omni** light casts light rays from its center evenly in all directions, much like a light bulb without a shade. **Omni** lights are effective when used to fill in and provide reflected light.
- A **Spot** light's rays converge at a single point. Spot lights are used to illuminate individual parts and to represent cast conical light from a point.

The *basic lighting method* uses a **Target Direct** light and one or more **Omni** lights to provide sources of reflected illumination. **Spot** lights may be added later for more specialized illumination needs.

STEP 1

Establish a Camera view. If you desire an orthogonal view, select the camera and choose **Modify|Parameters** and choose **Orthographic Projection**. Make sure that lights and cameras have not been turned off in **Display|Off by Category**.

STEP 2

Establish a strong **Target Direct** light (Figure 14.14). If you are looking from the right, place this light above and in front of the object to the left. Start with a light intensity **Modify|General Parameters|Multiplier:1**. Increase **Modify|Directional Parameters|Hotspot** and **Falloff** values so that the entire scene is illuminated.

Note: If you are looking from the left, just reverse the placement of **Target Direct** and **Omni** lights.

Tip: You don't want the **Target Direct** light to act like a spotlight so the **Hotspot** and **Falloff** parameters need to be very large to encompass the entire scene. If this light is acting as the sun, its **Hotspot** and **Falloff** values should be nearly identical.

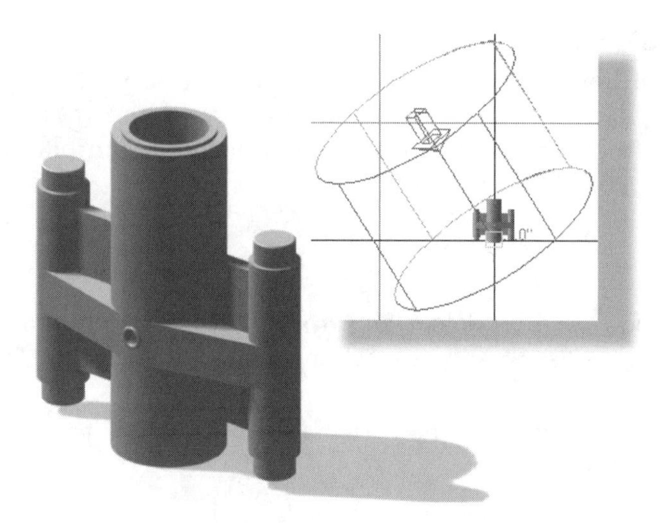

Figure 14.14 *Target Direct light provides strong illumination.*

STEP 3

Place **Omni** lights. Omni lights represent secondary light sources and illumination from the **Target Direct** light bouncing off the environment. Choose **Create|Lights|Omni** and in the Top view place one light to the right and in front of the object (light 1) and another behind the object to the left (light 2). Figure 14.15 shows this placement.

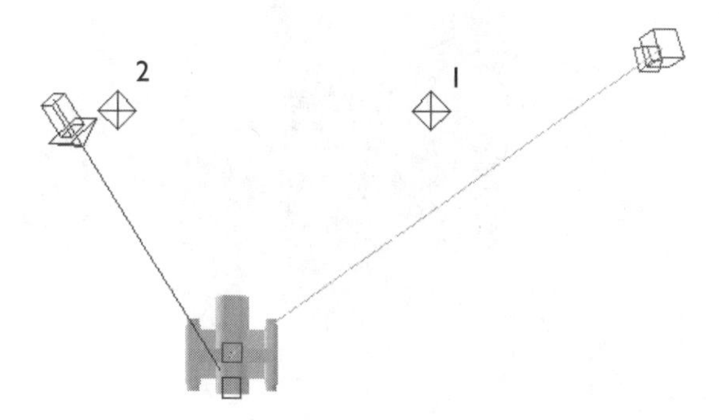

Figure 14.15 *Omni light placement in basic lighting.*

BASIC LIGHT ADJUSTMENT

By default all lights begin with white light, casting shadows, and a **Multiplier** (strength) of **1.0**. Because we want our light source to act like the sun, we should turn of casting shadows and diminish the Multiplier so that there are no conflicts with the sun's shadows.

STEP 1

Turn off shadow casting. Select an **Omni** light and right–click. In the pop-up menu toggle **Cast Shadows** off. Repeat this for the other **Omni** light.

Note: You cannot select multiple lights and affect their properties simultaneously.

STEP 2

Decrease light intensity. Select the **Omni** light in front and to the camera side **Modify|General Parameters|Multiplier** and reduce the value to **0.7**. Select the **Omni** light behind and to the left of the object and reduce its **Multiplier** to **0.5**. The result of these adjustments is shown in Figure 14.16. When you compared with Figure 14.14 you can see the impact of the secondary lights.

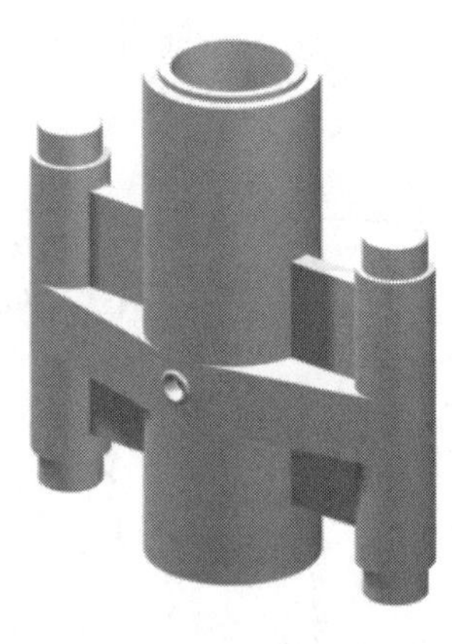

Figure 14.16 *Omni lights do not cast shadows and are less bright.*

Omni lights, because they illuminate both the object and the ground, have obliterated the shadow cast by the **Target Direct** light. To correct this we will exclude the ground plane from the list of objects the **Omni** lights illuminate.

STEP 1

Select one **Omni** light. Choose **Modify|General Parameters|Exclude..** and select the ground from the list on the left. Move the ground from the *include list* (on left) to the *exclude list* (on right) by clicking the **>>** button (Figure 14.17). Click **OK**. Repeat this for the other **Omni** light. When the scene is rendered again there is no interference with the target direct shadow (Figure 14.18).

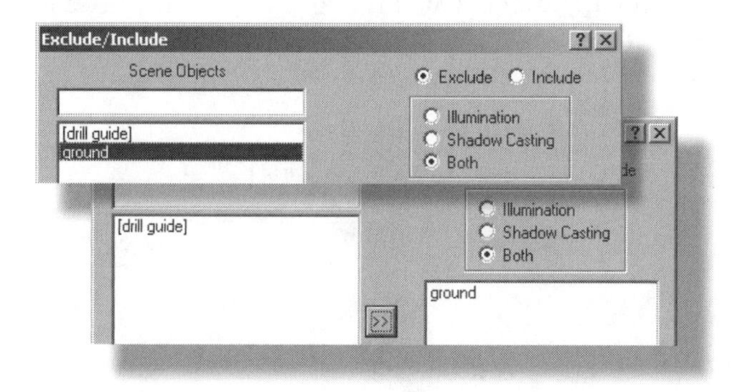

Figure 14.17 *Excluding an object from illumination.*

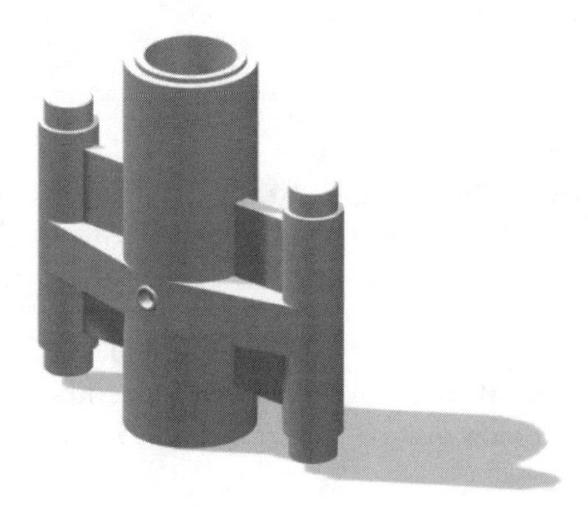

Figure 14.18 *Shadow from Target Direct light is visible.*

EXAMPLE: CREATING AN EFFECTIVE SCENE

In Chapter 8, Modeling Assemblies, you modeled the Pulley Assembly. Let's see how a camera and basic lighting can improve visualization.

STEP 1

Establish a camera. In the case of the Pulley Assembly, a **Target Camera** is placed above, in front, and to the left.

STEP 2

Increase Ambient light. In the **Environment** dialogue box, increase the **Ambient** color, keeping **Level** at **1.0**. Figure 14.19 shows default (a) and increased (b) ambient light.

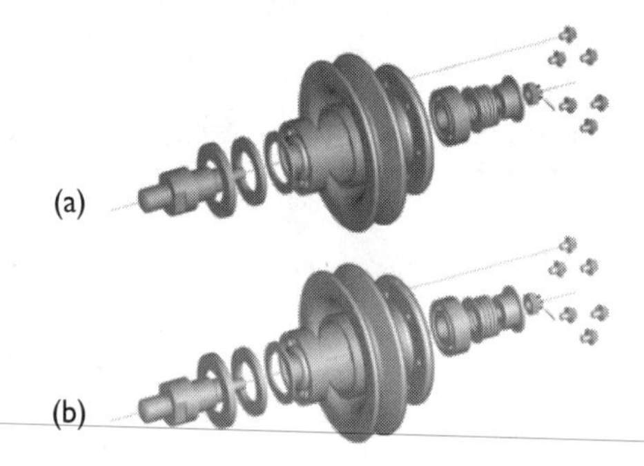

Figure 14.19 *Increasing ambient illumination.*

STEP 3

Create a strong **Target Direct** light. Bring the **Target Direct** light in from the right and in front of the assembly. Choose **Create|Lights|Target Direct** and in the Top view drag from the light to the target. Switch to the Front view and raise the light to shine down on the assembly. Place the **Target** in the middle of the assembly. Set **Multiplier** to **1.0** and **Falloff**, **Cast Shadows**, **Hotspot**, and **Falloff** to include the entire assembly. Figure 14.20 shows this light in the Front view and its effect in the Camera view.

STEP 4

Create secondary **Omni** lights. Place one **Omni** light above, in front, and to the left. Place another **Omni** light to the right, above, and behind. Start **Multiplier** values at **0.7** and **0.5** as was previously discussed. Keep **Cast Shadow** on to push shadows away from the camera. These **Omni** lights fill in the shade formed by the **Target Direct** light (Figure 14.21).

This is a departure from the first example of basic lighting given in this chapter but points out an important aspect about lighting: there are no lighting designs that will work in all instances. Because manufactured and constructed goods have widely varying shapes and materials, basic cameras and lighting will almost always be adjusted.

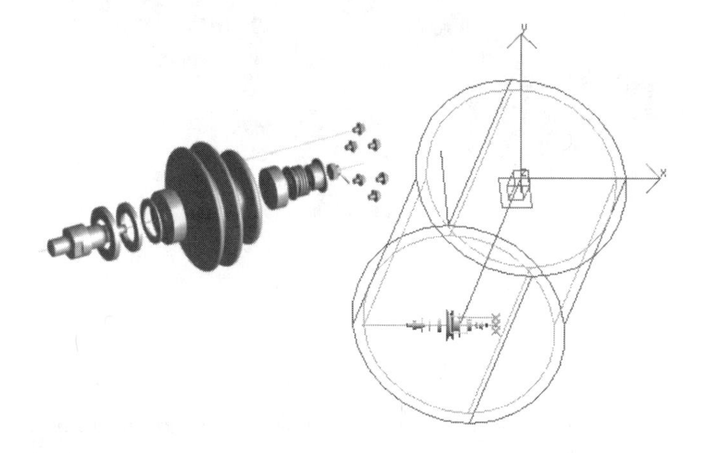

Figure 14.20 *Target Direct light provides general illumination.*

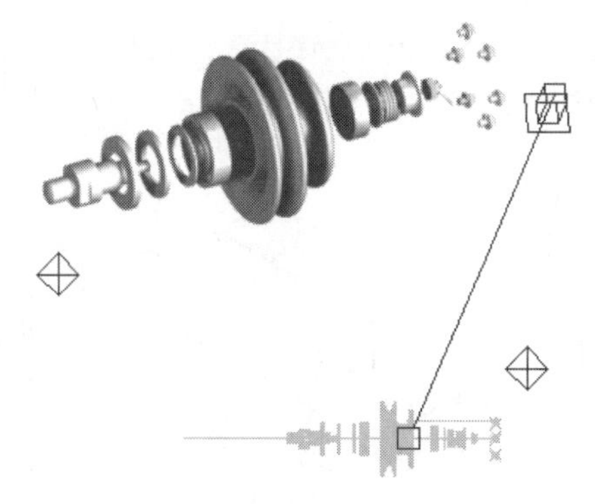

Figure 14.21 *Omni lights fill in the shadow.*

STEP 5

Exclude the centerlines. The lights wash out the centerlines (renderable **Spline Shapes** of a set thickness). **Exclude** these from all three lights. The final illuminated Camera view is shown in Figure 14.22.

Figure 14.22 *Final scene.*

Tip: The effect of a light is impacted by both its strength and its proximity to the subject. By moving a light away from the object you spread the light out and diminish its effect.

PROBLEMS

A set of basic objects can be found on the CD-ROM in *chapters\chapter_14*. Apply the camera and lighting instructions in the following problems.

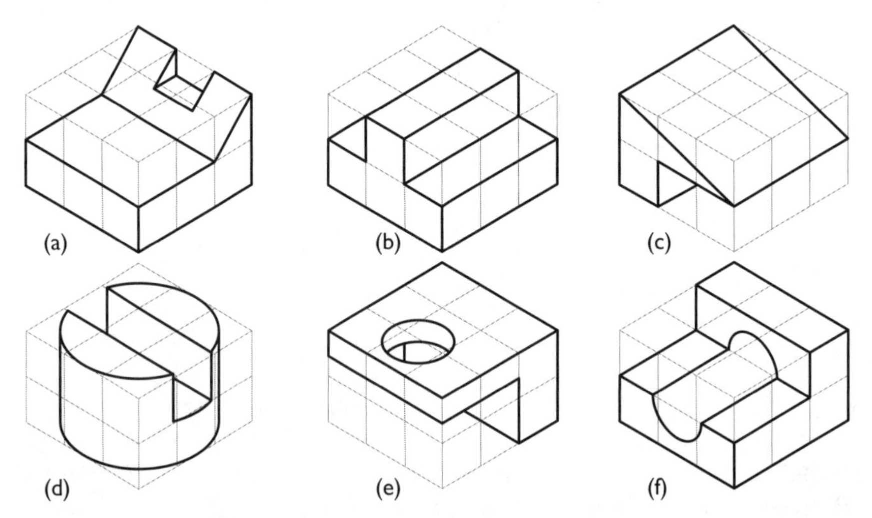

Problem 14.1 *Objects for lighting studies. In each case establish lights that fully explain geometric features. Place the camera in a favorable isometric position.*

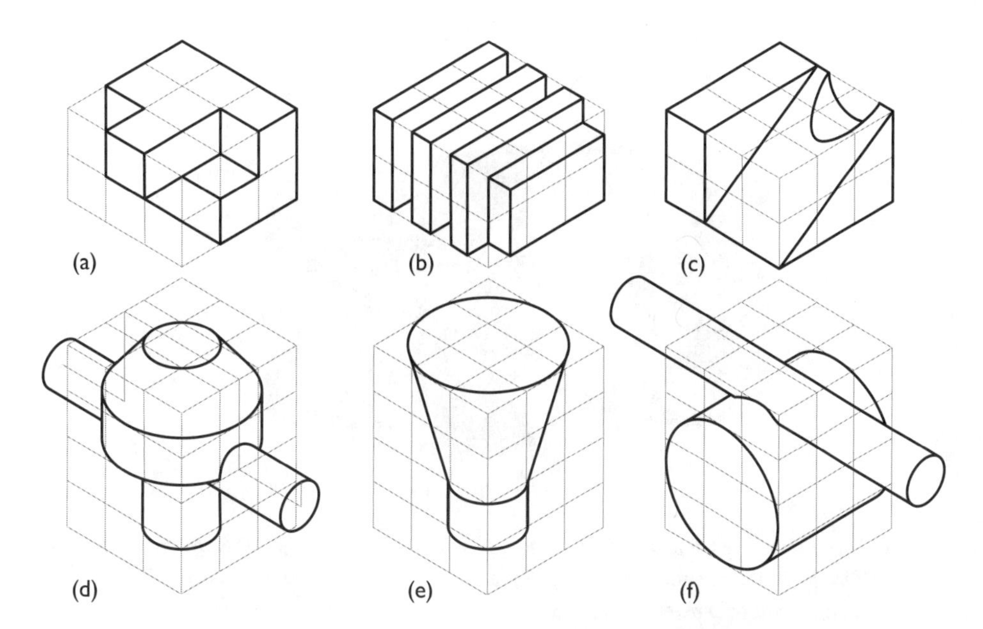

Problem 14.2 *Objects for lighting studies. In each case establish lights that fully explain geometric features. Place the camera in a favorable isometric position.*

Problem 14.3 *Piping study. Load problem14-3.max from the CD-ROM and establish lights that provide the strong overhead lighting you would encounter on the factory floor. The camera has been provided for you in this file. Make sure you set sources of reflected light to illuminate the sides of the object against a black background.*

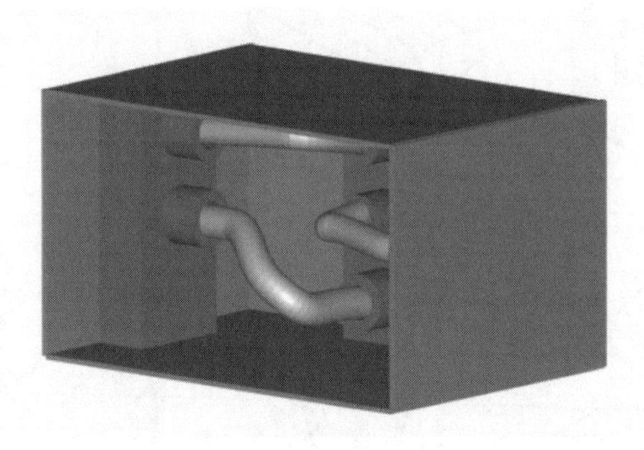

Problem 14.4 *Perspective camera problem. Load problem14-4.max from the CD-ROM and establish a perspective camera with a FOV sufficient to see all five inside walls of the Hydraulic Junction Box. Experiment with camera parameters until an acceptable level of distortion is achieved..*

Problem 14.5 *Confined lighting. Establish lights in the previous problem that effectively illuminate the Hydraulic Junction Box. You will have to exclude certain objects.*

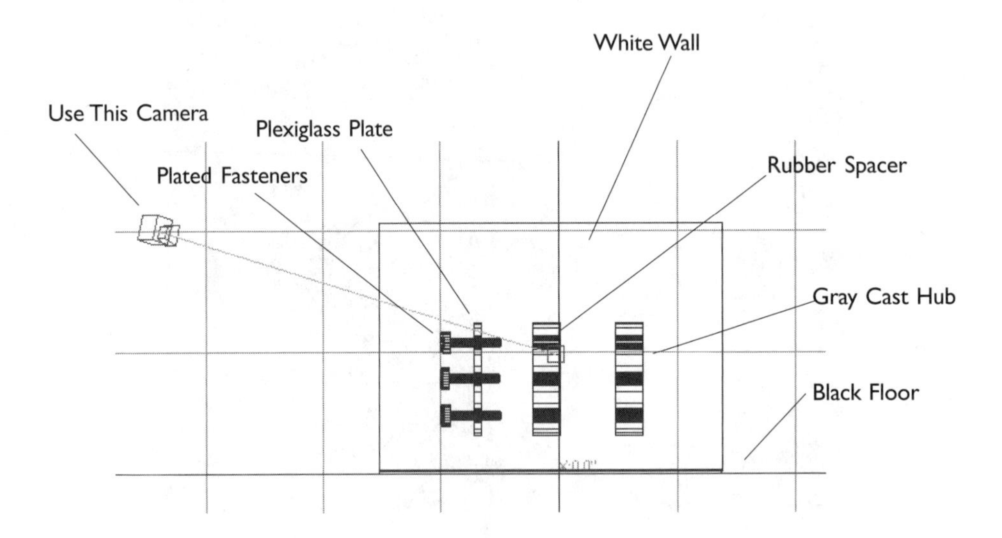

Problem 14.6 *Excluding objects from lights. Load problem14-6.max from the CD-ROM. This small assembly contains very light, very dark, and transparent parts. By establishing lights that exclude certain objects, the scene can be illuminated so that all parts are clearly visible.*

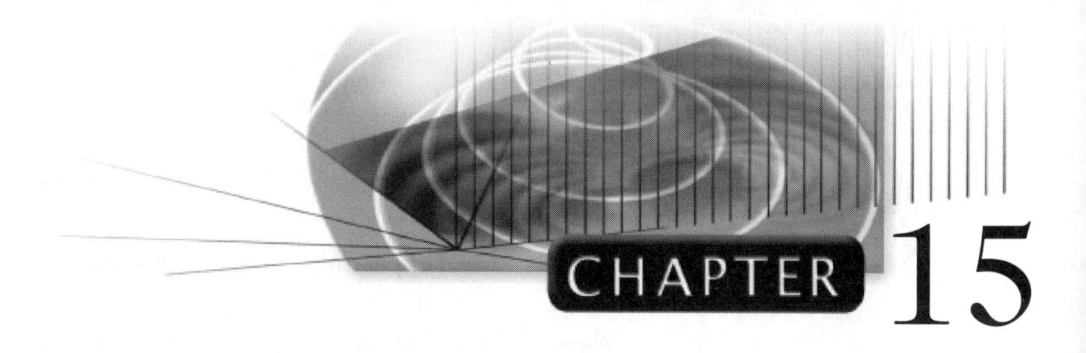

Camera and Light Applications

CHAPTER OVERVIEW

Once you have gained experience in creating Camera views and positioning lights you'll realize that there is so much more that these 3D Studio objects can do to make your technical presentations and designs more effective.

Because there are an infinite number of possible views, it becomes important to settle on a number of standard views that can be easily replicated. This facilitates complimentary pictorial views from a number of graphic applications. For example, viewpoint pictorials from AutoCAD can be matched with camera orthographic projections in 3D Studio. It allows you to match perspectives or axonometrics made at different times or by different people.

Because cameras can omit geometric objects that are too near or too far (clipping) sectional views (of sorts) can be created without actually cutting the parts.

There are numerous adjustments that can be made to lights. Directly adjusting the face normal to reposition a highlight is an effective way to get a light into exact position. You can even adjust where a light starts and stops and fill a light with haze or fog.

KEY COMMANDS AND TERMS

- **Angular Perspective**—a perspective view formed by rotating the subject about the vertical axis.
- **Attenuation**—the fading in and out of illumination generated by a light object.
- **Attenuation Ring**—the visual indicator showing where near and far attenuation begins and ends.

- **Camera Clipping**—a function that clips or truncates the display of geometry within a camera's field of view.
- **Cone of Vision**—the field of view in perspective within which acceptable distortion exists.
- **Dimetric View**—an orthogonal view formed when the line of sight makes an equal angle with two of the three axes.
- **Dolly Camera** (standard view)—the **Camera View** transform that allows very small numbers to be used for axonometric standard views.
- **Foreshortened**—reduction in size by viewing a feature at an angle.
- **Isometric View**—an orthogonal view formed when the line of sight makes an equal angle with each of the three axes.
- **Oblique Perspective**—a perspective view formed by rotating the subject about the vertical and horizontal axes.
- **Parallel Perspective**—a perspective view formed when the subject remains aligned to the vertical and horizontal axes.
- **Perspective Picture Plane**—the view plane; the plane on which the perspective image is projected.
- **Principle Orthographic View**—one of six orthographic views formed by taking a direction of sight parallel to one of the world axes.
- **Restore Active View**—the command that switches the current view to the principle or User View saved with **Save Active View**.
- **Save Active View**—the command that discards the currently saved active view (if one exists) and replaces it with the current active view.
- **Standard Views**—views with known camera (orthographic viewpoint) positions.
- **Use**—the toggles in **Modify|Attenuation Parameters** that activate near and far attenuation.
- **User View**—any principle orthogonal view whose viewpoint has been translated.
- **Viewpoint**—the camera position in a User view; the camera position in a Camera (perspective) view that has been orthogonalized by selecting the **Orthographic Projection** option.

CAMERA VIEWS AND SAVED VIEWS

A camera is a permanent way of recording a view and 3D Studio has six saved orthogonal views: Front, Top, Bottom, Rear, Right, and Left. Orbiting the orthogonal viewpoint so that the line of sight is no longer parallel to one of the world axes creates a User view. A User view can be saved by choosing **View|Save Active User View**. Then when you want to go back to this view choose **View|Restore Active User View**. The limitation to this method is that only one active view can be saved. Because camera views can be orthogonalized, they are independent of saved and restored views. You can always retrieve a Camera view and as such, orthogonalized camera views are more useful than saved User views.

A CAMERA APPROACH TO STANDARD AXONOMETRIC VIEWS

When principle orthogonal views (Top, Front, Side, etc.) are rotated, a User view is created. This is 3D Studio's description of an *axonometric view*. Unfortunately, 3D Studio provides no mechanism for either entering coordinates for a specific User view or recording its viewpoint location in order to get a certain axonometric view. But because Camera views can be orthogonalized, you have a method for arriving at predictable axonometric views.

Tip: Axonometric User views have a distinct advantage over perspective Camera views. Because axonometric views are orthogonal views, parallel dimensions can be directly compared. If you want to check two vertical dimensions, you can. You can compare two horizontal measurements (as long as they are parallel), or two parallel angular measurements...anywhere in the scene. In perspective, this is nearly impossible.

There are three categories of axonometric views, depending on the position of the viewer: isometric, dimetric, and trimetric. Isometric may be the most common axonometric because of its ease of construction and predictability of pictorial display. Most engineers and technologists are used to making isometric sketches. Dimetric and trimetric views have been avoided historically but because computer graphics automates a once-difficult task, they should not be dismissed. Dimetric and trimetric views can solve problems of symmetrical features aligning in isometric, or exceptionally long and narrow objects appearing to *spread* as they recede. The only difference between the three views in 3D Studio is the position of the camera.

An *Isometric View* (Figure 15.1) is formed when the camera's position is an equal distance along the X, Y, and Z-axes with **Orthographic Projection** chosen in the **Camera|Modify** roll-out. The camera's target is at 0, 0, 0. This results in a view where features in planes perpendicular to these axes are equally inclined (foreshortened). This means that distances along X, Y, and Z world axes can be directly compared, something we often do with technical drawings. Figure 15.1 shows the camera at X=1000, Y=–1000, Z=1000.

When **Orthographic Projection** is checked in the **Modify|Parameters** rollout, an isometric view is displayed. By changing value signs (+ or –) you will arrive at isometric views from the left, rear, or below.

Because an orthographic projection camera is actually at infinity, the actual numbers you enter are immaterial. For example, you could enter 10, 10, 10 as the three values for an isometric. Then you can **Dolly Camera** to zoom the Isometric view out so that the image is at a comfortable scale.

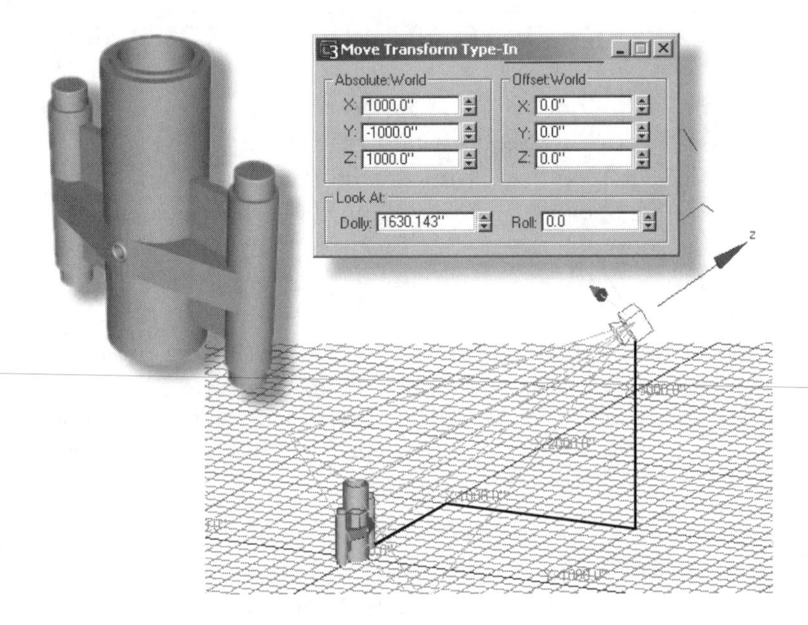

Figure 15.1 *Isometric view by camera position.*

Tip: VIZ displays the option for the more common isometric views in its **View|Views** menu. For example, a view from the SE produces the view shown above. Choosing SW moves the viewer in the negative X direction, negative Y, positive Z; 3D Studio MAX does not provide this feature. However, a file can easily be created containing orthogonalized cameras at SE, SW, NW, and NE positions. These cameras can then be **Merged** with the current scene as needed.

A *Dimetric View* (Figure 15.2) is formed when the camera's position is an equal distance along two of the three axes. Two of the principle planes are equally foreshortened. This view is more realistic than an isometric view but because one of the axes is unequally scaled, distances along this axis cannot be directly compared to the other two. Figure 15.2 displays a 20-40-40 dimetric view. It is referred to as such because the exposure of ellipses in principle planes is 20 degrees on top, and 40 degrees on the sides.

The dimetric values of 1000, 1000, 510 could have as easily been 1.0, 1.0, and .510. As with an isometric view, the relationships of the camera's positional numbers are critical, not their magnitude.

Note: Max files with camera positions for standard axonometric views can be found on the CD-ROM in *tools\axon_cameras.* You can **Merge** the cameras from these files into your scenes for predictable axonometric views.

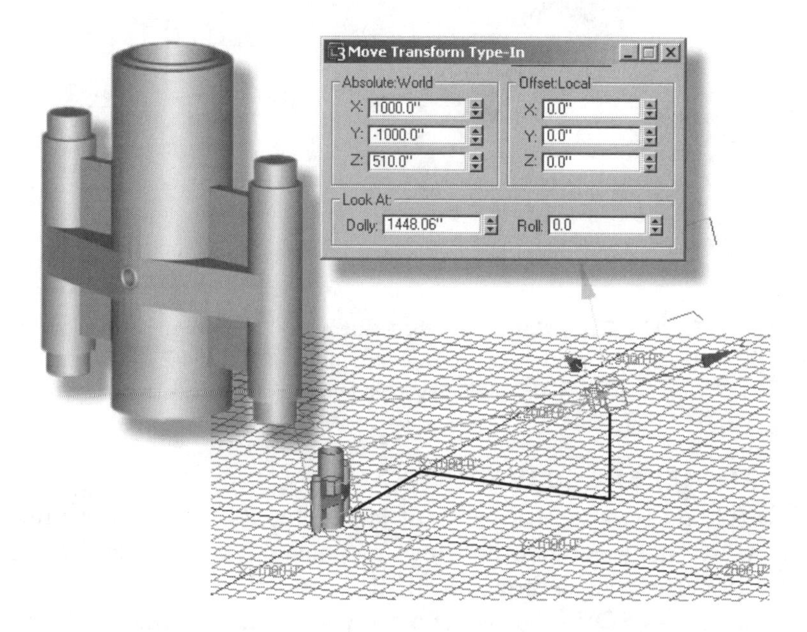

Figure 15.2 *Dimetric view by camera position.*

A *Trimetric View* (Figure 15.3) is formed when the camera is located at a position of unequal distance along each world axis. This results in the most realistic axonometric view, nearly perspective in its realism, but with the benefit that distances along an axis can be directly compared (although distances along any two axes cannot be compared). A 70-15-10 trimetric view has been created by placing the camera at X=1.00, Y=1.15, Z=3.88 (actually 1000, –1115, 3880). As you might expect with the Z value more than three times the other two, the view is more from above. You see very little of the Z-X or Z-Y planes.

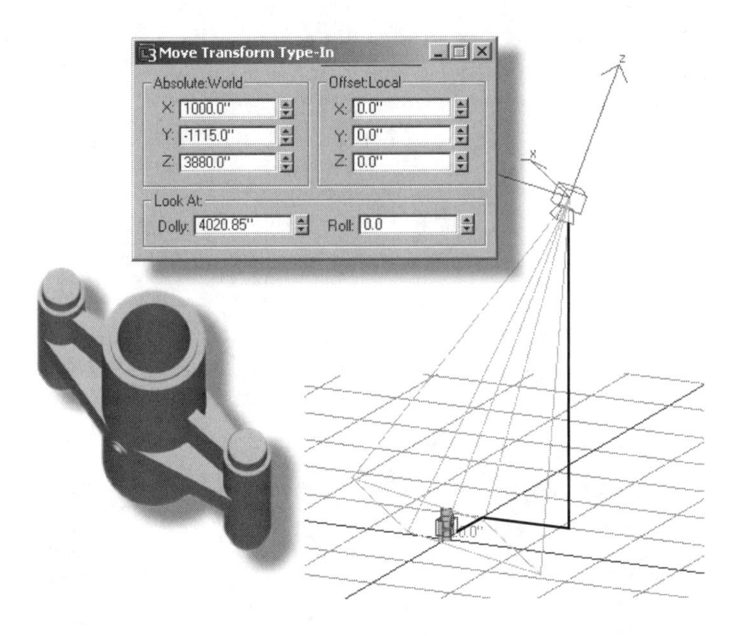

Figure 15.3 *Trimetric view by camera position.*

Tip: Use **Dolly Camera** in the Camera view to move the camera closer or farther away from the subject. The relationship of X, Y, and Z distances remains the same because the camera moves along the central vision line.

STANDARD PERSPECTIVE VIEWS

Those who rely on perspective views of large objects also have a number of standard views. These perspective views are based on a standard cone of vision of 30 degrees in the horizontal direction. By entering 30 into the **Modify|Parameters|FOV** field you can see you'll need a camera with roughly a 67 mm lens (Figure 15.4). By choosing the vertical icon in the **FOV Direction** fly-out menu, a 30 degree horizontal **FOV** results in approximately a 22–degree vertical **FOV**. This is very close to the numbers for technical perspectives with acceptable distortion.

The view plane should match the 30 x 22 aspect ratio (1.363) of the 67 mm camera's cone of vision. By checking the **Render Design** (**Render Scene**) dialogue box, you'll see that the 800 x 600 viewport's aspect ratio is 1.333. This is close enough to 1.363 to accept the standard display monitor aspect ratio. What this means is that objects within the viewplane of a 67 mm camera will have acceptable distortion.

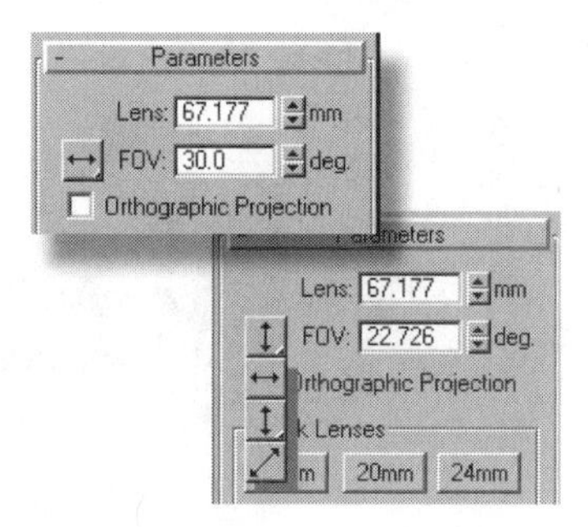

Figure 15.4 *Camera for standard perspective views.*

Figure 15.5 shows a 67 mm camera with its target and view plane on the world Z-X plane. This Z-X plane corresponds to the perspective picture plane; the object has been tilted forward 30 degrees, resulting in an oblique perspective. It usually isn't advisable to tip the model from its orientation to the X-Y ground plane (it makes subsequent constructions problematic) so the model is left alone and the camera is tipped 30 degrees (Figure 15.6), resulting in the same view.

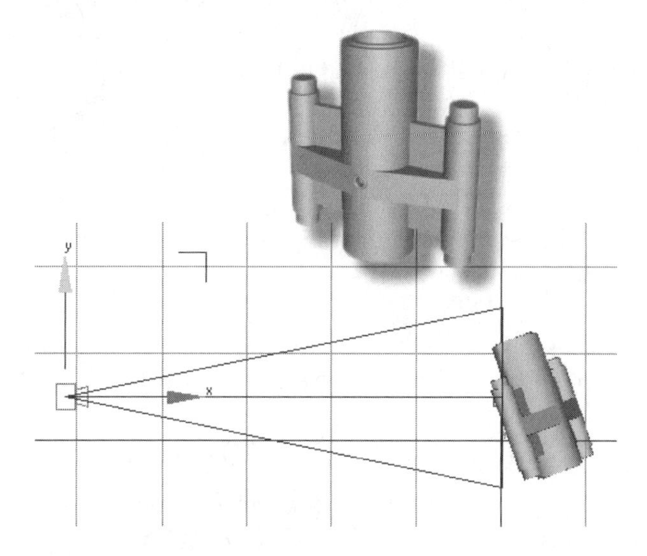

Figure 15.5 *Camera-view plane-object relationship.*

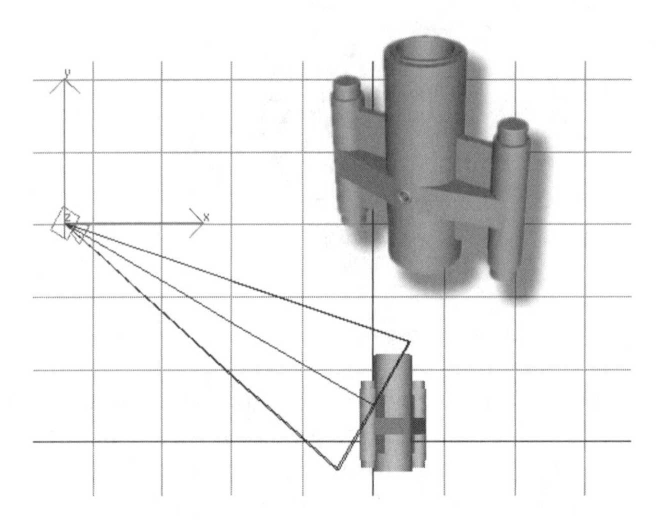

Figure 15.6 *Camera is tipped in a 30-60 perspective.*

In Figure 15.6 the object has been rotated 30 degrees about world Z. This is called a 30-60 perspective (angular perspective) because one face of the box that encloses the object is at 30 degrees, and the other at 60 degrees. This gives one long face, and one short face. Two other perspective views are commonly encountered. In a parallel perspective the object isn't rotated (Figure 15.7), but remains parallel and perpendicular to the picture plane (a 0-90 perspective). In a 45-45 perspective (also angular), the object is exposed equally on the left and right (Figure 15.8). An oblique perspective includes both tipping and rotation.

Tip: By choosing parallel, angular, or oblique perspectives of 0-90, 45-45, or 30-60, subsequent perspectives can easily be matched.

SECTIONAL VIEWS WITH CAMERA CLIPPING

Each camera object has two planes that can be used to exclude geometry from the view or rendering. The **Near Clipping Plane** excludes any geometry between the clipping plane and the camera. The **Far Clipping Plane** excludes geometry beyond a certain distance from the camera. One would think that this could be used to create sectional views without actually slicing the geometry. But as you will find, these clipping planes are more suited for selectively omitting unwanted geometry from a scene rather than for enabling technical sections.

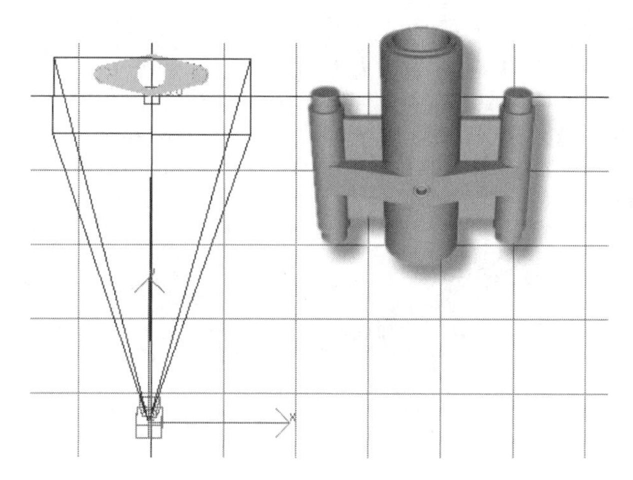

Figure 15.7 *Parallel perspective.*

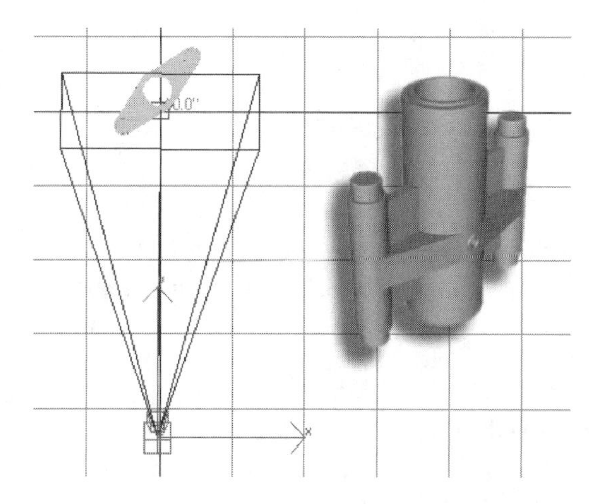

Figure 15.8 *45-45 perspective.*

Note: No mention is made of one-point, two-point, or three-point perspective because with 3D Studio, vanishing points are irrelevant. The real terms are parallel, angular, and oblique, making reference to the relationship of the geometry to the plane of projection.

The problem is that clipping does not produce sectional views in the normal engineering sense because visibility of clipping plane intersections with geometry is not determined. Remember, 3D Studio is a surface modeler, not a solid modeler so no material exists on the inside of the model with which to form sectioned surfaces. Figure 15.9 shows the effect of using a clipping plane to remove the front half of a collar and pin assembly (a). With the objects centered at 0, 0, 0 and the near clipping plane at Y=0, the front half of the collar and pin are removed. If carefully constructed "filler" planes are positioned slightly to the rear of the near clipping plane (b), a conventional sectional view results.

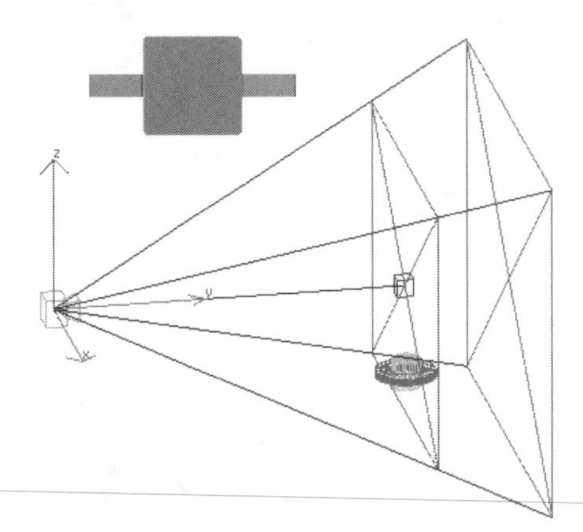

Figure 15.9 *Sectional view with camera clipping.*

Bringing the camera up to position looking down at the assembly only exacerbates the problem (Figure 15.10). The clipping plane and the sectional plane are now not aligned. With the clipping planes angular to the objects' axes, a standard sectional view is impossible. Plus the clipped portion runs diagonally across the collar and pin, extending in front of the section. So while sections are possible with clipping planes, the missing geometry must be explicitly created along with a careful positioning of "filler" planes. Other sectional methods would probably be more efficient.

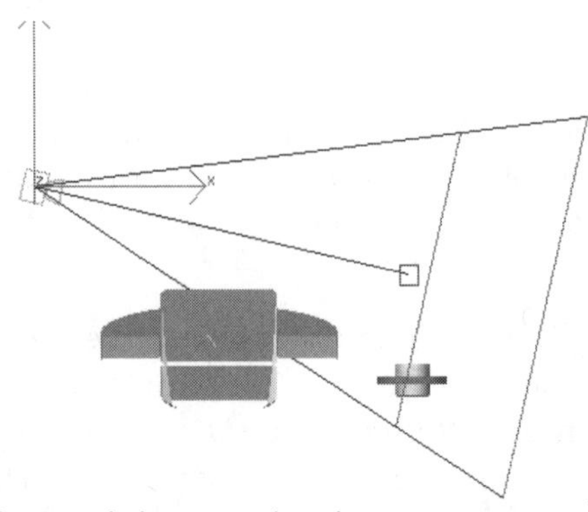

Figure 15.10 *Clipping and sectional planes not aligned.*

PLACING A HIGHLIGHT

Quite often a highlight placed at a specific location makes the difference in how a surface is interpreted. Take Figure 15.11 for example. The Storage Tank is comprised of cones, cylinders, boxes, and a plane; the tank appears evenly lit with standard lights and ambient illumination. You can experiment with a light source, moving the light around manually, until you arrive at the hot spot at the desired position. Or you can do the opposite: move the hot spot around.

STEP 1

Establish sources of light. Figure 15.11 shows a standard lighting arrangement with a **Target Direct** light and two **Omni** lights.

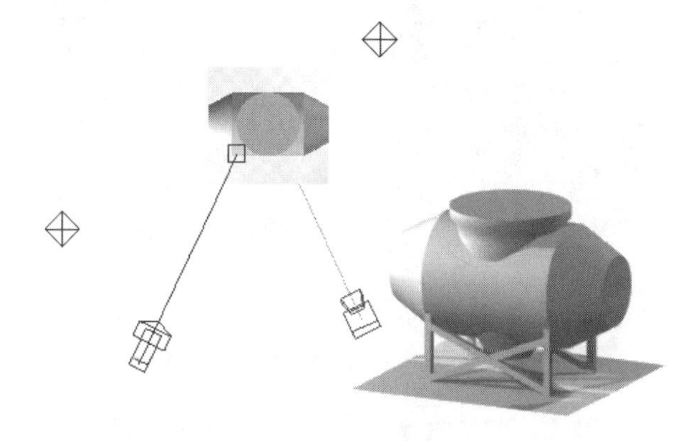

Figure 15.11 *Standard lighting of Storage Tank.*

STEP 2

Select the left **Omni** light. Any light object can have its highlight adjusted, even Omni lights that don't actually have hot spots. Lock this selection by pressing the **Space Bar**. Switch to the Camera view.

STEP 3

Move the hotspot. The highlight is defined as falling on that surface whose normal (perpendicular to the surface) is concurrent with the shortest light ray to the source. In VIZ, choose **Modify|Place Highlight** or in Max choose **Tools|Place Highlight**. Or, choose the **Place Highlight** tool in the **Align** fly-out menu in the main toolbar.

When you drag the cursor over the object, the selected light is aligned with the surface normal at that point on the object (Figure 15.12). A small line appears representing the normal and as you move the cursor, the light changes position to maintain alignment with the new normal. In (a) the light is aligned with a vector on the left cone. This alignment does not illuminate the central cylinder. When the cursor is moved over the cylinder (b), the vector allows both cone and cylinder to be illuminated.

Note: Because edges don't have normals (only planes do), you can't use this **Place Highlight** method to position a highlight at the intersection of two planes. This highlight must be positioned by manually moving the light to the desired position.

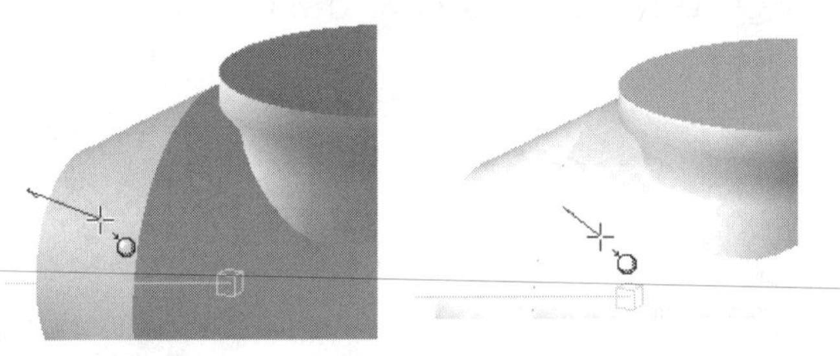

Figure 15.12 *Vector normal is used to align light.*

USING LIGHTS CREATIVELY

There is a broad set of lighting parameters available for fine-tuning illumination. In engineering and technology we are less concerned with these nuances than those in entertainment, but two bear mentioning: Attenuation and Atmosphere.

Attenuation is the characteristic of a light to diminish over distance and functions very much like graduated clipping planes. **Far Attenuation** determines the distance at which illumination drops off to zero. **Near Attenuation** determines where the light begins to fade in. Both are turned on or off by a toggle called **Use**. Figure 15.13 shows a lathe under standard lighting. The **Target Spot** illuminates only the transparent lid; all other objects are excluded from illumination and casting shadows. Two **Omni** lights provide general illumination.

Assume that you want to feature the part that is being machined. You could zoom up on the section of interest but in doing so you lose the context,

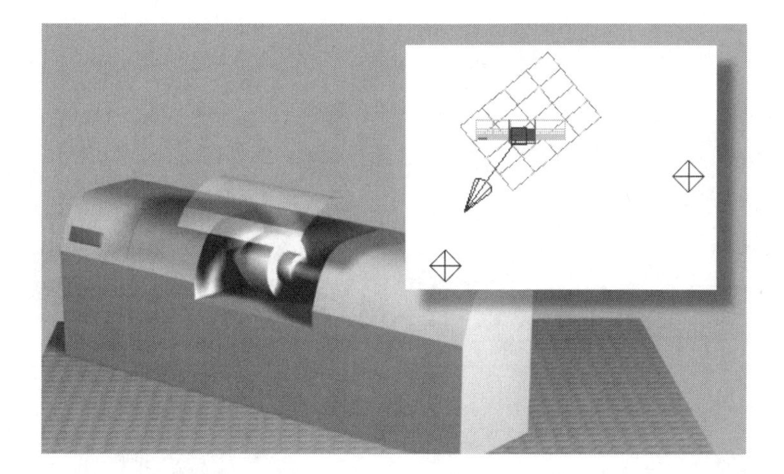

Figure 15.13 *Attenuation not in use.*

where the part is on the lathe. The solution is to adjust the lights so that the section of interest is illuminated, and the rest of the scene is in relative darkness.

STEP 1

Select light. Because we have two lights in our scene, each has to be adjusted individually. Select the **Omni** light on the left and choose **Modify|Attenuation Parameters** and turn on the **Use** toggle.

STEP 2

Adjust Attenuation. Open the **Attenuation Parameters** roll-out. Enter values into the four fields so that the work piece is bracketed between the second ring (**Near Attenuation|End**) and the third ring (**Far Attenuation|Start**). This is the area of full illumination as set by **General Parameters|Multiplier** (Figure 15.14). The Spot doesn't have to be attenuated because it only illuminates the lid.

STEP 3

Set other light. Repeat the steps for attenuating the first light with the second **Omni** light. When completed, the two lights' attenuation rings cross to bracket the work piece (Figure 15.15) and the work piece is featured by illumination (Figure 15.16).

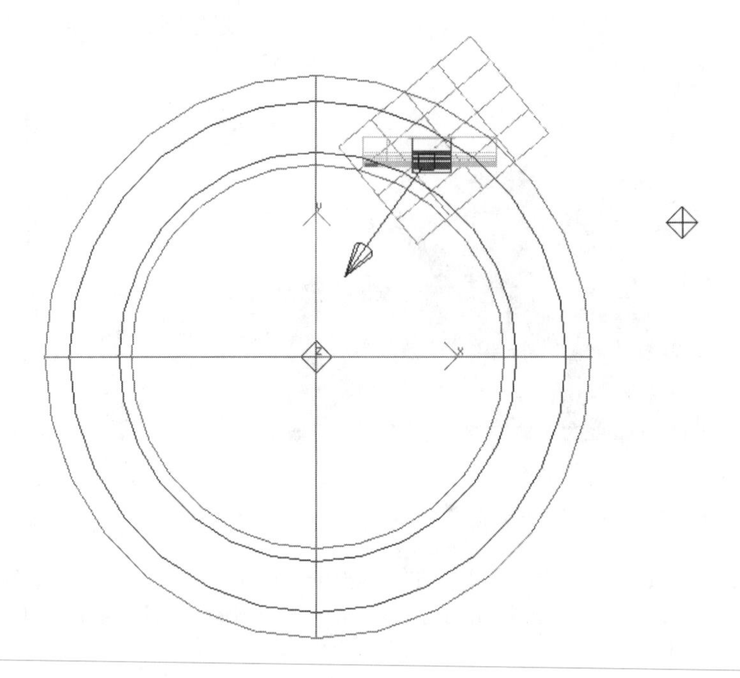

Figure 15.14 *Attenuation parameters shown by rings.*

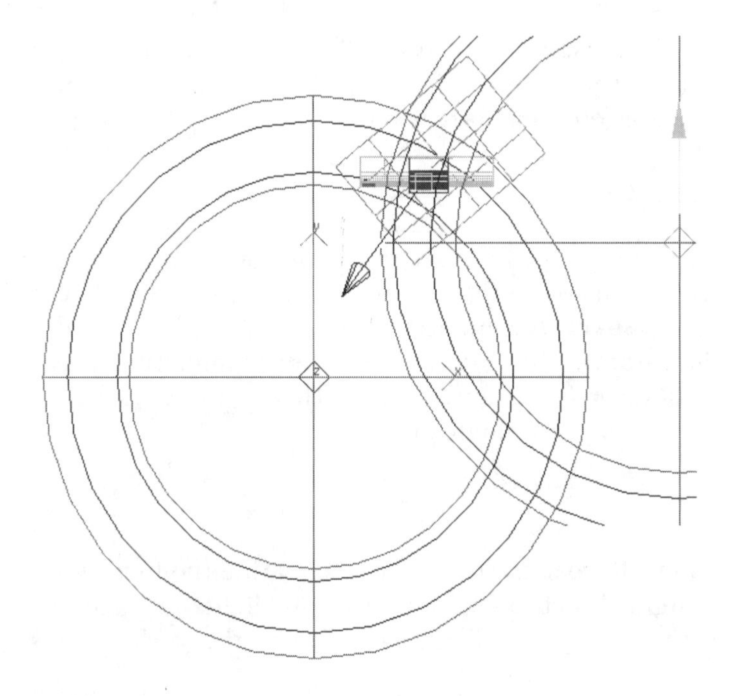

Figure 15.15 *Attenuation rings bracket the subject.*

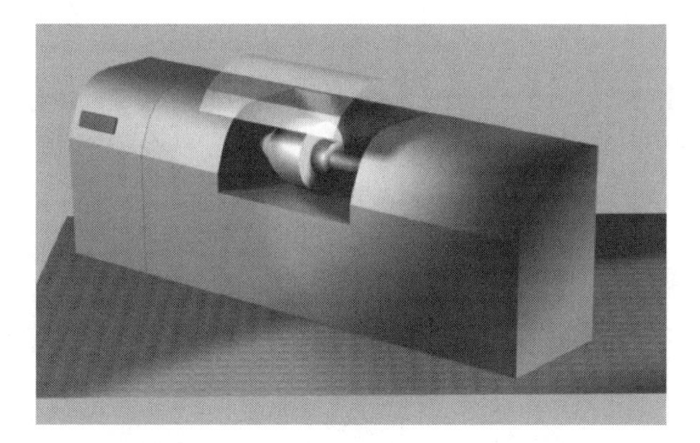

Figure 15.16 *Effect of lighting attenuation.*

Atmospheric effects can be effective in industrial environments where the air may contain particulates that illuminate the volume of light. Figure 15.17 shows the lathe of our previous example illuminated with atmospheric spotlights.

Figure 15.17 *Atmospheric lighting.*

STEP 1

Establish the lights. Atmospheric lights are effective when they contrast with background values. The darker the background, the more the atmospheric lights will show. Combining **Attenuation** of general lights with **Volume Light** is effective when attenuated lights keep their illumination from reducing the atmospheric effect.

STEP 2

Assign atmosphere. With the light selected, choose **Modify|Atmospheres & Effects|Add** and **Volume Light** from the list of **Atmosphere**. Click **OK**. The strength of the atmospheric effect is controlled by the value of **Multiplier**.

Tip: If you want an atmospheric light to be seen in the air but not contribute to the illumination of the scene, Exclude all objects.

STEP 3

Set attenuation. To make the volumetric lights more effective, choose **Attenuation Parameters** and enter values that cause the lights to fade out as they approach the object. In Figure 15.18 you can see that the far attenuation begins above the subject and ends slightly below the top surface of the lathe. The final atmospheric effect is shown in Figure 15.19.

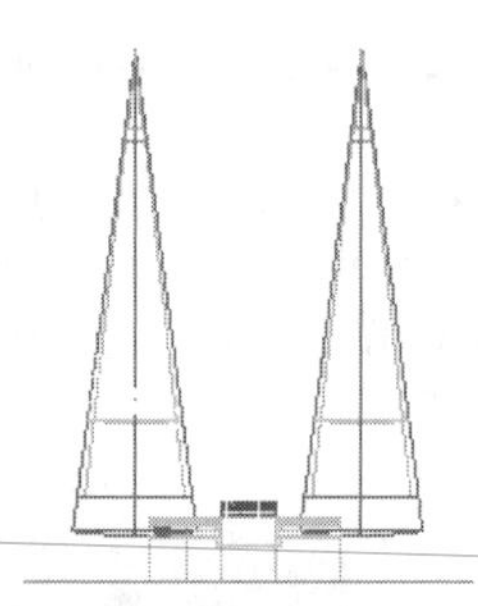

Figure 15.18 *Atmospheric volumetric lighting with attenuation.*

Figure 15.19 *Final atmospheric rendering.*

PROBLEMS

Interpret the verbal descriptions below into effectively illuminated scenes. e prepared to document your solutions through sketches and screen grabs. The problems that follow can be used to practice first modeling and then lighting.

Problem 15.1 *Industrial environment. You are looking upward inside a huge factory with a complicated steel superstructure supporting catwalks, ladders, tanks, and pipes. The room is illuminated by hanging lamps arranged in a grid near the ceiling. Though the air is smoky, the room is also illuminated by skylights. Using a storyboard sheet found on the CD-ROM in the tools\ directory, plan the factory environment and its lights.*

Problem 15.2 *Attenuated lighting. Pick an assembly you have already modeled and animated. Practice the light attenuation technique described in Figures 15.13 through 15.16. Create a lighting setup that illuminates a central part as other parts are brought into position. Assure adequate ambient lighting to illuminate parts before assembly.*

Problem 15.3 *Load the file tank..max from chapters\chapter_15 on the CD-ROM. The orthographic projection camera creates a parallel projection. A parallel light source would place a highlight on each of the tanks at exactly the same spot. Use the* ***Place Highlight*** *command to put a highlight on each tank to indicate a parallel light source.*

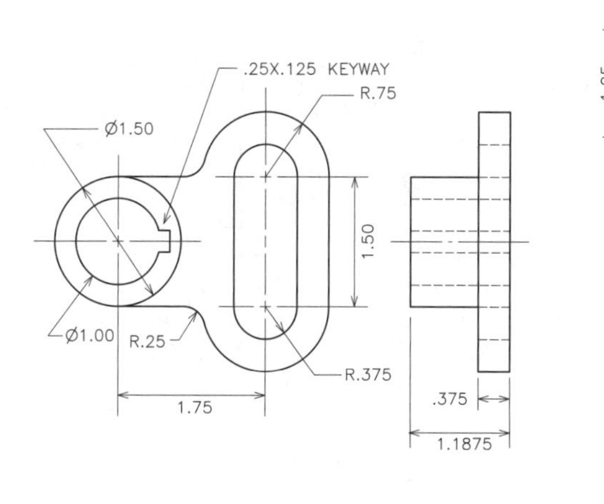

Problem 15.4 *Adjuster Bracket.*

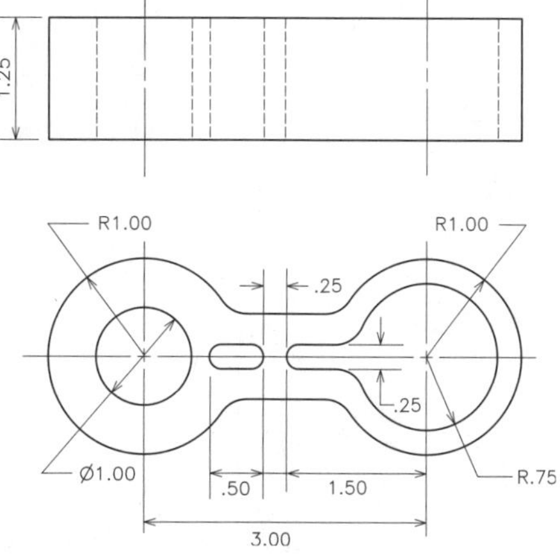

Problem 15.5 *Connector Link.*

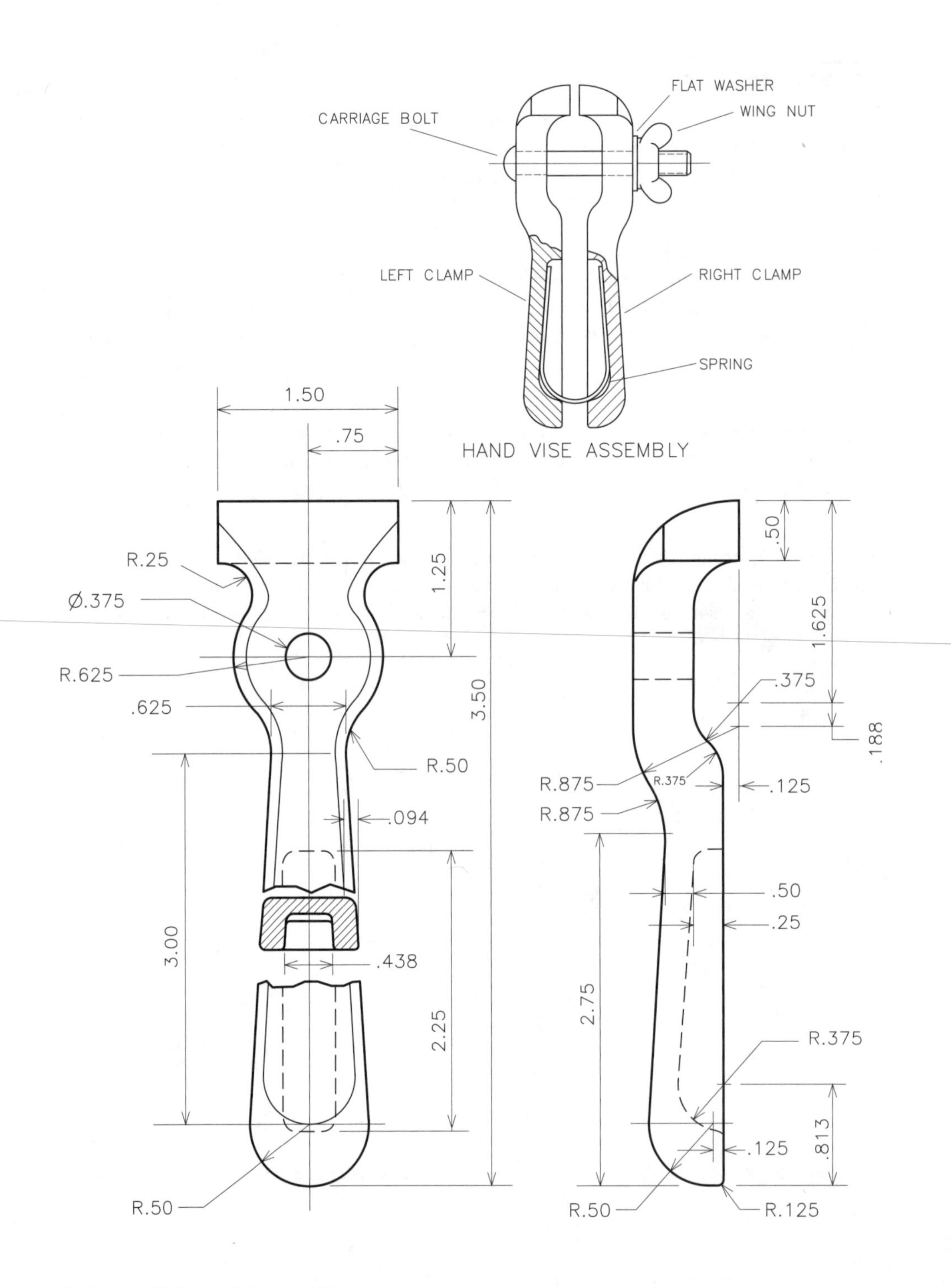

Problem 15.6 *Hand Vise Assembly.*

Animation Basics

CHAPTER OVERVIEW

Technical animations have become an important part of product development. Where in the past physical models had to be constructed of wood, plaster, foam, or cardboard to check alignment, fit, and assembly, computer models can be animated to perform a variety of functions. In sales and marketing, traditional product renderings made before actual products were produced are replaced by graphics made from the same geometry on which molds, patterns, or machine tool instructions are based. All in all, technical animations have made a positive contribution to modern manufacturing.

The same phenomenon that allows us to perceive a series of static images as continuous movement in a motion picture is the key to animation in 3D Studio VIZ and MAX. The difference between technical animation and entertainment animation is that while entertainment animation is based on individual images or frames, technical animations are more closely tied to time. In a technical animation it may be more important that an event start at a certain time and last for a given number of seconds; 3D Studio can time events to 1/4,800th of a second. Because of this small internal time increment, animations can easily be scaled or converted to various formats.

Note: 3D Studio's internal clock measures time at 4,800 "ticks" per second. This is the smallest unit of time that animations can address.

Almost any object, its transforms, or its parameters can be animated. With such power and flexibility the temptation is to animate everything. However, technical animations are usually created to show specific relationships. Extraneous movement must be reduced so that the technical condition of interest can be observed. Only by intelligently combining geometry, materials, cameras, and lights can effective technical animations be produced. For this reason, animation can be broken into two general sections: *geometry animation and camera* and *other object animation*. In this chapter you will learn techniques for putting objects in motion while keeping the camera static. In the next chapter you will discover how moving objects, lights, and cameras can produce even more effective animations.

KEY COMMANDS AND TERMS

- **Anchors**—the white boxes at the beginning and end of animation tracks in the **Track View**.
- **Active Viewport Only**—the playback parameter that suppresses animation playback in all but the active viewport.
- **Animate**—the button in the bottom menu that toggles the recording of keyframe changes on and off.
- **Animation**—the display of static images at a speed sufficient to perceive changes from one image to the next as continuous.
- **AVI**—Microsoft's Audio Video Interleave format for time-based images and sound.
- **CODEC**—Compression Decompression: a method for reducing (compressing) and uncompressing (extracting) data.
- **Exploded Assembly**-the display of individual parts such that their position communicates the manner in which the parts are assembled.
- **External Compression**—the reduction of data by a utility not included in the data file itself.
- **Frame**—an image at a unique time point in an animation.
- **Frame Rate**—the rate, in seconds, that frames are displayed in an animation.
- **Frames Per Second**—see **Frame Rate**.
- **Hierarchical Link**—the logical association of parts for the purpose of coordinated movement during an animation.
- **Internal Compression**—the reduction of data by a utility included in the data file itself.
- **Key**—a graphic indicator that a change has occurred at a given time point or frame in an animation.
- **Lossless Compression**—compression that loses no data in the compression-decompression process.
- **Lossy Compression**—compression that loses data in the compression-decompression process.

- **MOV**—Apple's QuickTime movie format for time-based images and sound.
- **Output Size**—the physical dimensions (in pixels) of the frames in an animation (for example 800 x 600).
- **Parent-Child**—the hierarchical relationship that causes children parts to be transformed in relation to the parent part.
- **Real Time Playback**—the **Time Configuration Option** that allows frames to be dropped from the playback to maintain the animation length.
- **Re-Scale Time**—the changing of the animation's length (or a portion of the animation) or the movement of a section of animation from one time point to another.
- **Tick**—the internal timing interval used by 3D Studio (1/4800 second)
- **Time Configuration**—the dialogue box in which animation parameters are entered.
- **Time Slider**—the controller in the **Track Bar** used to advance the animation to a specific frame or time point.
- **Track Bar**—the area of the bottom menu in which the **Time Slider** and **Keys** for the selected object are displayed.
- **Track View**—a view of all objects in a scene and their **Tracks** and **Keys**.
- **Tween**—the process of interpolating between **Key Frames**.
- **Video Compression**—the reduction of data in an animation file.

IMPORTANCE OF ANIMATION

Computer graphics encourages designers to think about their products in 3D space from first conceptualizations. Animation frees designers to think about how their products move and interact in 3D space. Some products may never need to be animated while animation may be critical with other products, especially involving assemblies.

In early stages animation allows a designer to study a design from various vantages. Before computer graphics orthographic views had to be "read", assembling a mental picture of three-dimensionality by comparing features view-to-view. Unfortunately, efficiency in this mental process has always varied greatly from designer to designer. Pictorial (axonometric or perspective) views could be drawn but each was static. If you wanted to see something on the other side, you had to create another drawing. Animation allows you to put a product in motion so that it can be observed from all directions. In effect you have access to an infinite number of views.

BASIC ANIMATION PROCESS

The part in Figure 16.1 contains a number of features that cannot be seen in a single principle view. You have to "read" the views, comparing features from one view to another. Even with the addition of a pictorial view, it may still be difficult to visualize. Were the part to be animated, revolving about world Z-axis, the relationship of these features could more easily be evaluated. The part Link and the finished animation can be found on the CD-ROM in *chapters\chapter_16*.

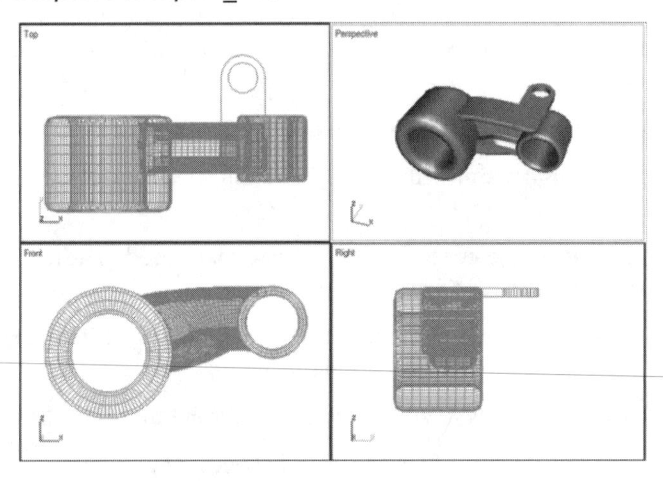

Figure 16.1 *Visualization by static views.*

STEP 1

Establish the camera. We will use an isometric **Target Camera** set at **X=1000**, **Y=–1000**, **Z=1000** with its **Target** set at **0, 0, 0**.

STEP 2

Set up basic animation parameters. Switch to the Camera view. Right–click on any of the VCR-type controls in the bottom menu. The **Time Configuration** dialogue box is displayed. Use these settings:

- **Frame Rate**. Use **Custom** and enter **12** in the **FPS** (frames per second) field. For most design decisions, 12 FPS is adequate for realistic motion.
- **Time Display**. Check **MM:SS:Ticks.** This displays the animation in terms of time rather than frames.

- **Playback.** Accept the defaults of **Real Time, Active Viewport Only**, and **Speed 1X**.
- **Animation.** These fields read time due to the **Time Display** parameter. Set the **End Time** to **0:20.0,** an animation length of **20** seconds. The **Length** field changes appropriately.
- **Key Steps.** Accept the **Use Track Bar** parameter.

Note: There is a direct relationship between frames per second, length of animation, time required to render the frames, and storage size needed for the animation. The greater the former, the larger the latter. The remaining variable—frame size and compression—can be used to reduce the storage size of the animation.

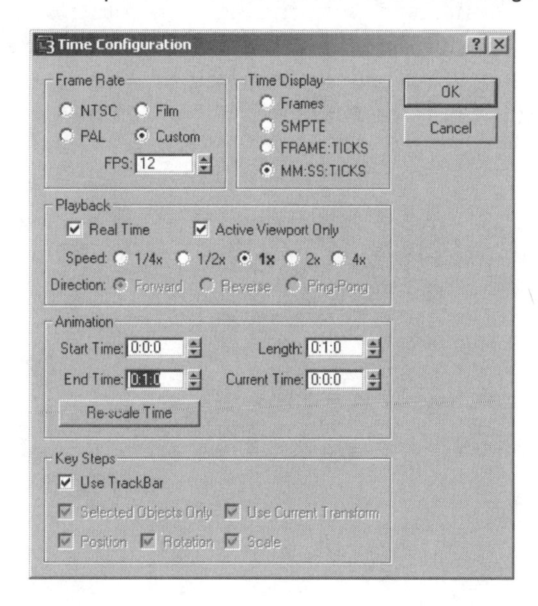

Figure 16.2 *Time configuration.*

STEP 3

Design the animation. A *frame planning sheet* has been provided for you on the CD-ROM inside the *tools* subdirectory. This planning sheet is shown in detail in Figure 16.10. For our animation, the Link is going to complete 90 degrees of Z-axis rotation approximately every 6.3 seconds. These four positions are marked by **Keys** and correspond to specific points in time or **Key Frames**. The frames between the key frames are *tweened* or interpolated from the **Key Frame** positions.

STEP 4

Animate the Link. Switch to the Top view where the Z-axis rotation can be affected. Click on the **Animate** button in the bottom menu. It turns red signifying that all transforms will be recorded; a red frame surrounds the active view signifying that any changes will be recorded at the current **Key** position.

Move the **Time Slider** until it reads **6.3**...and rotate the selected object –90 degrees about the screen Z-axis by right–clicking the **Select and Rotate** tool. Enter the value in the **Absolute World Z-Axis** field. Move the slider to **12.6**... seconds and perform another 90 degree Z-axis rotation. Move the slider to **20.0** seconds and perform the final rotation. Figure 16.3 shows the position of the **Time Slider** and the object at each **Key Frame**. Notice that a **Key,** signifying the presence of a **Key Frame**, has been deposited at each time increment. Click the **Animate** button again to stop recording scene changes.

Note: This rotation could have been performed by a single 270-degree rotation at the 20-second Key position. But then you wouldn't have had the fun of moving the Time Slider and rotating the part.

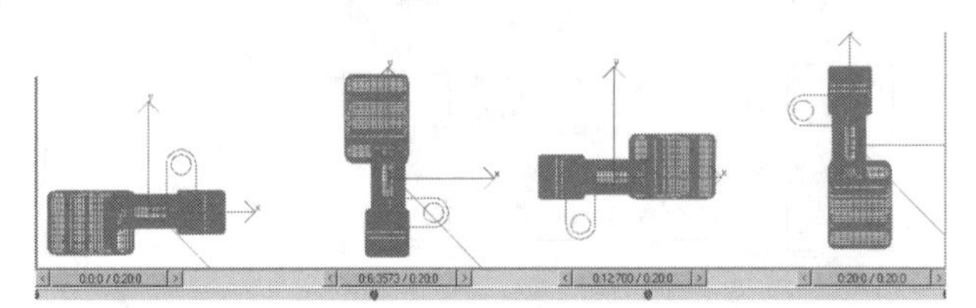

Figure 16.3 *Rotations at key frame intervals.*

STEP 5

Test the animation. Switch to the Camera view and display in **Smooth + Highlights**. Manually drag the **Time Slider** to the right across the animation track. The animation is played in the Camera view at the speed you move the slider.

Many times this is sufficient in making design decisions. The slider can be moved forward and backward at will, fast or slow. Or, you can use the VCR controllers to play, stop, and rewind the animation. At this point, the animation exists only as settings in the 3D Studio file.

STEP 6

Create an animation movie. Manually playing the animation is fine if you have the 3D Studio software but what if you don't? What if you want to share the animation with someone over the Internet? Or include the animation in an electronic presentation or as part of multimedia training? To do any of these you have to create an *animation movie file* that captures each frame (20 sec x 12 FPS = 240 total frames). On the PC Windows side, Microsoft's AVI Audio/Video Interleave format is probably a first choice. For Macintosh applications, QuickTime (MOV) is appropriate.

Click on **Render Design** (**Render Scene** in MAX). In the dialogue window (Figure 16.4), use these settings:

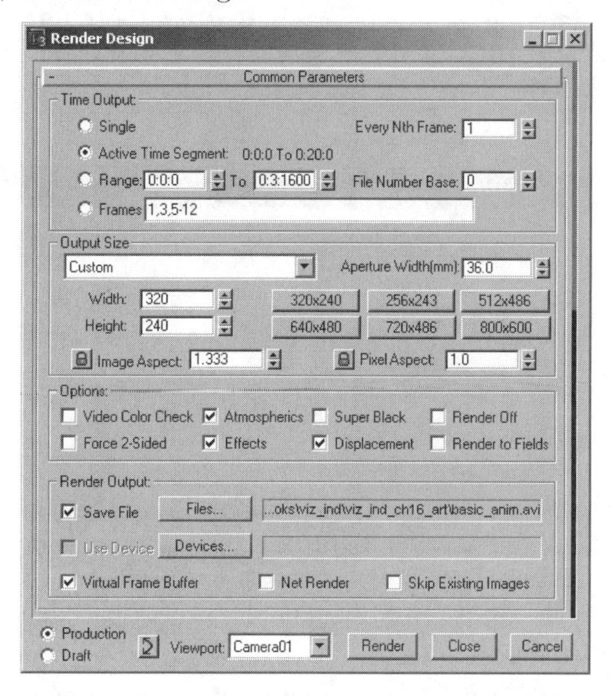

Figure 16.4 *Rendering settings.*

- **Time Output.** Use **Active Time Segment.** This will create a movie equal in length to your full 20-second animation.
- **Output Size.** Use the required output size. If you are just testing the animation, use a small size such as 320 x 240.
- **Options.** Accept the defaults.

- **Render Output**. Click on Files and browse to the directory in which you want to save the file.
- **Save as Type**. Choose AVI unless you have specific output needs. Use the **Virtual Frame Buffer**.
- **Click Render.** In 3D Studio, the scan line renderer renders each frame in succession in AVI format to the file you specified.

The first time you save the video file you will be presented with the **Video Compression** dialogue box (Figure 16.5). Compression is an important aspect of creating animation files. An analysis of our subject reveals a solid white background, effective in lossless compression. Compression algorithms either look for unchanging pixels within a frame, unchanging pixels between frames, or by analyzing groups of pixels that can be reduced to a smaller number. Either way, the difference in file size can be considerable. In our example, consider the data in Table 16.1.

Video Compression Comparison		
Animation	Uncompressed	Cinepak @ 50% Quality
AVI Animation 240 frames @ 320 x 240	54 Mb	1.2 Mb

Table 16.1 *Compression comparison.*

Because you may be making dozens of animations in the course of a project, pick a frame size, compression algorithm, and compression ratio (quality) that results in the smallest file you can stand. In Table 16.1, you can see that a 50% Cinepak compression results in a dramatic reduction in file size, with acceptable quality.

Pick a *compression CODEC* that is supported by your target system. Though QuickTime may produce superior results to AVI (especially when sound is required), most Windows computers do not have QuickTime installed. Because most Windows users will view videos on their desktops via the *Windows Media Player*, make sure that the CODEC you choose is compatible with the latest version of this product.

Note: A CODEC is a Compression Decompression algorithm or method. CODECs can be internal, expanding without any other resources, or external, requiring specialized hardware or software. Motion Picture Experts Group (MPEG) is a highly effective compression method that requires an external MPEG decoder, either in software or hardware.

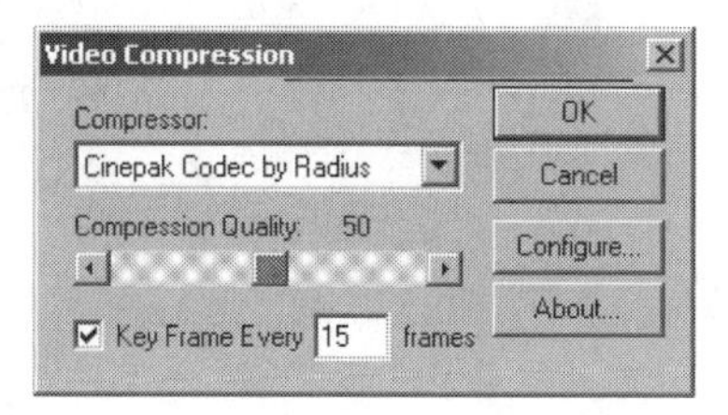

Figure 16.5 *Compression dialogue box.*

Tip: Compression works either by substituting a token for repeating pixels (lossless) or by considering a group of pixels and calculating a mathematically derived substitute (lossy). The higher the compression percentage in these lossy methods, the smaller the file size but more data is lost when the compressed file is uncompressed.

INTRODUCTION TO THE TRACK VIEW

At every point in time or frame at which you make a change in an object, in transforms (position, scale, rotation) or materials, a **Key** is placed to record the action. When you activate **Use Track Bar** in the **Time Configuration** dialogue box, you allow these **Keys** to be shown below the **Time Slider** in the bottom menu. They can be adjusted there.

All of these **Keys** are also displayed in the **Track View**, a special view of all objects in a scene and what has happened to them over the time range of the animation. Figure 16.6 shows the **Track View** of the Link animation. The length of the animation is represented by a black bar; the beginning and end marked by repositionable white anchors. You can see the **Rotation Keys** at the four time positions. You can also see that the Link has not been moved or scaled because there are no **Keys** along those transforms. You can switch to the **Track View** using the shortcut **E** key but once a window displays the **Track View**, you must use **View|Layout** to return the viewport to a standard view in VIZ or **Customize|Viewport Configuration|Layout** in MAX.

The **Track View** gives you direct control over all objects and their **Keys**. For example, dragging a time bar anchor to the desired position can scale the animation length; all **Keys** are scaled proportionally (Figure 16.7). The **Track View** will be explored in detail in the next chapter.

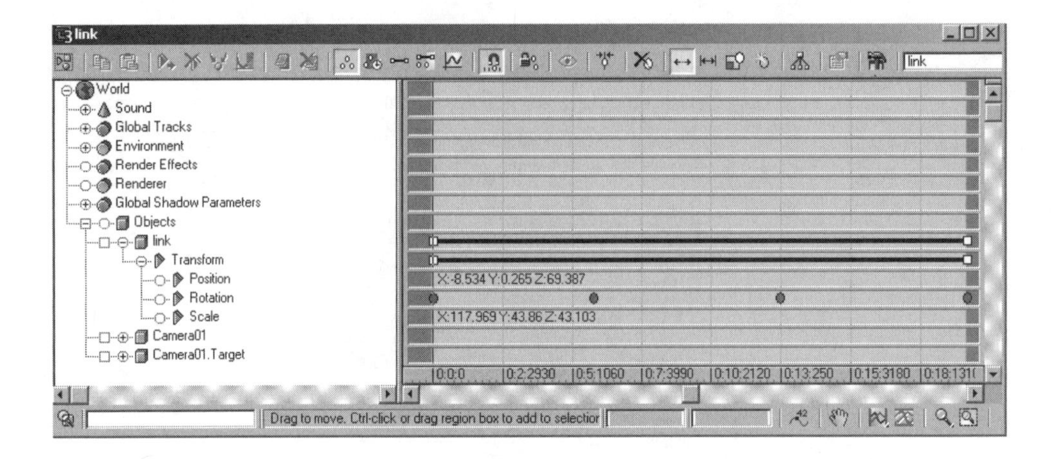

Figure 16.6 *The Track View.*

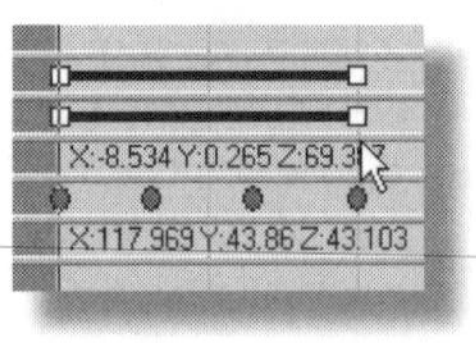

Figure 16.7 *Adjustment in the Track View.*

ANIMATION GUIDELINES

- A **Key** is placed at **Frame 1 Time 0** only after another **Key** is created. You can then go back and modify or delete the first **Key** and then insert a new one.
- Changing an object parameter with the **Animate** button on changes the object at that key and tweens between these new settings and the settings at the previous key.
- A change in the object with the **Animate** button off is applied equally and fully across all **Keys**.
- A change in an object with the **Animate** button off that has already been animated will be applied equally across all animation **Keys**.
- For long animations you can specify an active time segment in the **Time Configuration** dialogue box. This restricts the time segment the Time Slider can access and limits playback to those frames.
- Use **Re-scale Time** to move a segment from one place in the animation to another or to shorten or lengthen the segment.

- Clicking on the **Time Slider's Direction Arrows** moves the display one tick per click. Drag the slider for interactive movement. Click on the track to the left or right of the slider to jump the animation to that spot.
- Changing output format (NTSC to PAL, for example) automatically changes the frame rate to maintain the same time length.
- If you check **Real Time Playback** in the **Time Configuration** dialogue box, playback in 3D Studio drops frames as necessary to play the animation back at the specified elapsed time length. If your animation includes complex geometry and you are asking for more than your system can deliver, playback may be jerky.
- By having **Active Viewport Only** checked in the **Time Configuration** dialogue box, you can assure the best chance of smooth playback in the selected viewport. Switch viewport with shortcut keystrokes and the animation continues from the current frame in the new viewport.

EXAMPLE: THE GEAR PULLER

Figure 16.8 shows the assembly view of a Gear Puller assembly. The components can be found as Problem 1 at the end of this chapter and in PDF format on the CD-ROM in *chapters\chapter_16\gear_puller*. Also in this directory is *gear_puller.max*, the finished assembly ready to animate. However, you are encouraged to model the assembly yourself and then follow the animation instruction.

Figure 16.8 *Gear Puller assembled.*

STEP 1

Analyze the parts for assembly order. Orient the parts spatially to facilitate the assembly. This might be the position on a workbench a technician might be expected to use for assembly. Or, the starting position might mimic how the parts are held in the hand as they are assembled. Figure 16.9 shows the parts correctly aligned for an animated assembly. The parts have been "exploded" along their assembly axes (parallel to world X, Y, and Z) and positioned within the camera view so that each part "floats" free of other parts. The one exception is the right Spring Pin that overlaps the Yoke. This was done to keep the overall area of the arrayed parts small, assuring that the parts are displayed at a maximum scale.

Tip: Rendering times increase dramatically with the addition of lights and materials. For this reason, work out your animation before assigning materials and adding more than a standard lighting arrangement.

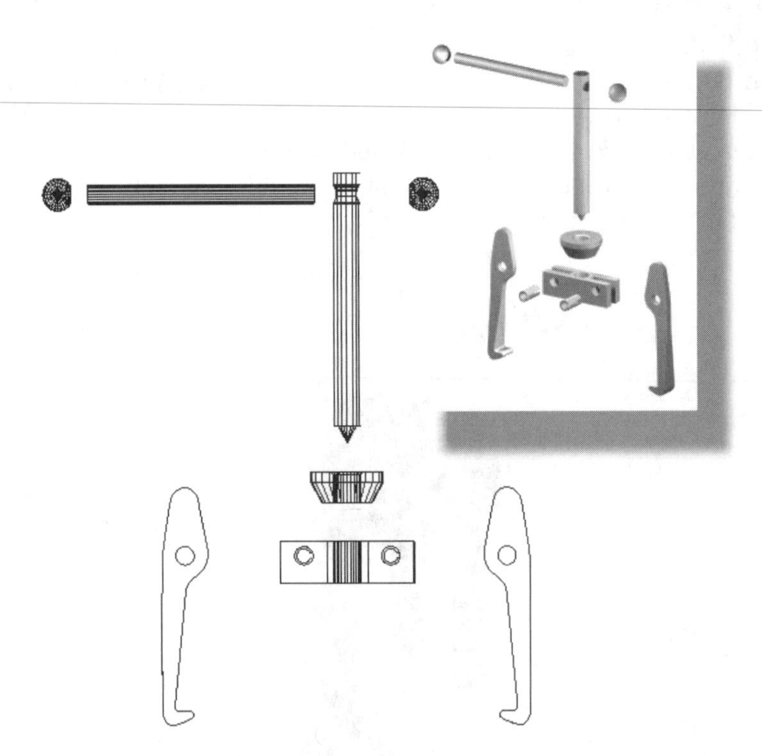

Figure 16.9 *Gear Puller components positioned for assembly.*

STEP 2

Establish standard lights. These lights must start out distributed sufficiently to illuminate all parts. What you don't know right now is whether or not the illumination will decrease as the parts are assembled, moving away from the lights. In other words, you may have to move lights along with parts to maintain consistent illumination.

STEP 3

Establish a static camera. The camera position must include all the parts in their unassembled position. Make this an **Orthographic Projection** camera.

STEP 4

Plan the animation. Figure 6.10 shows the *Frame Planning Sheet* used to animate the Gear Puller. This sheet is used to plan the length, frame rate, order of assembly, and concurrency of transforms. The length of a transformation is shown by an arrow and the particular transform by a letter: T for transform (move) and R for rotate.

The animation is centered about the Yoke, the part that doesn't move. It is the first part on the Frame Planning Sheet. The Yoke determines the final position of all other parts. There may be deviations from the Frame Planning Sheet as you complete the animation but you shouldn't begin the animation by just moving components around. The success of a technical assembly depends both on order of assembly and alignment, elements that if done without planning almost always bring disaster.

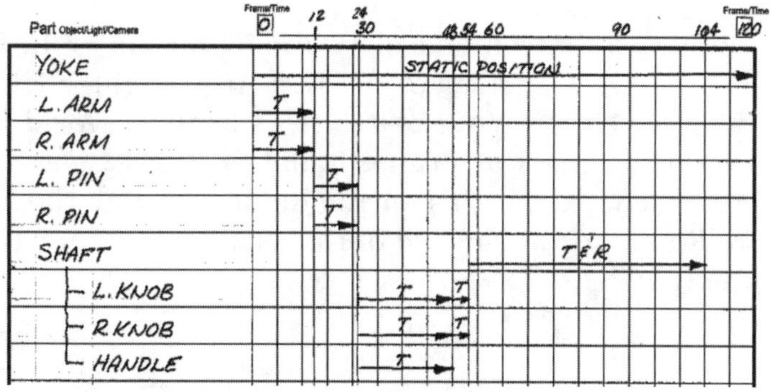

Part Object/Light/Camera
Frame/Time 0 12 24 30 48 54 60 90 104 120 Frame/Time
YOKE — STATIC POSITION
L. ARM — T
R. ARM — T
L. PIN — T
R. PIN — T
SHAFT — T & R
L. KNOB — T T
R KNOB — T T
HANDLE — T

Figure 16.10 *Frame planning for the Gear Puller assembly.*

STEP 5

Set animation parameters. The Yoke is centered at 0, 0, 0 such that the holes with which the Arms and Spring Pins are aligned are at Z=0 (Figure 16.11). The Arms and Spring Pins have pivots that are centered on these holes. This makes precise alignment a matter of expressly moving the parts to locations derived from the detail drawings.

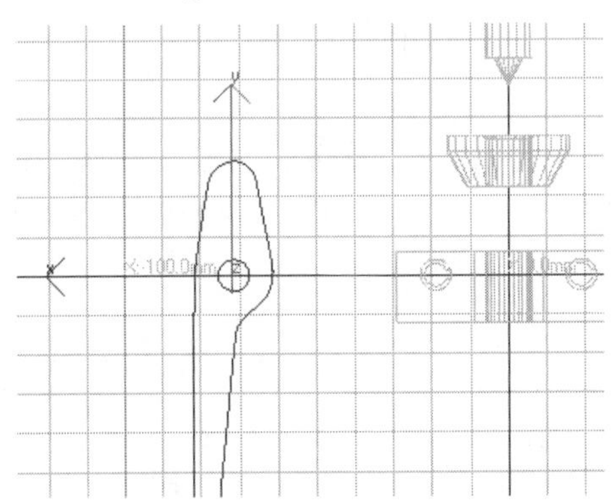

Figure 16.11 *Position of parts and pivots encourage precise alignment.*

Begin by setting **Custom Frame Rate** of **12 FPS** and 120 total frame **Length** in **Time Configuration** dialogue box. This corresponds to your timing decisions on the Frame Planning Sheet.

STEP 6

Begin the animation. Complete simultaneous movement of arms at **Frame 12** (Figure 6.12).

Activate the **Animate** button and move the **Time Slider** to **Frame 12**. Switch to the Front view and select the Left Arm. Right–click on **Select and Move** and enter **X=-19**, **Y=0**, **Z=0** in the **Absolute World** fields. **Keys** are inserted onto the track at **Frame 0** and **Frame 12** and all transformations are recorded. Repeat this for the Right Arm with the X-axis value of **19** (refer to Problem 1 at the end of this chapter). Figure 16.12 shows this transformation.

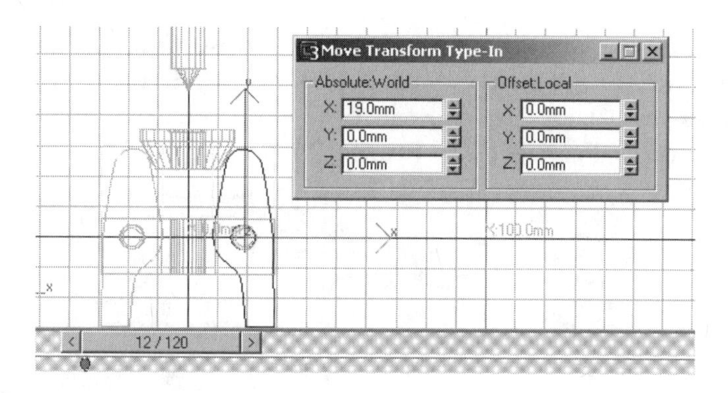

Figure 16.12 *Arms are in position at **Frame 12**.*

STEP 7

Complete simultaneous movement of Spring Pins at **Frame 24** (Figure 6.13). Move the **Time Slider** to **Frame 24**. Switch to the Top view and select the Left Spring Pin. Right–click on **Select and Move** and enter **X=–19**, **Y=0**, **Z=0** in the **Absolute World** fields. **Keys** are inserted onto the track at **Frames 0** and **24** and all transformations are recorded. Repeat this for the Right Spring Pin with the X-axis value of **19**. Figure 16.13 shows this transformation.

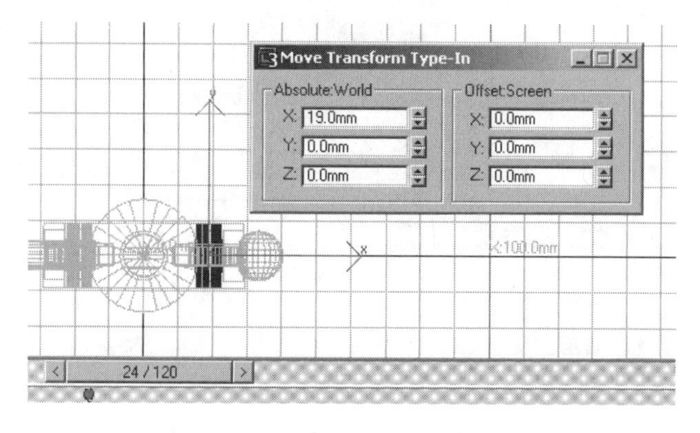

Figure 16.13 *Spring Pins are in position at **Frame 24**.*

STEP 8

Adjust **Starting Keys** in the **Track View**. Refer to the Frame Planning Sheet and you'll see that the Spring Pins don't move until the Arms have completed their movement at **Frame 12**. Choose **Open Track View** from the **Main Menu** bar. Open the Spring Pins and their transforms. Drag the **Left Anchors** until they are at **Frame 12**. The pins now move from their original position at **Frame 12** to **Frame 24** (Figure 16.14).

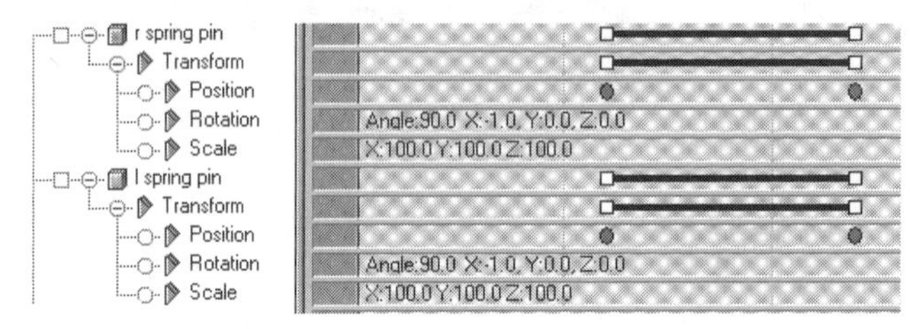

Figure 16.14 *Spring Pin tracks adjusted to begin at **Frame 12**.*

STEP 9

Move the **Time Slider** to **Frame 48**. Move the Handle to **X=0**, centered on the Threaded Shaft (the Z-axis value is only reflective of where it is above the Yoke). Move the Knobs at the same time so at **Frame 48** the handle is in position to accept the Knobs (Figure 16.15). In the Track view, adjust the **Track Anchors** for the Handle and both Knobs to begin at **Frame 24** (Figure 16.16).

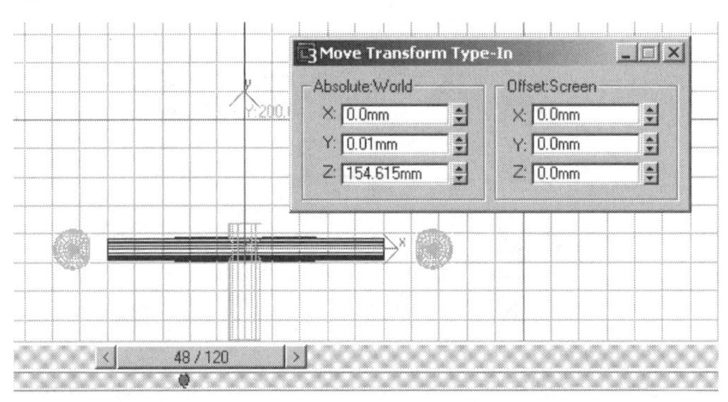

Figure 16.15 *Handle and Knobs in position at **Frame 48**.*

Note: Remember, a **Key** can't be placed at **Frame/Time 0** without another **Key** being previously recorded. A zero **Key** is inserted automatically when the first **Key** is created. This is why we always have to go back and adjust the point at which animation starts for all parts but those starting at **Key 0**.

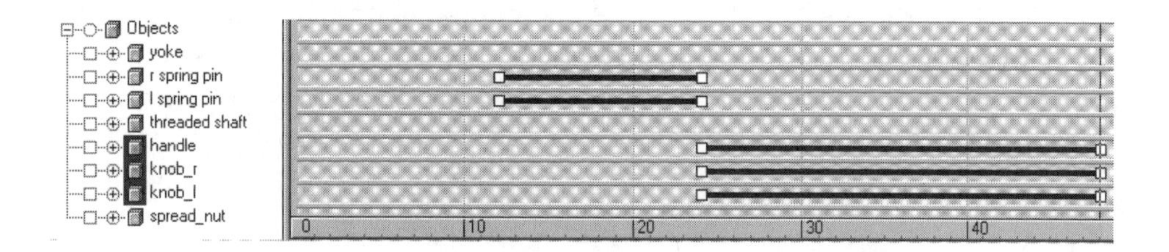

Figure 16.16 *Handle and Knob tracks adjusted to begin movement at **Frame 24**.*

STEP 10

Install Knobs. The Knobs move onto the Handle between **Frame 48** and **Frame 54**. These parts actually rotate as threads are engaged but because there is no detail on the Knobs that would show this rotation, they are simply moved into position (Figure 16.17). Their position isn't critical so move them manually. Because there was a **Key** at **Frame 48**, you don't have to adjust the **Track Anchors** in the Track view.

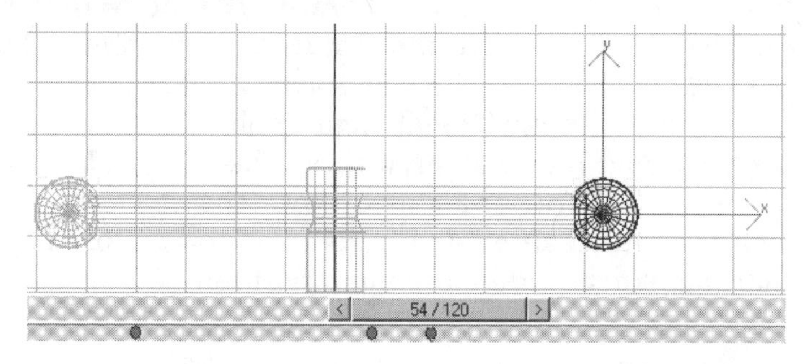

Figure 16.17 *Knobs in position in **Frame 54**.*

STEP 11

Link the handle subassembly. The Handle subassembly (Handle, Knobs, and Threaded Shaft) moves down and screws into the Spread Nut and Yoke from **Frame 54** to **Frame 104**. Because these four parts all move and rotate together, we need to **Link** the *children* (Left Knob, Handle, and Right Knob) to the *parent*, the Threaded Shaft. Transforms applied to the parent are translated to the children so that all parts move together.

Choose the **Select and Link** tool from the **Main Toolbar** and select the Left Knob. Drag to the Threaded Shaft and release. The Shaft flashes and a parent-child relationship is established. Do the same for the Handle and the Right Knob. When finished, the parts will display their hierarchical relationship when **Display Subtree** is checked in a **Select** dialogue box (Figure 16.18). You'll notice that the parts are also rearranged in the Track View to show this linked relationship.

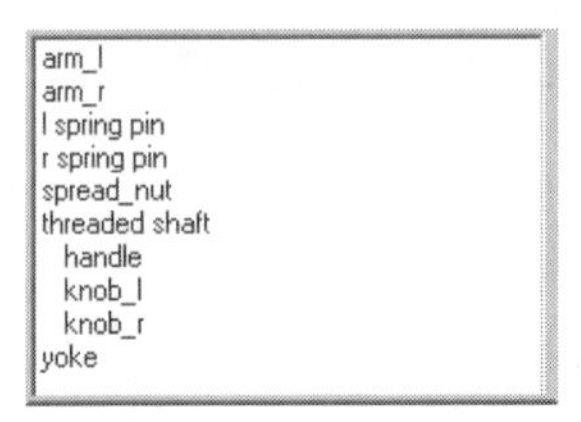

Figure 16.18 *Parent-child relationship.*

STEP 12

Transform the Handle subassembly. Select the Threaded Shaft and switch to the Top view. Move the **Time Slder** to **Frame 104**. Activate the **Animate** button. Right–click on the **Select and Rotate** tool and enter **3600** in the **Screen Z-axis** field. When the Threaded Shaft rotates ten times between **Frame 54** and **Frame 104**, the Handle and Knobs maintain their position relative to the shaft and rotate along with it. This is the parent-child relationship. Switch to the Track view and move the starting point of the Threaded Shaft track to **Frame 54**. The rotation now begins at **Frame 54**, after the handle and knobs have moved into position.

Note: Depending on the technical accuracy needed in the animation, you can determine how much vertical movement results from one full rotation (based on thread density). You can then match rotations and screw thread spacing.

Switch to the Front view. With the **Time Slider** at **Frame 104** move the Threaded Shaft downward into final position. (It threads into the Spread Nut and Yoke and it is the independent nature of these two parts on the shaft that determines the width of the arms at their bottoms.) Because of the parent child relationship the Handle and Knobs move and rotate with the shaft.

STEP 13

Test the animation. Switch to the Camera view. With the **Animate** button unselected, drag the **Time Slider** slowly and observe the animation (Figure 16.19). It should exactly follow the Frame Planning Sheet.

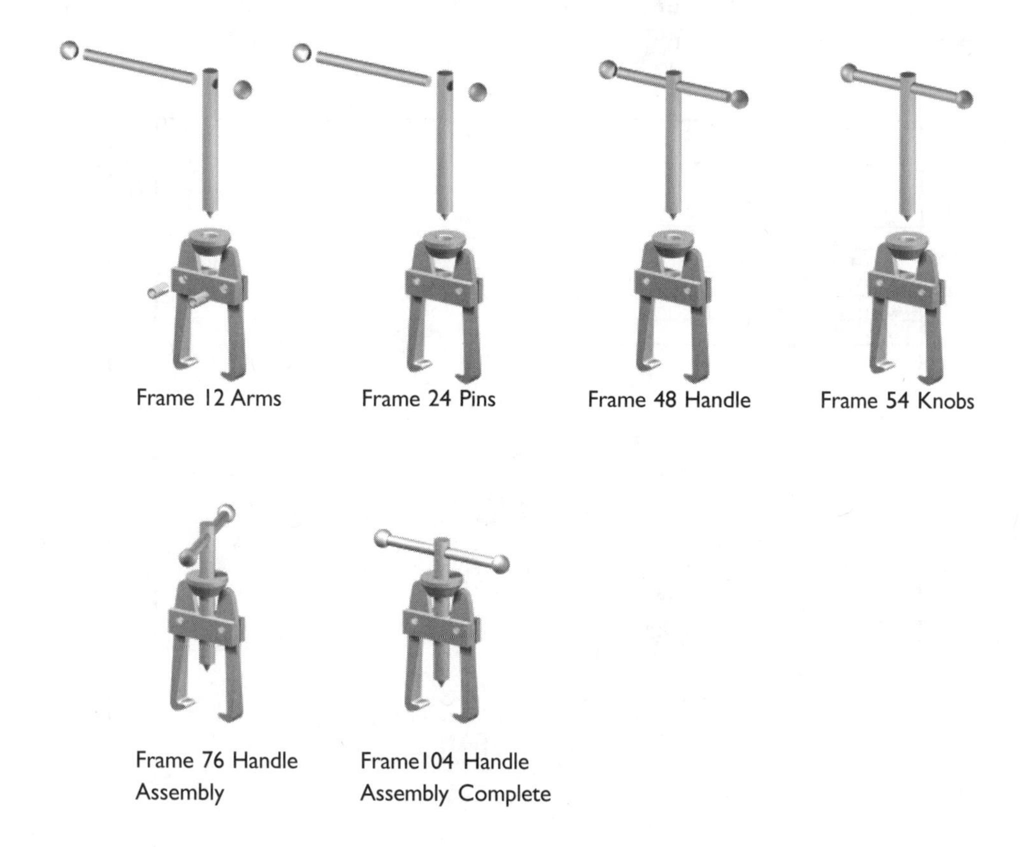

Figure 16.19 *Final animation follows the Frame Planning Sheet.*

STEP 14

Save the animation. When you save the 3D Studio file, all transforms affected while the **Animate** button was active are saved with the file. You can always reload the file and play the animation using the VCR controllers or by dragging the **Time Slider**. Save the movie file only when necessary to save both rendering time and hard disk space.

PROBLEMS

Detail and assembly views of the Gear Puller appear on the following pages. Also, a completed assembly for animation purposes can be found on the CD-ROM in *chapters\chapter_16\gear_puller*. Assemblies suitable for animation can be found in many of the technical and engineering texts in Referencesfor Further Study at the end of this book.

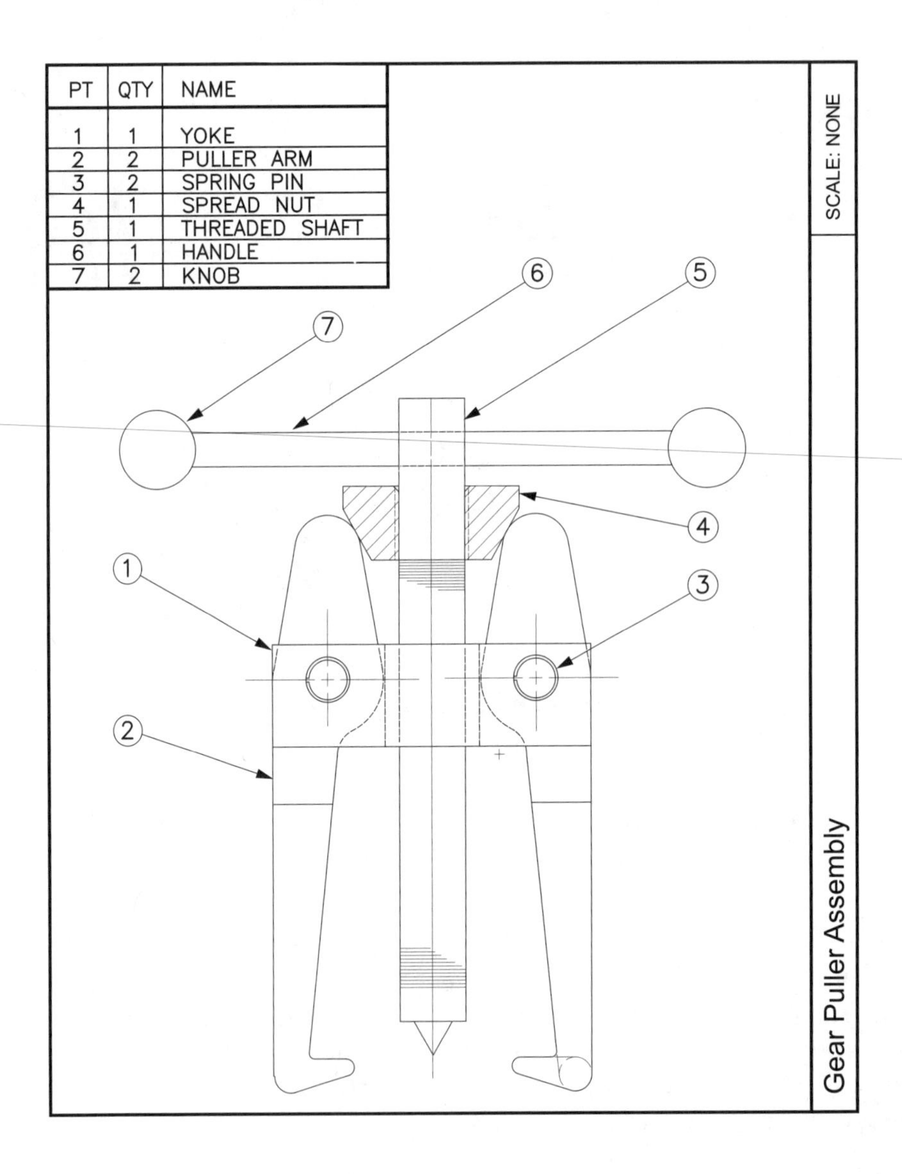

Problem 16.1a *Gear Puller Assembly (Earl:* Graphics for Engineers*).*

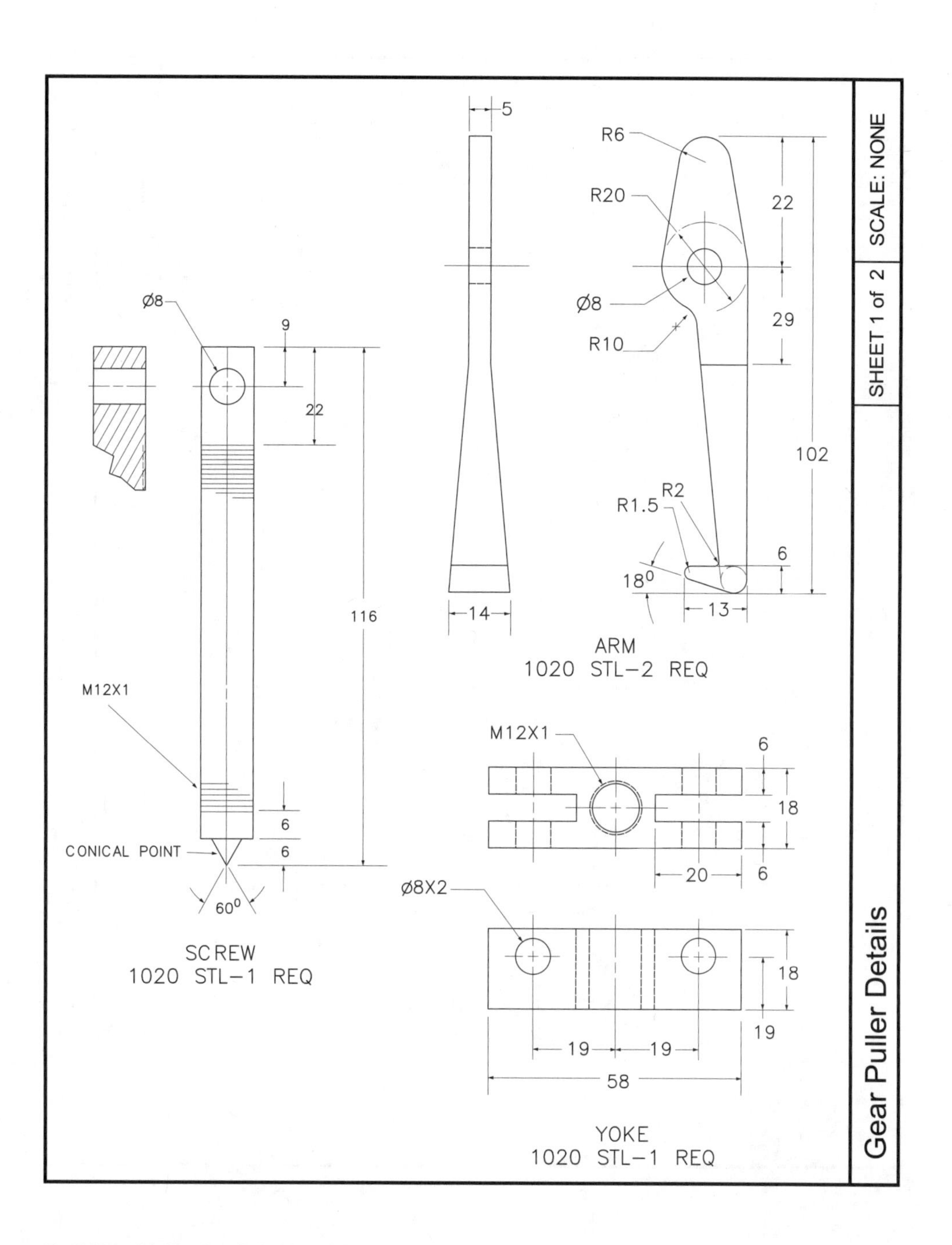

Problem 16.1b *Gear Puller Details.*

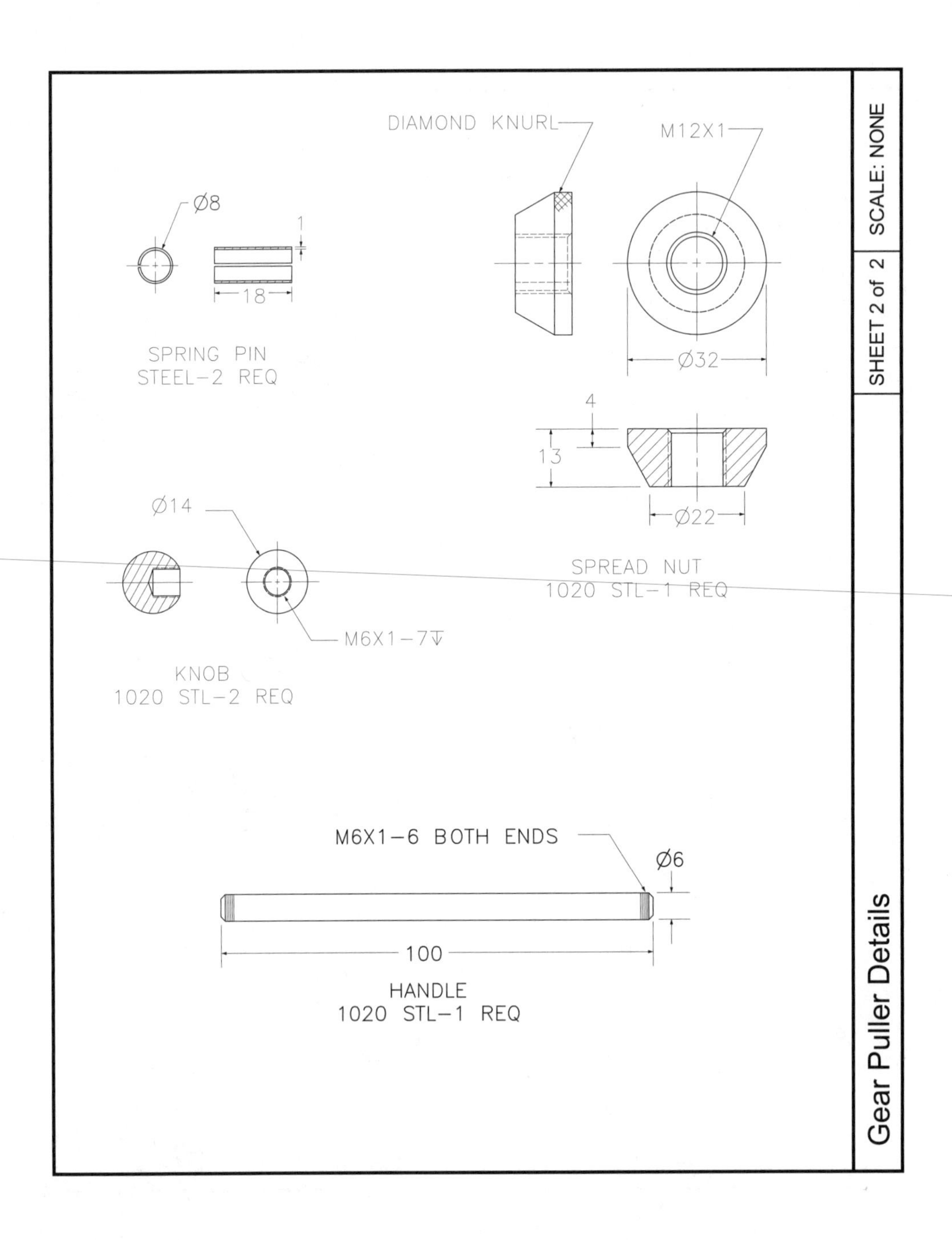

Problem 16.1c *Gear Puller Details.*

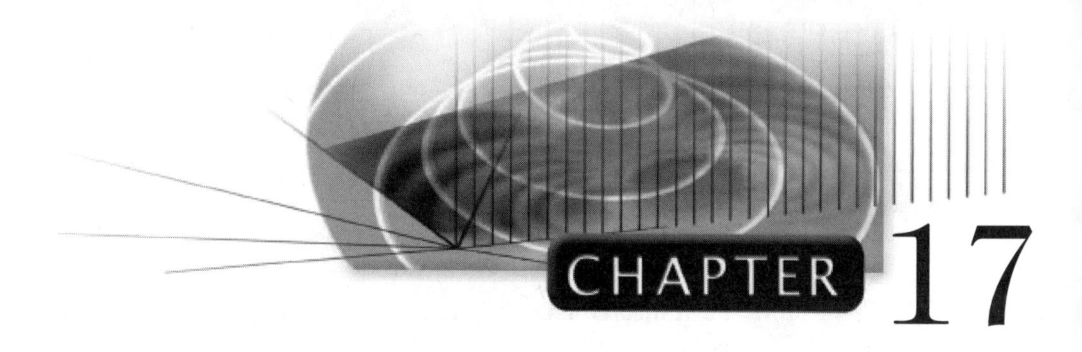

Camera Animation/The Fly-Around

CHAPTER OVERVIEW

In the previous chapter you learned how objects can be animated to show assembly order and position. The Gear Puller that was used as an example could be animated with a static camera position because no parts became hidden during assembly. But what if a static camera position doesn't allow all the parts to be seen? Two strategies could be used to solve this problem: moving the camera to multiple static Camera views or by smoothly moving the camera around as the parts assemble. The first method produces instantaneous *camera cuts*. The second is referred to as a *fly-around*. And as you will see, several methods exist for this fly-around technique.

Smoothly moving a camera's vantage can be an effective method for adding realism to an animation. Very seldom do we have experience watching something for very long without moving our eyes. But we avoid rapidly looking around because most people become disoriented. Plan slower camera movement over longer periods of time to avoid this disorientation.

KEY COMMANDS AND TERMS

- **Assign Controller**—the button in the **Motion|Assign Parameters** rollout that allows a **Path** to be assigned as the **Position Controller** for a camera.
- **Bezier Controller**—controller that bases the position of interpolated **Keys** on Bezier curves.
- **Camera Cut**—an immediate change in camera position. See **Pop**.
- **Camera Path**—a spline object used to define the motion of a camera.
- **Camera/Position**—the dialogue in which options are given for how **Interpolated Keys** are distributed between **Key Frames**.

- **Free Camera**—a camera having no separate target object.
- **Intermediate Key**—a **Key** inserted between **Key Frames** to explicitly control the movement of an object.
- **Path Into/Out Of**—options for interpolating path shape between **Keys** for positional change.
- **Pick Path**—the option in the **Motion|Path Parameters** roll–out that allows a spline to be picked as the position controller for a selected object.
- **Pop**—the practice of moving a camera from one location to another in the span of one frame; a camera cut.
- **PRS Parameters**—the **Position/Rotation/Scale** options in the **Motion|PRS Parameters** roll-out that insert appropriate **Keys** for the selected object.
- **TCB Controller**—the **Tension/Continuity/Bias** controller dialogue that provides a method for shaping and distributing interpolated keys either side of the selected **Key Frame**.

POPPING A TARGET CAMERA

Figure 17.1 shows a small Housing assembly that is suitable for changing camera position during the course of an assembly animation. The Face Plate and Pin are assembled on the front side of the Housing while the O-ring and Retainer are assembled at the rear. Without changing camera position, the assembly of the rear parts would be obscured by the Housing. Our first solution is to move the camera to a second position in the span of one frame so that all parts can be observed. We call this "popping" a camera and produces a "camera cut."

Figure 17.1 *Housing assembly.*

The completed animation can be found on the CD-ROM in *chapters\chapter_17\pop_anim.max* and a file ready to animate in *chapters\chapter_17\housing_assy.max*.

STEP 1

Plan the animation. Using a copy of the *Frame Planning Sheet* found on the CD-ROM in the *\tools* directory, plan the animation. Figure 17.2 shows this planning sheet.

The general strategy is to have the camera in front to record the Face Plate and Pin and then pop to the rear to record the O-ring and Retainer.

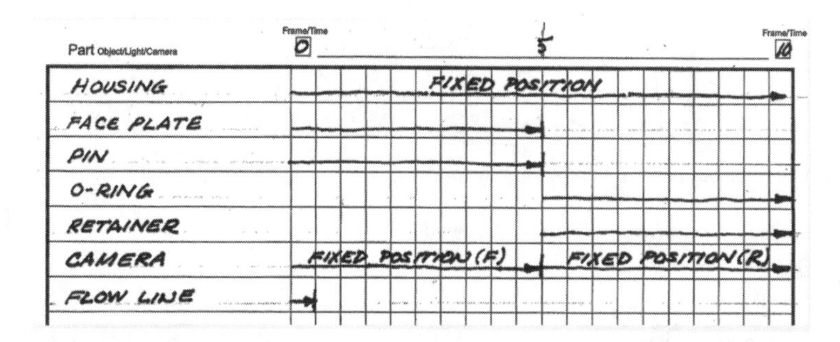

Figure 17.2 *Frame Planning Sheet for Housing assembly.*

STEP 2

Set up the animation. Right–click on the VCR controllers and in the **Time Configuration** dialogue box set **Frame Rate Custom**, **FPS:12**, **Time Display:MM:SS:TICKS**, and **Length=10** seconds. (We will switch back and forth between seconds and frames as necessary.) This gives you a 120–frame, 10-second animation. Because you have two parts in the front and two parts in the rear, we will split the animation in half: the first 5 seconds (frames 0–59) are spent on the Face Plate and Pin while the last 5 seconds (frames 60–120) record the assembly of the O-ring and Retainer.

STEP 3

Set a **Target Camera** at its initial position (Figure 17.3). This was the camera that produced Figure 16.1.

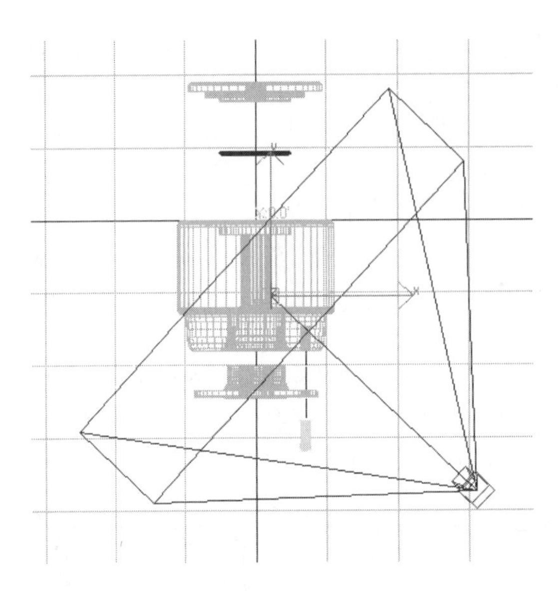

Figure 17.3 *Camera position at 1 through 5 seconds.*

STEP 4

Animate the movement of the Face Plate and the Pin. Switch to the Right Side view and turn on the **Animate** button. **Move** the **Time Slider** to the **5-Second** position and assemble the Face Plate and Pin (Figure 17.4). (The flow line is popped off-screen at **Frame 1** so it's out of the way. A **Key** is inserted into the **Face Plate** and **Pin Tracks** at **0** and **5** seconds.

Figure 17.4 *Pin and Face Plate assembled at 5-second position.*

STEP 5

Pop the camera. Open the **Time Configuration** dialogue box and switch to **Time Display|Frames**. Switch to the Top view, turn on the **Animate** button, and **Move** the **Time Slider** to **Frame 60**. Move the camera and its target to a position similar to that shown in Figure 17.5. Because you are moving the camera in the Top view, there is no change in world Z-axis position; the camera moves in a horizontal plane. **Keys** are inserted for both the camera and its target at the **0** and **5** seconds.

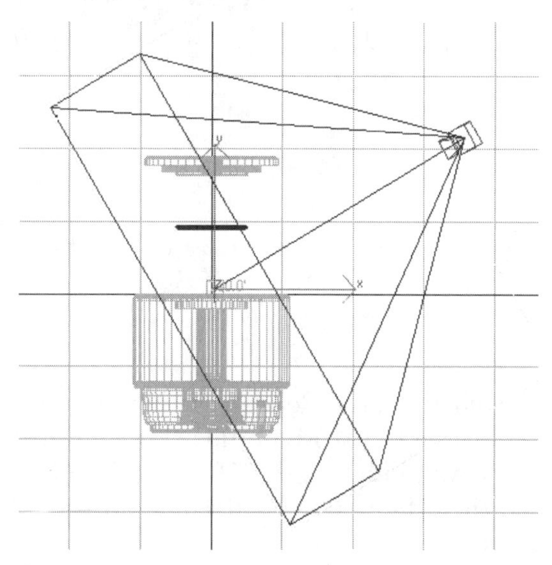

Figure 17.5 *Camera position for second half of animation.*

Note: Each **Target Camera** object has two parts: the camera and its target. Each of these have their own **Keys** during an animation and must be adjusted to assure the desired camera movement.

STEP 6

Adjust the **Keys**. Because this is the first recorded movement of the camera, its change of position is **Tweened** between frames **0** and **59**. You want its original position to hold steady until the frame right before it pops to the rear at **Frame 60**. To get the camera to hold its position between **Keys**, open the **Track View** and the **Camera** and **Camera Target Transforms**. Right–click on the camera's first key, bringing up the **Camera/Position** dialogue box. This dialogue is used to instruct 3D Studio how to move the camera between **Keys**.

Choose the option displayed (Figure 17.6) and the path into and out of the **Key** will be straight. Do this for both camera and target. **Duplicate** the **Frame 1 Key** by dragging with the **Shift Key** depressed, depositing the **Keys** at **Frame 59**. Again, do this for both the camera and its target.

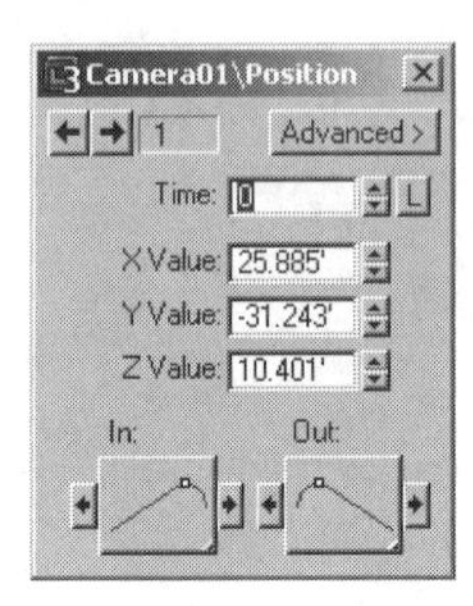

Figure 17.6 *Camera Position controls object movement into and out of the Key.*

STEP 7

Animate the O-ring and Retainer. Repeat the animation procedure for the O-ring and Retainer. Switch to the Right Side view and turn on the **Animate** button. Position the **Time Slider** at the **10-second** mark (**Frame 120**) and move the O-ring and then the Retainer into final position. **Pop** the flow line off screen at **Frame 61**.

These parts' **Keys** must be adjusted in the same manner as were the **Keys** for the front two parts, so that movement begins at **Frame 60**. Adjust **Key Curves** for **Key 0** in the **Position** dialogue box and duplicate **Key 0** to **Frame 60**. A sample frame of the second half of the animation is shown in Figure 17.7 and the completed animation is found on the CD-ROM in *chapters\chapter_17*.

ANIMATING A TARGET CAMERA

Our first example popped the camera and its target to a new position in a "camera cut" method. We cut from the front camera position to the rear camera position. This has the advantage of not wasting frames moving between camera positions. You get right to where the action is without looking at sections of the assembly where there are no parts. Additionally, you have the advantage of avoiding disorienting rapid camera movements.

An alternative method is to smoothly move the camera from its initial position, through an intermediate position, to its final position. Because we have already animated the parts and established the beginning and ending camera positions, all we have to do is **Delete** and **Edit Keys** to produce animation with continuous movement. This completed animation can be found on the CD-ROM in *chapters\chapter_17\target_anim.max*.

Figure 17.7 *O-ring and Retainer during animation.*

STEP 1

Delete unnecessary **Keys. Delete** all **Keys** for the camera and its target except those at the beginning and ending of the animation. Switch to the Top view and turn on **Animate**.

STEP 2

Create **Intermediate Keys**. Move the **Time Slider** to **Frame 60** and position the **Camera** so that its central vision line is perpendicular to world Y-axis (Figure 17.8). Go to the Track view and adjust this new **Key** (select and then right–click) for in and out path curves, as you did in Figure 17.6. Do this for the **Target** also. This assures that the **Target Camera** exhibits straight motion out of the first key, into and out of the second key, and into the third.

Note: By moving a camera and its target over time, you define an invisible path that the camera follows.

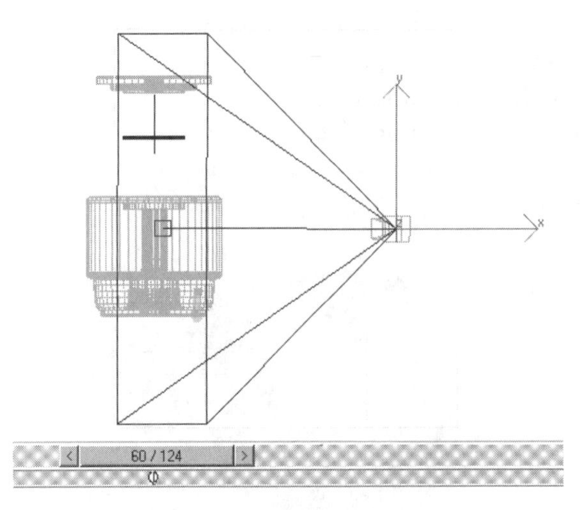

Figure 17.8 *Camera and target in intermediate position.*

This new animation accomplishes the same goal as did the first, but in this case by moving the camera smoothly between the first and last **Keys**, through an **Intermediate Key**. However, moving the camera and parts at the same time can be problematic. Not only can parts become hidden behind other parts, the camera can move such that parts disappear into the shade.

ANIMATING A FREE CAMERA

You have seen how a camera can be moved by popping it and its target to new positions in the span of a single frame. You have also seen how a camera can be smoothly animated by letting 3D Studio **Tween** between **Key Frames**. The third method involves attaching a camera to a spline path that defines its movement.

Tip: A **Free Camera** is more difficult to aim than a **Target Camera.** You can get the camera into a general position in orthographic views (move-rotate-move-rotate) but to fine-tune the aim, use the **Camera View** and its camera transforms in the **Bottom Menu**.

We will use the same assembly as done previously. A completed path animation can be found on the CD-ROM in *chapters\chapter 17\path_anim.max*.

Another camera object is available for this technique: the **Free Camera**. A **Free Camera** does not have an explicit target. Instead, it is pointed toward its target by rotating about **Screen**, **World**, or **Local Axes**. Be prepared to switch between axis systems as needed.

STEP 1

Create a **Free Camera** in initial position. Switch to the Top view and choose **Create|Cameras|Free**. Click in the approximate location for the initial position. The camera is inserted with its central line of vision parallel to the screen Z-axis (you're looking at the back of the camera).

Rotate the camera about screen Z until it points at the assembly. Switch to the Front view and move the camera to a position above the assembly.

Select **Local Axis** from the drop–down in the **Main Menu**. This aligns local Z with the camera's central vision line. **Rotate** the camera about its local X-axis so that the assembly is in the field of view (Figure 17.9). In the Top view, the Free Camera is pointed in at the assembly. In the Front view, the camera is pointed down at the assembly. Because the local Z-axis is concurrent with the central vision line, constraining movement to local Z moves the camera in and out, changing the field of view. Switch to the Camera view to fine–tune this initial view with the **Camera Transforms**.

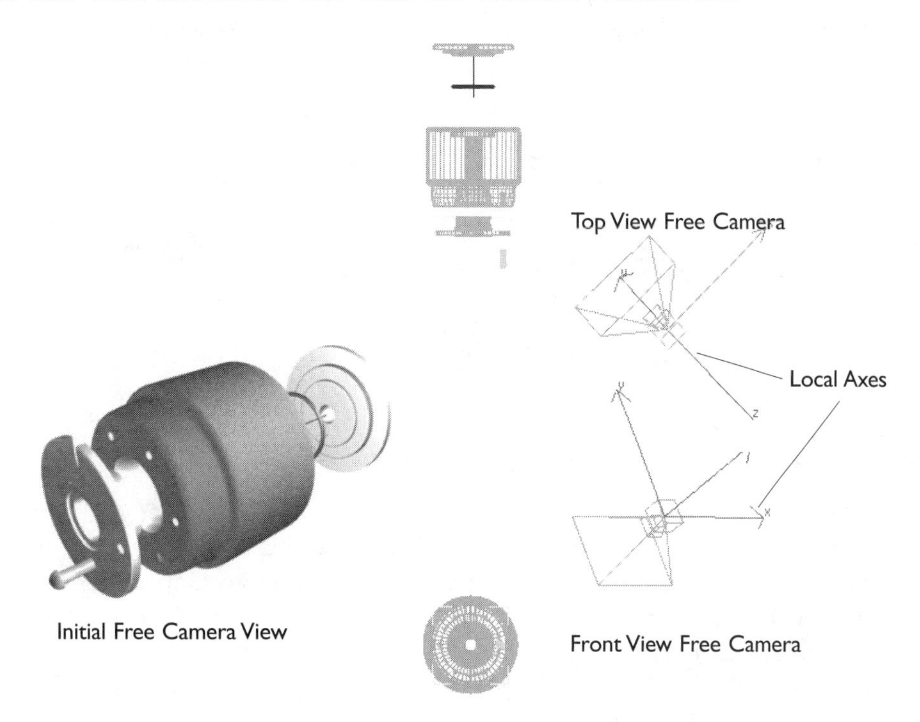

Figure 17.9 ***Free Camera*** *in initial position.*

Tip: One reason to use a **Free Camera** is avoid the "flipping around" characteristic of a **Target Camera** as it approached true vertical. By pointing the **Free Camera**, you essentially move its target.

STEP 2

Create the camera path. In the Top view **Create** a **Spline Shape** that describes the path you want the camera to follow (Figure 17-10). Because you are in the Top view, the path will be parallel to the world X-Y plane. In thr Front view, move the path vertically to align with the camera.

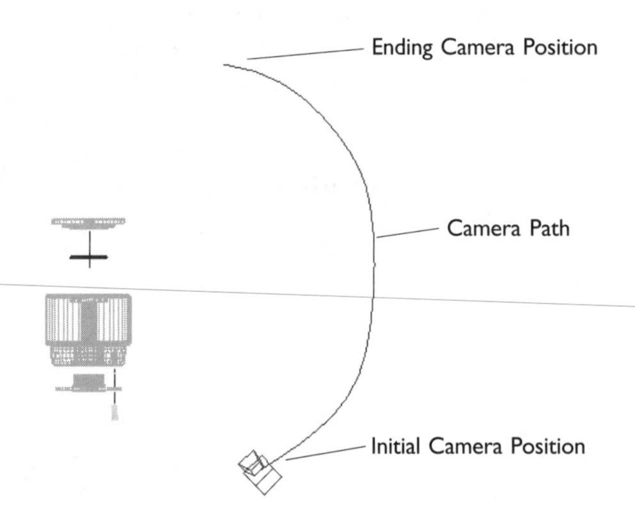

Figure 17.10 *Camera Path.*

STEP 3

Assign the path as the camera's position controller. Select the **Free Camera** and choose **Motion|Parameters|Assign Controller**. In the transform hierarchy window, choose **Position** and click the **Assign Controller** button left and above the hierarchy display. In the **Assign Position Controller** dialogue box select **Path**. This makes the path the position controller for the camera and the word "Path" now follows **Transform|Position** in the **Assign Controller** roll-out.

In the **Path Parameters** roll-out choose **Pick Path** and select the path. Manually move the **Time Slider** across the animation. The camera follows the path from beginning to end as the animation progresses. However, you'll notice that the camera keeps the same orientation.

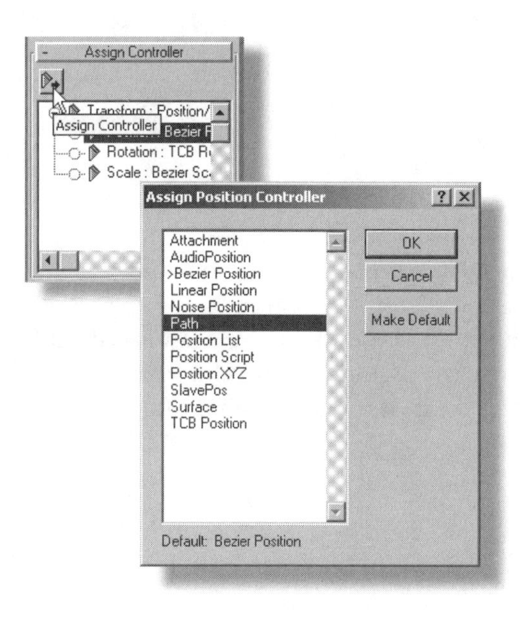

Figure 17.11 *Assign the path as camera controller.*

STEP 4

Retarget the camera. This is done differently than was done with a **Target Camera**. Move the **Time Slider** to **Frame 60** and turn on **Animate**. Select **Screen Coordinates** from the **Main Menu** drop–down, choose **PRS Parameters|Create Key|Rotation**, and **Rotate** the camera about the screen Z-axis until it points at the assembly (Figure 17.12). This creates a **Key Frame** at the middle of the animation.

Tip: It is easy to rotate a **Free Camera** so that the subject is out of the field of view. You may want to display plan (top), elevation (front or side), and camera views simultaneously so that the impact of a rotation can be immediately seen. Save often and keep the **Undo** button handy.

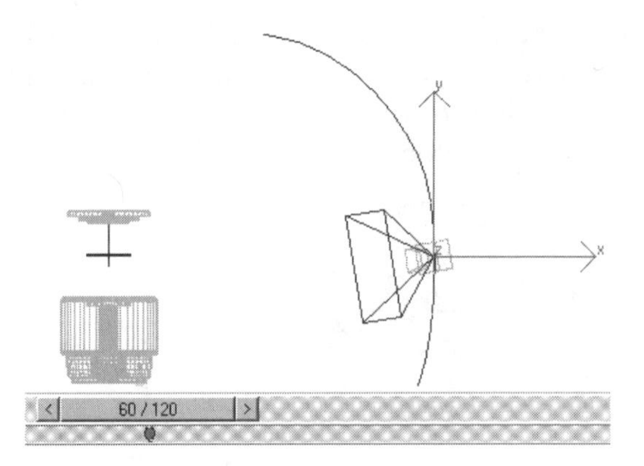

Figure 17.12 *Rotate* ***Free Camera*** *to keep assembly in field of view.*

Move the **Time Slider** to the end of the animation and repeat the process to point the camera at the assembly from the rear (Figure 17.13). The camera now moves along the path, its field of view remaining on the assembly.

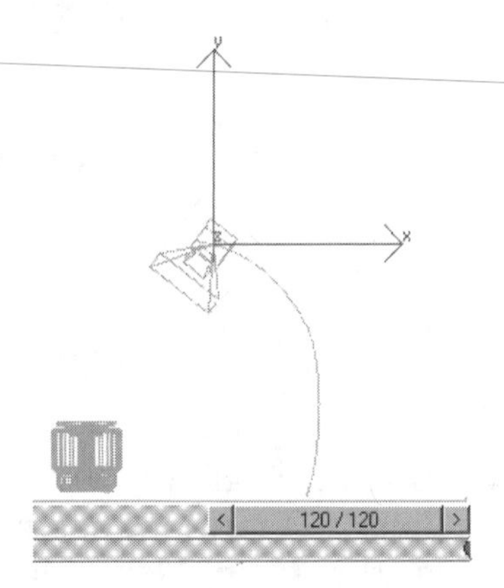

Figure 17.13 *Camera at* ***Frame 120*** *rotates to keep assembly in field of view.*

EXAMPLE: PULLEY ASSEMBLY ANIMATION

In Chapter 8, Modeling Assemblies, we completed the Fan Assembly. Refer back to Figure 8.26 to see that five parts are assembled from one end while eleven parts are assembled from the other. Keeping a static camera would compromise being able to accurately show all parts as they assemble. The solution is to move the camera from front to rear during the assembly. This completed animation can be found on the CD-ROM in *chapters\chapter_17* and the assembly, ready to animate, in *chapters\chapter_8*.

STEP 1

Plan the animation. Complete a Frame Planning Sheet to coordinate the assembly of all parts. **Key Frame** this animation as you normally would, over 120 frames.

STEP 2

Create a camera path. In this case we use a circular path that is edited. This makes a path that maintains a consistent distance from the center of the assembly (Figure 17.14).

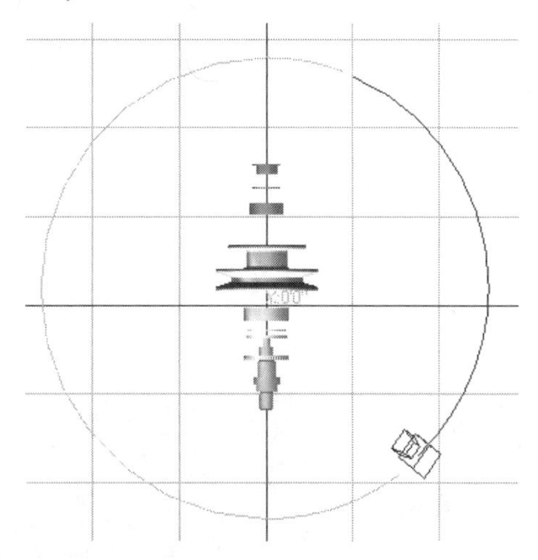

Figure 17.14 *Circular arc as path for camera.*

Select the circular path and choose **Modify|Edit Spline|Sub Object|Segment|Break** and break the circle at the beginning and end of the camera path. A small circular target appears at each break. Choose **Sub Object|Spline|Trim** and click on the part of the circle to remove. The portion of the circle to be used as the camera path remains.

STEP 3

Assign the path as the camera controller. Select the camera. Choose **Motion|Assign Controller|Position** and click the **Assign Controller Button**.

Choose **Path** from the **Assign Position Controller** dialogue box. Choose **Path Parameters|Pick Path**. The camera moves along the path from frame 0 to frame 120.

STEP 4

Point the camera. Move the **Time Slider** to **Frame 60**. Turn on **Animate** and choose **Rotation** from the **Assign Controller** roll-out. **Rotate** the camera about screen Z to point at the assembly (Figure 17.15). Move the **Time Slider** to **Frame 120** and **Rotate** the camera again. Turn **Animate** off.

Switch to the Camera view and test your animation. You should have smooth camera movement and the assembly of all parts should be visible.

Note: You want the camera to be viewing the rear parts before they have moved behind the pulley. Go back and adjust the starting track anchors for the rear parts so that their movement doesn't start too early, before the camera is in position.

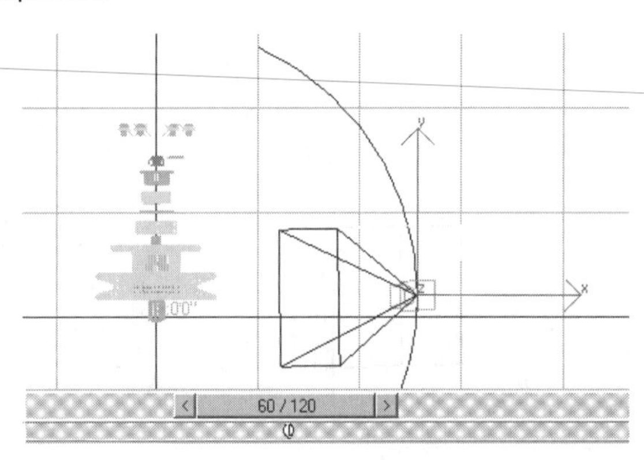

Figure 17.15 *Camera pointed at assembly at **Frame 60**.*

STEP 5

Adjust the camera **Keys**. The previous examples used **Bezier Controllers** as were shown in Figure 17.6. The **Keys** for the **Free Camera** use T**ension/Continuity/Bias** (TCB) controllers as shown in Figure 17.16. This gives even greater control over interpolated values between the keys using values between 0 and 50.

Default values of **25** provide smooth motion into and out of the key. **Tension** can override **Continuity** and **Bias** if set to **50**. The less **Tension** (approaching zero) the greater the effect of large (approaching **50**) continuity values. **Bias** skews the curve on one side of the **Key** or the other. Large **Continuity** values and small **Tension** values allow for large **Bias**. The **Ease To** and **Ease From** controllers effect how **Interpolated Keys** are distributed from halfway between adjacent **Key Frames**. For example, you can have more of the camera rotation happen between keys or at the key.

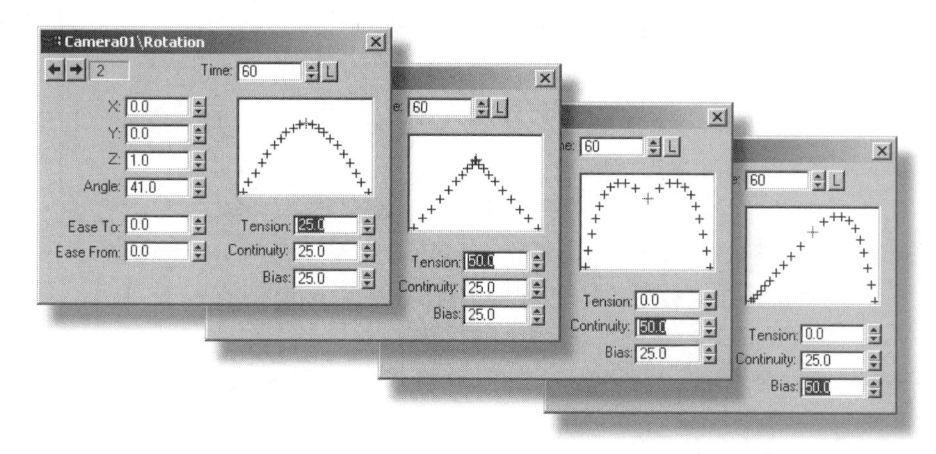

Figure 17.16 *The Tension/Continuity/Bias (TCB) Controller.*

PROBLEMS

Detail and assembly views of subjects appropriate for animated camera assemblies are on the following pages. These problems are based on problems found in several of the most popular texts.

DOUBLE ACTING CYLINDER			
PT	QTY	NAME	DESCRIPTION
1	1	SHAFT	STEEL
2	1	BUSHING	BRONZE
3	1	O-RING (SMALL)	RUBBER
4	1	PRESSURE END	STEEL
5	2	O-RING (LARGE)	RUBBER
6	1	O-RING (PISTON)	RUBBER
7	1	PISTON	STEEL
8	4	BODY	STEEL
9	1	WORKING END	STEEL
10	4	FLAT WASHER	.375X.875X.0781
11	4	HEX HEAD BOLT	.3125-24UNCX6.500

Problem 17.1a *Double Acting Cylinder (Bertoline:* Fundamentals of Graphics Comminication*).*

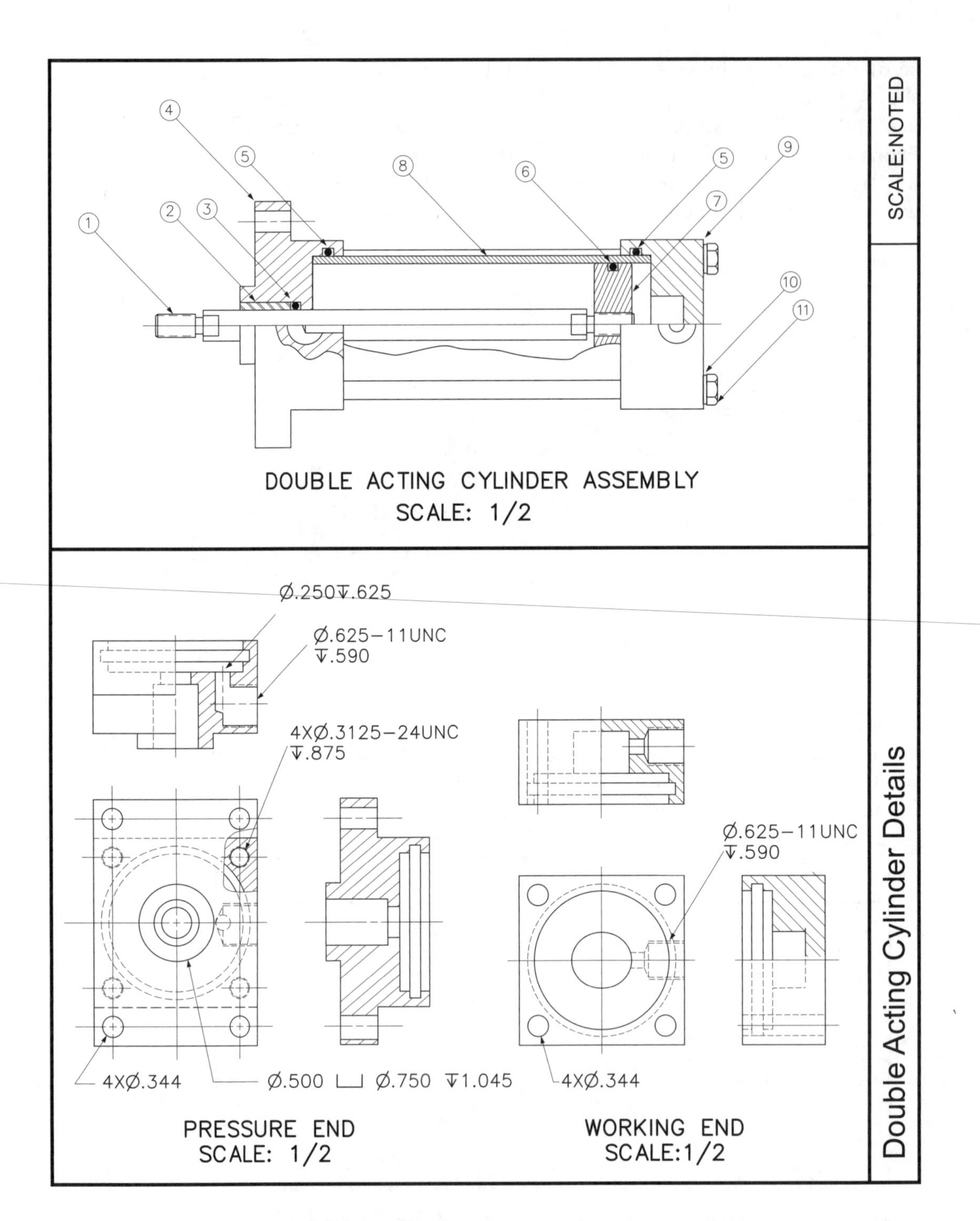

Problem 17.1b *Double Acting Cylinder (Bertoline:* Fundamentals of Graphics Communication*).*

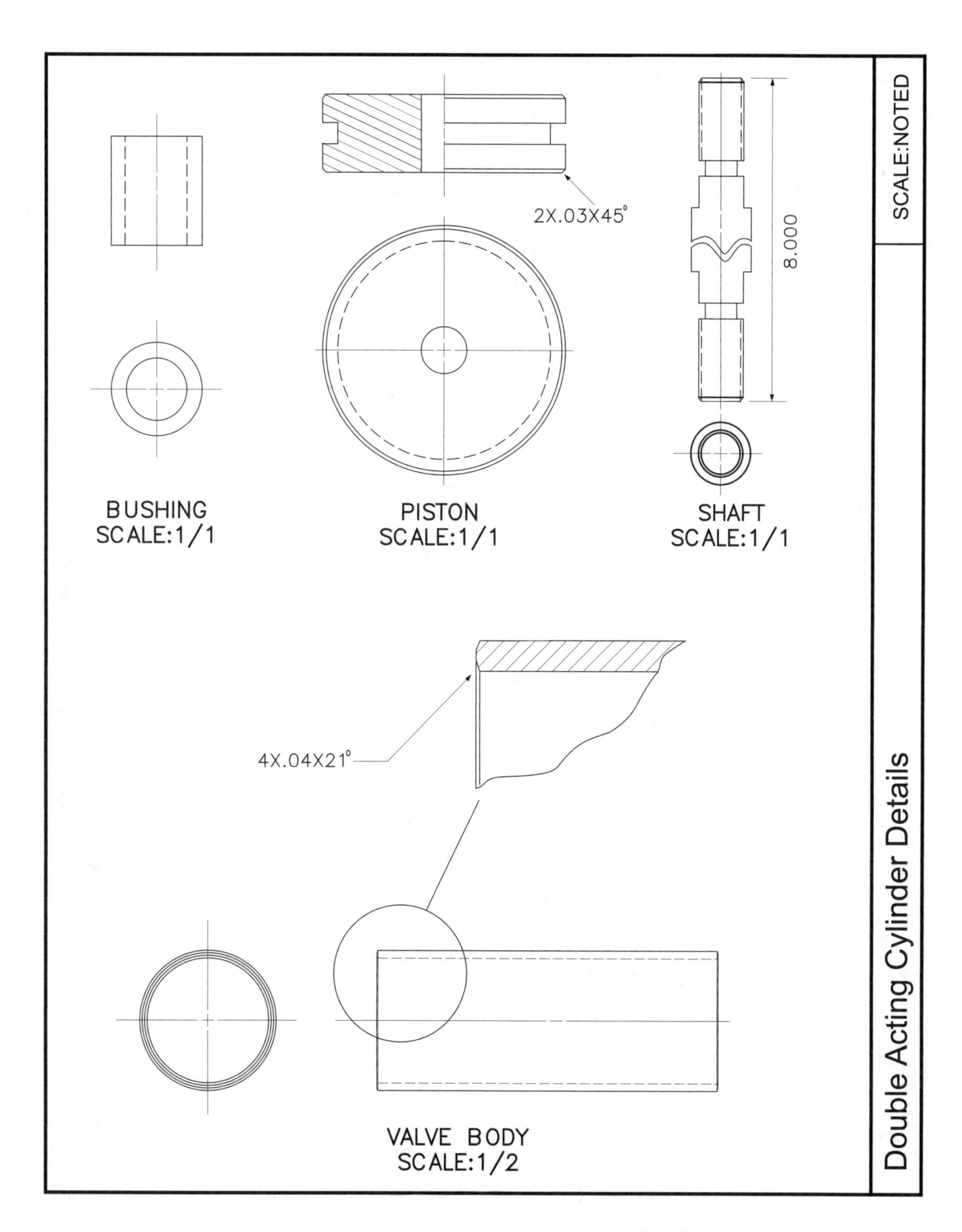

Problem 17.1c *Double Acting Cylinder (Bertoline:* Fundamentals of Graphics Communication*).*

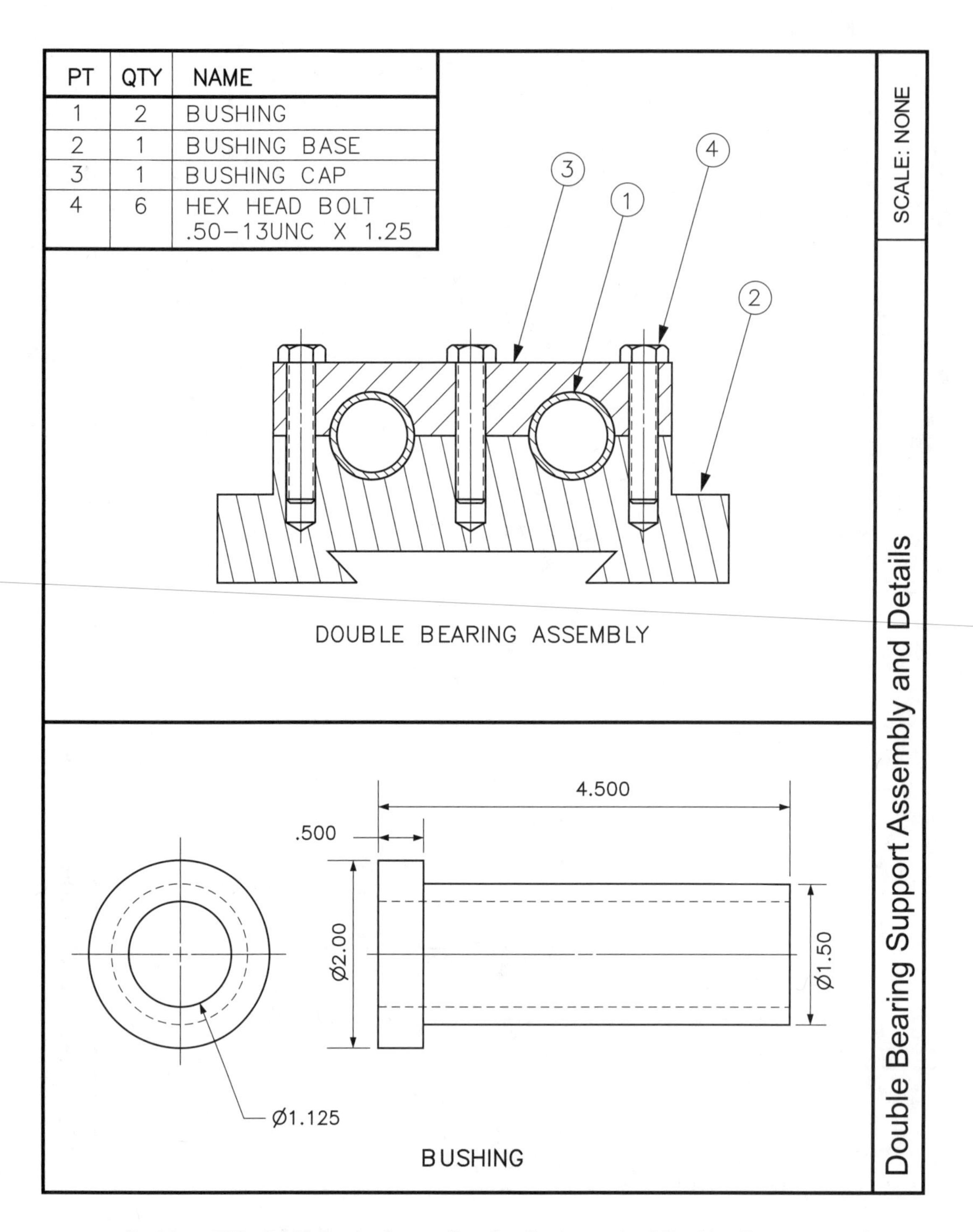

Problem 17.2a *Double Bearing Support (Bertoline:* Fundamentals of Graphics Communication*).*

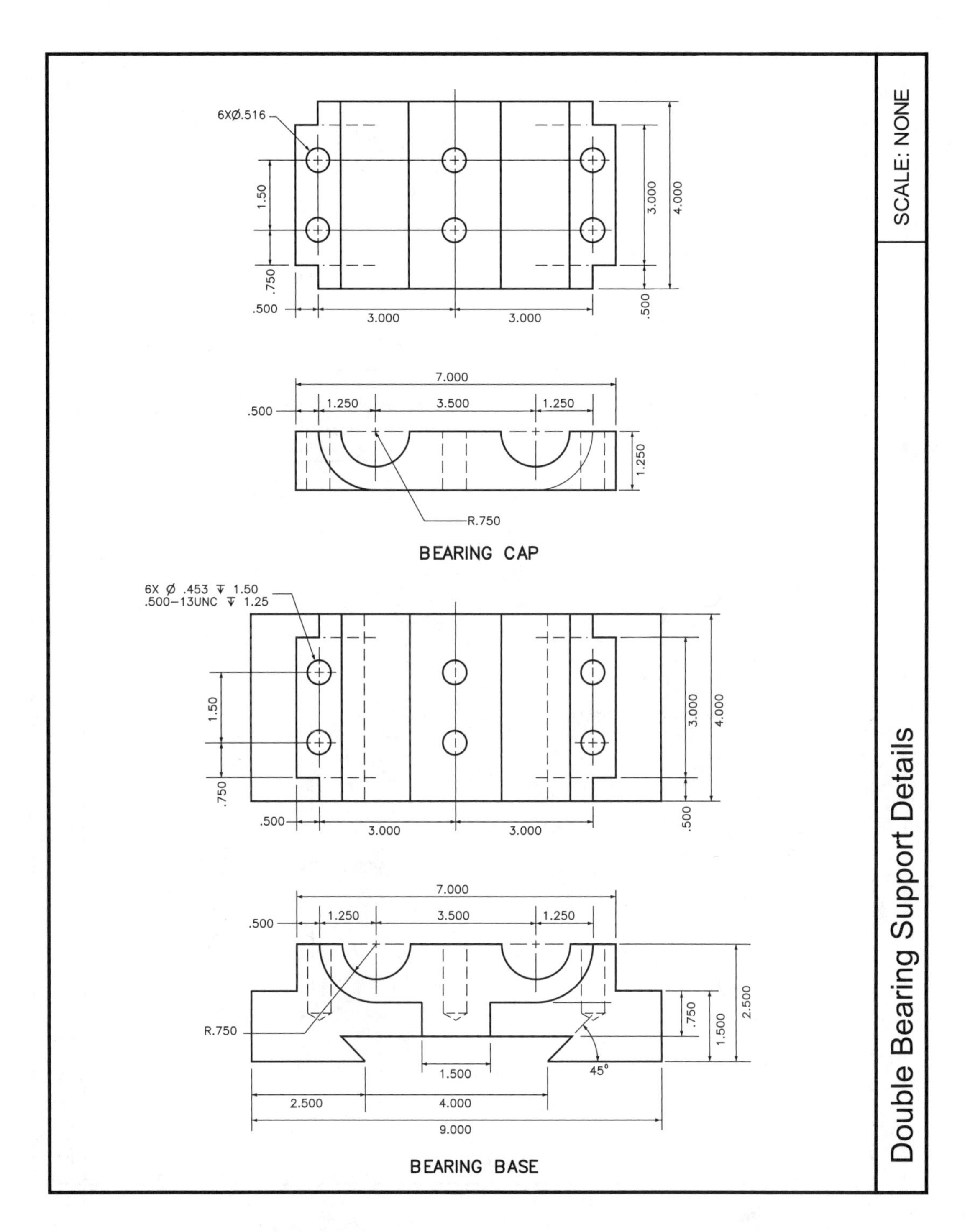

Problem 17.2b *Double Bearing Support (Bertoline:* Fundamentals of Graphics Communication*).*

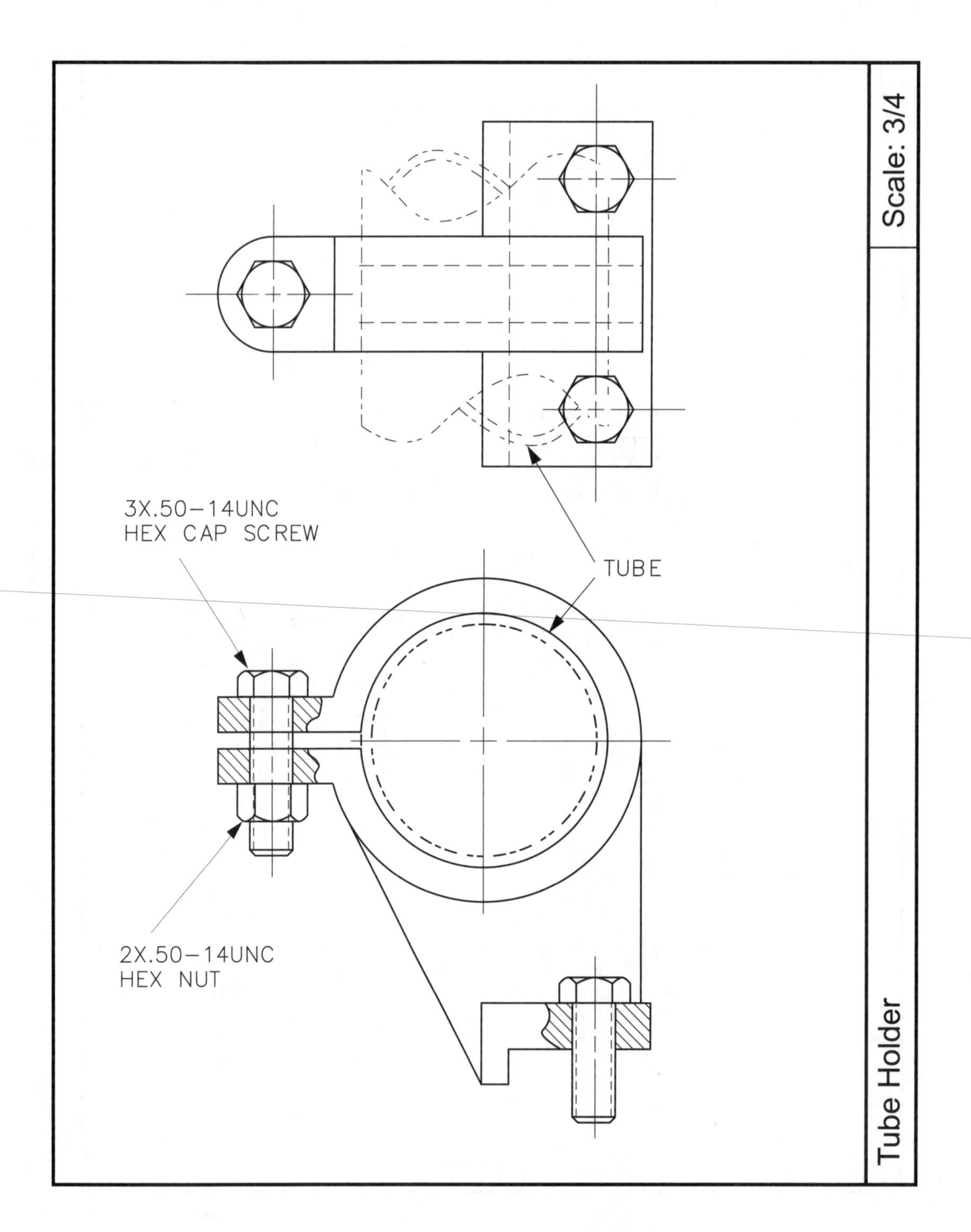

Problem 17.3 *Tube Holder (Luzadder & Duff:* Fundamentals of Engineering Drawing*).*

Programming in MaxScript

CHAPTER OVERVIEW

Both VIZ and MAX contain a robust programming language referred to as *MaxScript*. Because MaxScript is understood only by 3D Studio, it is called a *scripting language*. This is actually a peek at the way 3D Studio internally records your interactive modeling activities. Turning the process around, you can give 3D Studio MaxScript commands and it will execute them one after the other, just as if you had interactively modeled in one of the viewports.

There was a time when writing lines of code was the only way to communicate with the geometry engine of a graphics program. Thankfully, those days are gone. Modern computer graphics programs have developed intuitive graphical user interfaces to make it easier to get directly at the graphics, without programming. This is what you have been doing through the first seventeen chapters of this book.

There are tasks, however, that are more easily accomplished by writing a small program than by interactive modeling.

- Modeling tasks that are done over and over are candidates for MaxScript. For example, if you always collapse a model to an **Editable Mesh**, **Optimize** it, and then apply a **Smoothing Group**. You might want to create a MaxScript that does this to the selected object and add the script as a button to your command bar.
- If you have similar parts (same shape, different sizes) where dimensions can be expressed as *parametric* relationships, you can write a MaxScript that asks for sizes and automatically creates the model.
- If VIZ or MAX doesn't do something you repeatedly require (like placing cameras at predetermined positions) you can create *floaters*, windows that float above the viewports, providing these functions.

Even though MaxScript gives you tremendous power in directly addressing the 3D Studio geometry engine, it is foolish to spend time programming actions that are just as easily accomplished by interactive modeling. If you aren't extending the interface, simplifying repetitive procedures, or creating automatic parametric modeling scripts, you probably are better off modeling interactively.

KEY COMMANDS AND TERMS

- **Comment Line**—a line of MaxScript preceded by two hyphens (--) that is skipped and not interpreted.
- **Data Type**—the data form (integer number, floating point number, Boolean condition, etc.) in which a particular property must be expressed.
- **Determining Dimension**—a dimension that establishes a parametric relationship to other dimensions in an object.
- **Dot Notation**—the method of attaching a property to a named object; **mybox.height** attaches the property *height* to the named object *mybox*.
- **Editor Window**—a text editor internal to 3D Studio available when **Open Script** or **New Script** is chosen from the **Listener Window** or by choosing **Create|Utilities|MaxScript**.
- **Evaluate All**—the command in the **Editor Window** that sends its MaxScript to the **Output Pane** of the **Listener Window** where it is evaluated for correctness.
- **Floater**—a window that floats above the 3D Studio interface.
- **Listener Window**—the window containing **Recorder** and **Output Panes** where scripts can be written and evaluated.
- **MaxScript**—the internal programming language used to automate 3D Studio operations or extend its interface.
- **Mini Listener Window**—a window located in the lower left corner of the 3D Studio interface where the last line of both **Recorder** and **Output Panes** in the **Listener Window** are displayed.
- **Open Script**—the command that opens a MaxScript in the **Editor Window**.
- **Output Pane**—the lower (white) portion of the **Listener Window** where MaxScript is evaluated for correctness.
- **Parametric Modeling**—a modeling method whereby features are related to one another; change in one dimension is reflected in other dimensions due to parametric relationships.
- **Property**—a parameter of an object that can be manipulated in MaxScript.

- **Recorder Pane**-the upper (pink) portion of the **Listener Window** where interactive actions in the 3D Studio interface are recorded in MaxScript.
- **Run Script**-the command that runs a MaxScript.
- **ShowProperty**-the MaxScript command that displays properties associated with a named object.

IMPORTANCE OF SCRIPTING

Scripting is not a substitute for interactive modeling. You should think of it as an addition or enhancement to the 3D Studio graphical user interface. Indeed, the instances where scripting might increase productivity should literally scream out at you "there's got to be a better way of doing this!"

MaxScripts are interpreted, unformatted text files (.MS). That is, lines of instructions are evaluated one after the other starting with the first command. Some MaxScripts are executed only once. Other MaxScripts stay active and execute their instructions anytime they are run (chosen).

Note: Although MaxScript files carry the .MS extension, they are simply unformatted text files. You can check this by changing the .MS extension to .TXT and opening in WordPad. This means you can write MaxScript in any text editor or word processor as long as you save as unformatted text (.TXT). Change the extension to .MS so 3D Studio will recognize the file as a script.

When you choose **Utilities|MaxScript|Run Script**, 3D Studio loads the MaxScript text file, evaluates each line of the script, and performs the instructions. When you choose **Utilities|MaxScript|Open Script**, 3D Studio opens the MaxScript in the **Editor Window**. When evaluated, any errors appear in red in the **Output Pane** of the **Listener Window**.

RECORDING 3D STUDIO INTERACTIVE ACTIONS

Probably the easiest way to see what MaxScript can do is to see what has been going on behind the scenes as you have been modeling. Begin by opening the MaxScript **Listener Window**. It's called the **Listener Window** because that's exactly what it does: it listens to what is being done in the scene and displays these actions in the form (syntax) understood by the program.

STEP 1

Choose **Utility|MaxScript|Open Listener** to open the **Listener Window**. You can also right–click on the **Mini Listener Window** in the lower–left corner of the interface.

STEP 2

Create a Cylinder in any viewport. You will see instructions appear in the top (**Macro Recorder**) pane of the **Listener Window**. In fact, if you look at these instructions, you can probably make sense of them right away (Figure 18.1) because they address many of the properties in the **Create|Geometry|Cylinder** roll-out.

STEP 3

Create a box in the same viewport. Another line of MaxScript is added to the pink **Recorder Pane**. Again, many of the Box properties should be familiar.

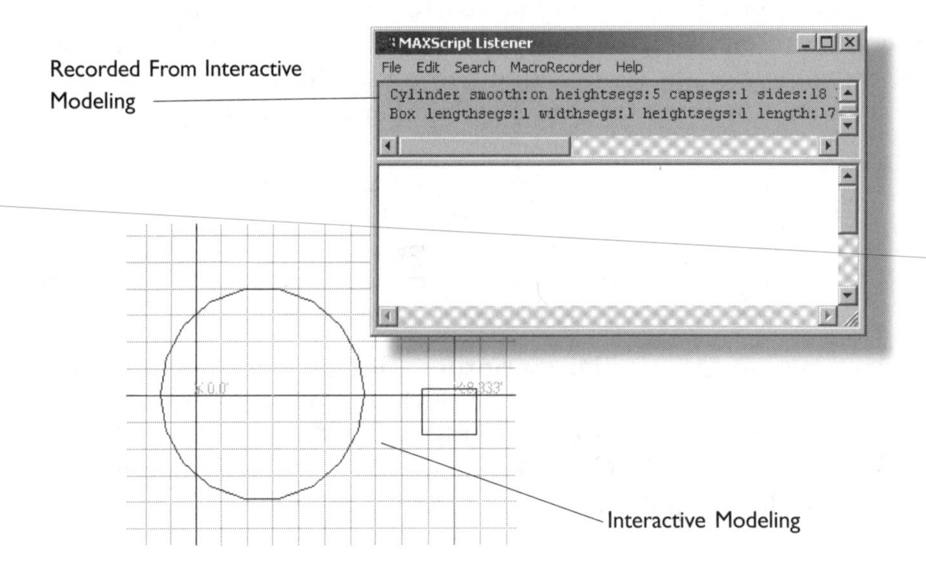

Figure 18.1 *Modeling actions in the **Recorder Pane** of the **Listener Window**.*

EDITING RECORDED MAXSCRIPT

The bottom section of the **Listener Window** is called the **Output Pane**. This section evaluates MaxScript for correctness that is either written, pasted, or passed to it. If the code is correct, it instructs 3D Studio to run the commands. If the code contains errors, red text will provide appropriate, although cryptic, error messages.

Note: MaxScript commands are interpreted and run. They are not executed in a computer programming sense because they are not compiled.

STEP 1

Delete all objects from the scene. Copy the cylinder command line from the **Recorder Pane** and Paste it into the **Output Pane**. Nothing happens because 3D Studio is waiting for the **Enter Key** to execute the command.

STEP 2

Press the **Enter Key** and the same Cylinder is created from the MaxScript code as was created interactively to initiate the first example. The command is in black. The output is in blue. Error messages are in red. Do the same for the Box (Figure 18.2)

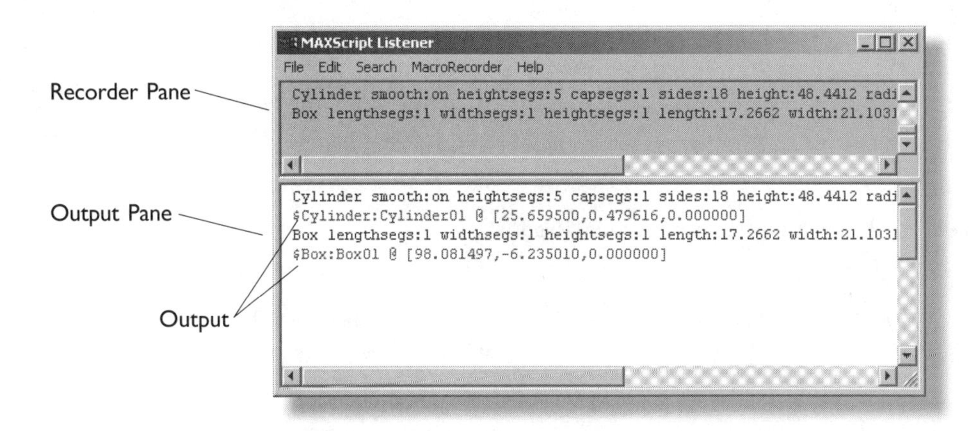

Figure 18.2 *MaxScript copied from the **Recorder Pane** and pasted into the **Output Pane**.*

You can directly edit the code in either pane. However, only code in the **Output Pane** will be evaluated for correctness. Let's move the Cylinder and Box to the world origin (0, 0, 0) and then place the Box on top of the Cylinder.

Tip: Double click to select one word (separated by spaces); triple click to select one line (separated by returns). This works in either pane of the **Listener** or **Editor Windows**.

STEP 3

Now that you have seen how to start a MaxScript from recorded actions, let's see how you can modify object properties. MaxScript uses *dot notation* to identify properties that may be associated with an object. If you enter the following into the **Output Pane** of the **Listener Window**:

myprism=box()

showProperties myprism

you'll be rewarded with a listing of the properties for a Box that MaxScript recognizes as *myprism* (Figure 18.3).

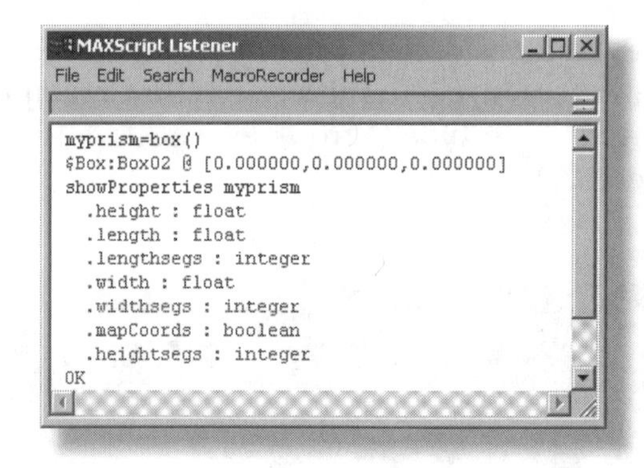

Figure 18.3 *Box properties recognized by MaxScript.*

Note: The X-Y local axes of MaxScript default objects will be matched with world X-Y axes. You'll be creating objects on the ground.

So **myprism.length** would refer to the length property of the object named *myprism* while **myprism.widthsegs** would refer to the number of segments in the width direction. To the right, after the colon, is the *data type* each property expects. *Float* expects whole or decimal numbers; *integer* expects only whole numbers; *Boolean* expects true or false.

STEP 4

Delete all objects from the scene. Note that 3D Studio will continue to add objects every time a line is executed so the scene can become rapidly cluttered. Highlight both lines of output (blue) and delete. Close up the two lines of script so there is no blank line between them.

Scroll to **pos:[xxx,yyy,zzz]** of each object and replace with **0, 0, 0**. You can continue working in the **Listener Window** (Figure 18.4), hitting **Enter** after each line. If correct, output lines are generated and the Cylinder and Box are created again, now centered on the world origin.

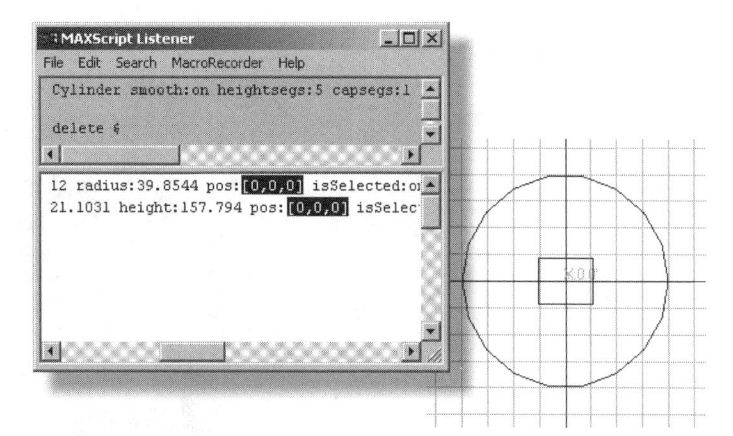

Figure 18.4 ***Cylinder*** *and* ***Box*** *origins edited.*

Note: The **Mini Listene**r is a place to enter a single line of MaxScript to see its affect. Program code can be copied and pasted between **Mini Listener** and either pane of the **Listener**, or the **Editor**.

STEP 5

Edit the Z-axis origin of the Box to equal the height of the Cylinder. Scroll the **Recorder Pane** to find the cylinder's height, **Height:XX.XXXX**. (This number depends on how you pulled out its height.) You know from previous experience that Cylinders and Boxes are drawn upward toward the observer. The origin of each object is centered on the bottom plane. This means that placing the origin of the Box at the height of the Cylinder will put the Box's bottom surface right on the Cylinder's top surface.

In the **Recorder Pane**, copy the Cylinder's height and paste it into the third coordinate position for the Box. Delete any output from the bottom pane. Copy the Cylinder line, paste it into the **Output Pane**, and press **Enter**. Copy the Box line, paste it into the **Output Pane** below the cylinder line, and press **Enter**. New output is generated with the edited coordinates. The Box is now centered on the Cylinder and sits on the Cylinder's top surface (Figure 18.5).

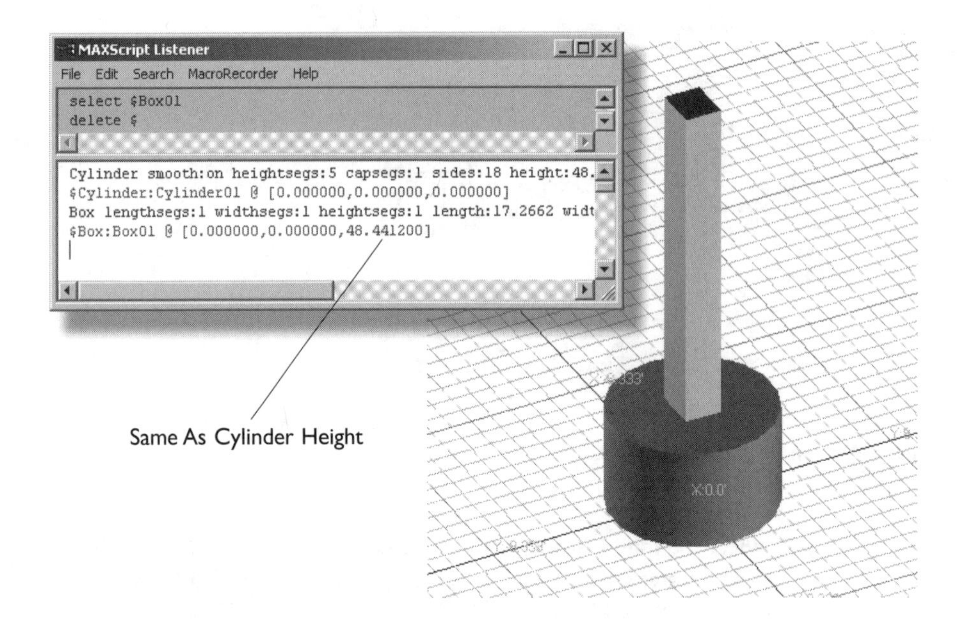

Figure 18.5 *Box origin coincides with Cylinder height.*

THE BASIC SCRIPTING PROCEDURE

In the previous example you saw how interactive modeling can be recorded as MaxScript and how that script can be edited into custom commands. You can also write MaxScript directly in the **Editor**, evaluate the correctness of that script in the **Output Pane** of the **Listener Window**, and save the script to be used later.

MaxScript has the common structures found in most programming languages: Functions, If-Then-Else, For-To, and Do-While. *Online Help* (**Help|MaxScript Reference**) can give you assistance in MaxScript commands, syntax, and grammar.

Figure 18.6 shows the planning sketch for a display of Rods that are dependent on Rod diameter (VIZ and MAX specify cylinders by radius, not the common industrial practice of using diameter). The *offset* between Rods and their heights are based on the radius (*rad*) of the Rod.

- If the Rod radius (*rad*) is smaller than .99, then the *offset* is twice the radius and the Rod height is equal to the radius.
- If the radius is 1.00 or larger, the *offset* is four times the radius and Rod height is 1.50 the radius.

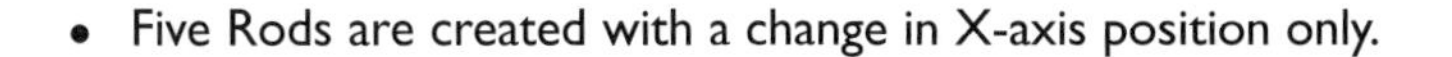

- Five Rods are created with a change in X-axis position only.

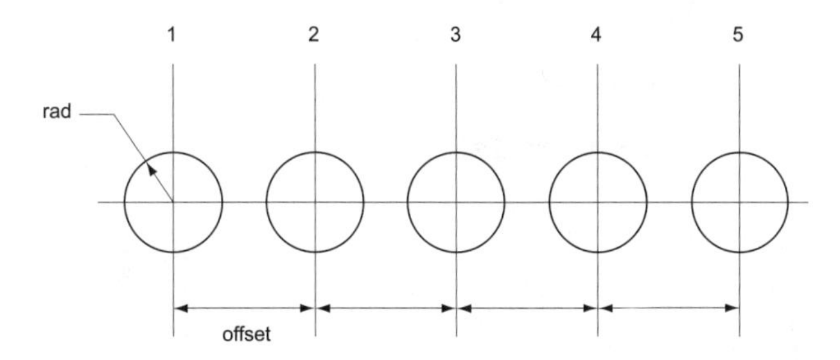

Figure 18.6 *Display of Rods based on radius.*

Note: A simple linear array can be created once you know this offset value. However, this MaxScript also automatically calculates the height of the cylinders as a function of its radius, something the Array command will not do.

STEP 1

Right–click on the **Mini Listener Window** and choose **File|New Script** or choose **Utilities|MaxScript|New Script** from the **Create Menu**. The **Editor Window** opens.

STEP 2

Create a test for the radius. We will begin by seeding the program with a known value.

```
rad=1.00
```

If *rad* is .99 or smaller, set **offset=rad*2** and **height=rad**. Otherwise, set **offset=rad*4** and **height=rad*1.50**.

```
rad=1.00
If rad<.99 then
(
    offset=rad*2
    height=rad
)
```

```
        else
    (
        offset=rad*4
        height=rad*1.50
    )
```

Right and left parentheses define the start and stop of the sections of the If statement. Notice that there is no need for an "End If" because of the second right parenthesis.

Note: MaxScript is not case sensitive. You may choose to use upper and lower case for ease of identification. Indenting is not required but may also be helpful in understanding MaxScript structure. Do not omit or add extra spaces. Spaces are used to delimit syntactical elements.

STEP 3

Save the script. In the **Editor Window** choose **File|Save** and give the MaxScript the name *rods.ms*. The script is saved by default in *3DSMAX\Scripts* or *3DSVIZ\Scripts*.

Test the script by choosing **File|Evaluate All** from the **Listener Window** menu. If you have a valid If-Then-Else statement the **Listener Output Pane** will display the value of each variable as control moves through the statements, followed by **OK**. If you get a red error message, check the format and characters used in the statement and reevaluate. Figure 18.7 shows the completed program and its correct evaluation in the **Output Pane**. Refer to this figure as you follow the development of the MaxScript in the following descriptions.

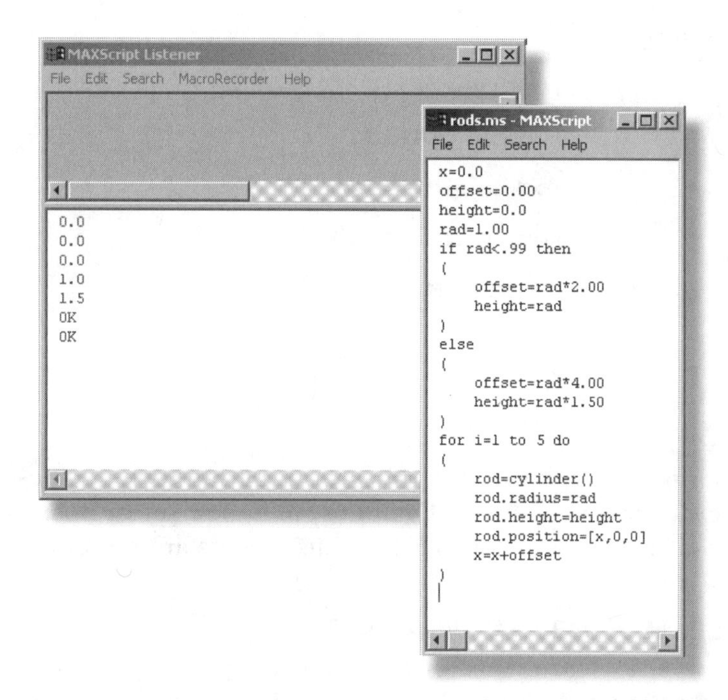

Figure 18.7 *MaxScript in the* **Editor Window** *is evaluated in the* **Output Pane** *of the* **Listener Window.**

STEP 4

Initialize the values of variables used in the MaxScript: the X-axis value of the initial position of the first Rod, the *offset*, and Rod height. These values must be initialized before they are entered into calculations.

```
x=0.0
offset=0.00
height=0.0
```

STEP 5

Write a For-To loop to create the display of five Rods based on the *offset* value and specified radius. The X position of successive Rods is incremented by the *offset* value each time through the For loop. Add the following lines to the MaxScript after the If-Then-Else statement.

```
for i=1 to 5 do
(
        rod=cylinder()
        rod.radius=rad
        rod.height=height
        rod.position=[x,0,0]
        x=x+offset
)
```

STEP 6

Save the script and rerun several times with varying radius values. You'll notice that each time you choose **Evaluate All**, a new set of Rods are created based on the current value of *rad*. Previous objects are not erased.

EXAMPLE: MOUNTING BRACKET

The best way to learn MaxScript is to apply it to a problem. Study Figure 18.8 where a Mounting Plate is described. Dimensions are expressed in generic units for simplicity. What is important is the relationship of dimensions to **body.length**, what we will call the *determining dimension*. This body length would be entered by the modeler with all other dimensions calculated from this value.

Assume you work for a company that makes a number of these plates, in many sizes, for various manufacturers. You wouldn't want to model the plates over and over. This appears to be a job for MaxScript. We will sneak up on the final solution by increasingly sophisticated solutions, where in the end, all you have to do is specify the *determining dimension*. All other sizes and positions will be calculated and the plate created accordingly. Let's start with the basics. Refer to Figure 18.8 as we work along.

STEP 1

Make a simple Box. Open the **Listener Window**, choose **File|New Script** and type the following *comment line* in the **Editor**.

```
--a MaxScript that creates a Mounting Plate from a determining dimension
```

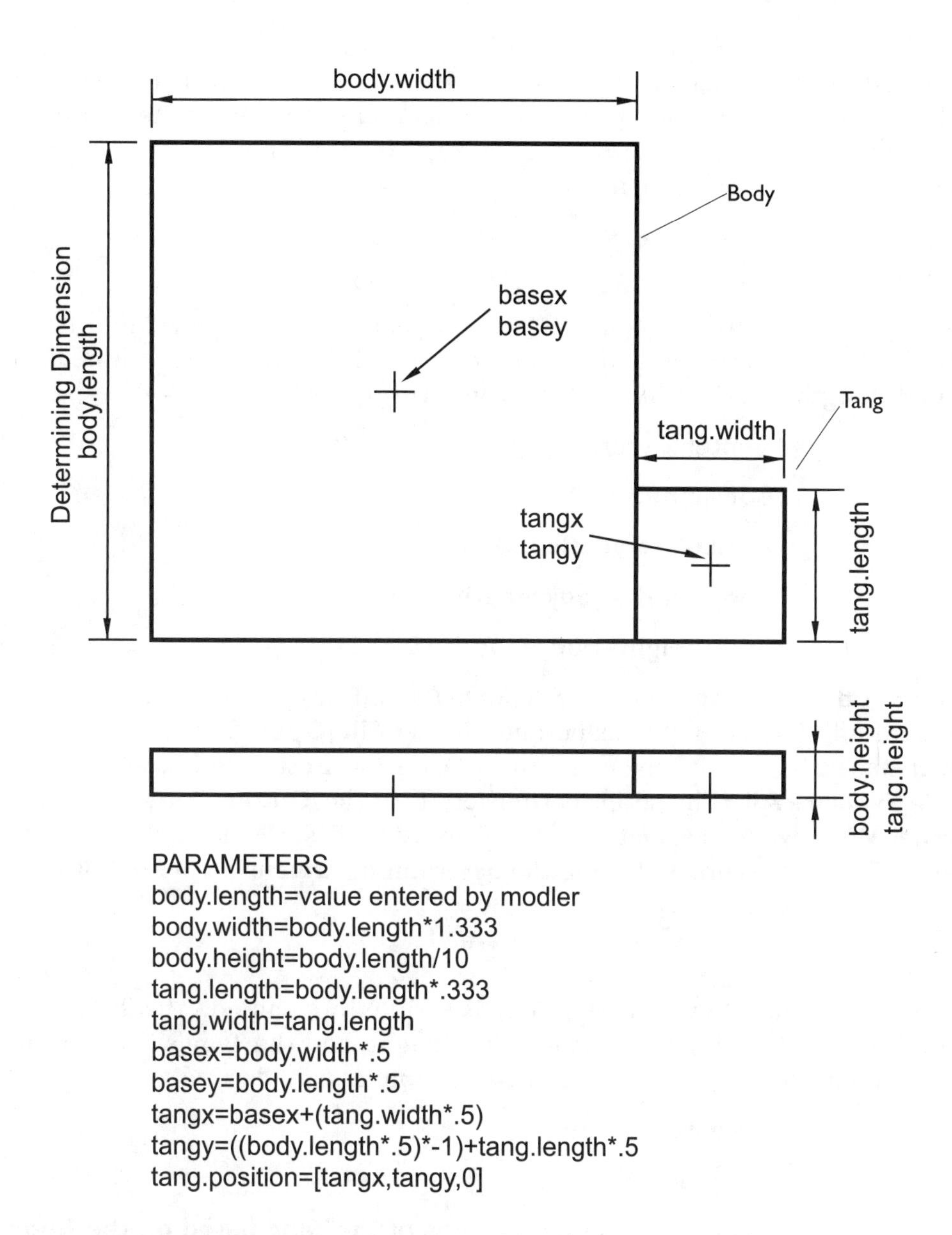

Figure 18.8 *Mounting Plate for MaxScript solution.*

A *comment line* (preceded by two dashes) is ignored by the MaxScript interpreter and is there for you and anyone else who might want to understand the script. Then enter on the next line:

```
body=box()
```

This establishes a default **Box** named “Body” at 0, 0, 0 world. The origin of the Box is in the center of its bottom plane and the **Box** rises above the world X-Y plane in the positive Z-axis direction. This knowledge allows us to use the information in Figure 18.8.

STEP 2

Make a flexible **Box**. In Figure 18.8 we see that the width and height of both the Body and the Tang are functions of the determining dimension (**body.length**). Let’s define the Body in terms of its length.

```
--enter determining dimension (length)
body.length=50
--calculate width and height
body.width=body.length*1.333
body.height=body.length/10
```

In the **Editor Window** choose **File|Save As** and save the script as *plate.ms*. Delete all objects in the scene and choose **File|Open Script** and choose *plate.ms*. The script opens in the MaxScript **Editor Window**. Choose **File|Evaluate All**. The script is transferred to the **Output Pane** of the **Listener Window** and evaluated. If there are no errors, the desired part is created. The Body is created using the determining dimension of 50 units.

STEP 3

Create the Tang. Now that the Body is scripted we can add the Tang. This involves two sets of data: length, width, height, and the Tang’s X,Y,Z origin. From Figure 18.8 you can see how these numbers are derived.

```
--create the Tang
tang=box()
--calculate the dimensions of the Tang based on the Body
tang.length=body.length*.333
tang.width=tang.length
tang.height=body.height
```

```
--calculate origin of the Body
basex=body.width*.5
basey=body.length*.5
--calculate Tang origin
tangx=basex+(tang.width*.5)
tangy=((body.length*.5)*-1)+tang.length*.5
tang.position=[tangx,tangy,0]
```

The Tang's width and length are calculated using **body.length**. Tang height the same as that of the Body (**body.height**). The X-Y origin of the Tang is calculated so that it is flush with the front and right edges of the Body.

The completed Mounting Plate and its script are shown in Figure 18.9. Go back and compare the MaxScript with the parametric relationships in Figure 18.8. You should be able to identify where each line of MaxScript is taken from the Mounting Plate drawing. Right now, to create a Mounting Plate you would have to bring up the script and edit the value and then manually run the script. In the next section we will automate this.

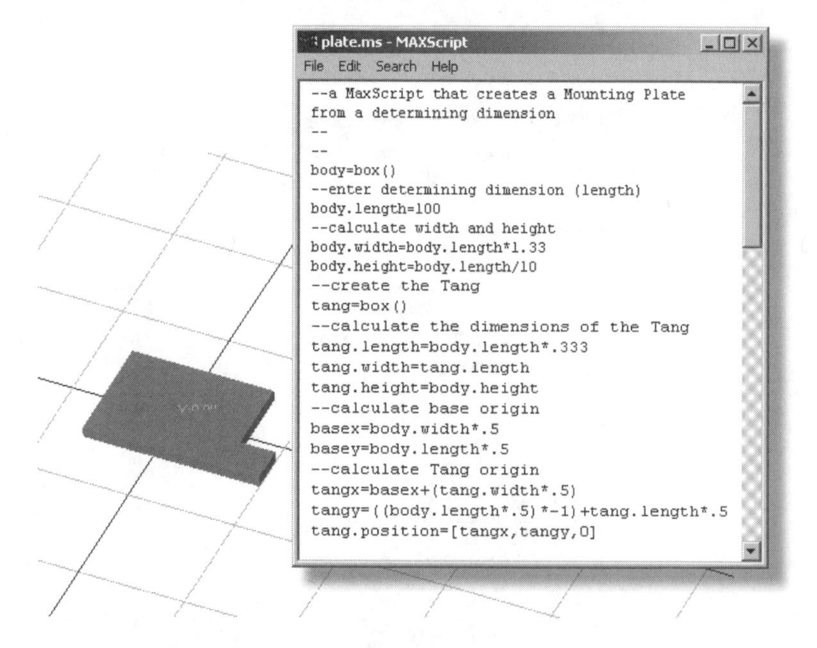

Figure 18.9 *Mounting Plate script and objects.*

EXTENDING THE INTERFACE

It would be better if our Mounting Plate script were more readily available. To do that, we will add a roll-out that automatically calls the script.

STEP 1

Insert the following lines of code at the beginning of *plate.ms*:

```
utility myplate "Mounting Plate"
(
    button btnCreate "Make Plate"
    on btnCreate pressed do
    (
```

and then follow the second left parentheses with the Mounting Plate script (Figure 18.10). Add two closing right parentheses at the end of the script and save again.

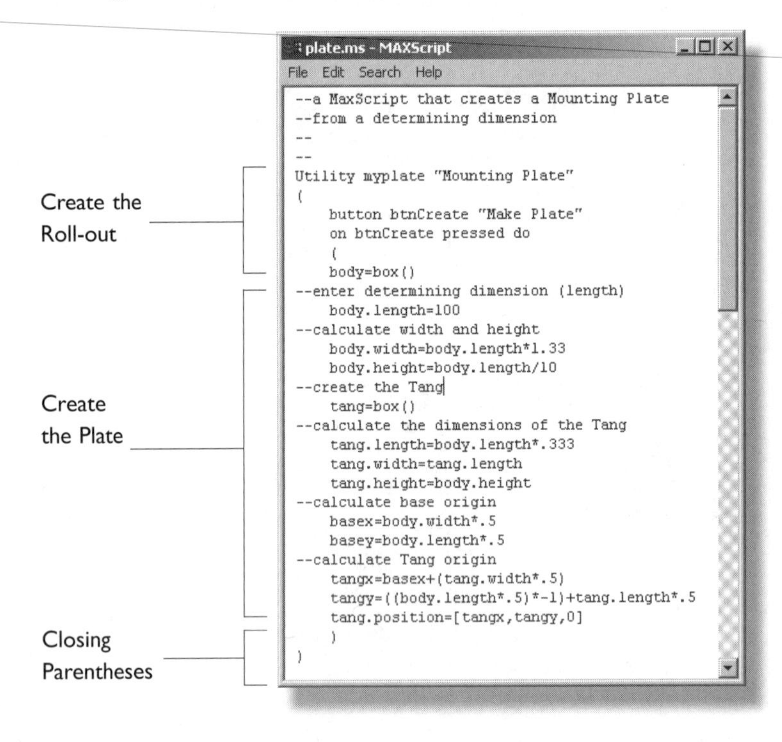


```
--a MaxScript that creates a Mounting Plate
--from a determining dimension
--
--
Utility myplate "Mounting Plate"
(
    button btnCreate "Make Plate"
    on btnCreate pressed do
    (
    body=box()
--enter determining dimension (length)
    body.length=100
--calculate width and height
    body.width=body.length*1.33
    body.height=body.length/10
--create the Tang
    tang=box()
--calculate the dimensions of the Tang
    tang.length=body.length*.333
    tang.width=tang.length
    tang.height=body.height
--calculate base origin
    basex=body.width*.5
    basey=body.length*.5
--calculate Tang origin
    tangx=basex+(tang.width*.5)
    tangy=((body.length*.5)*-1)+tang.length*.5
    tang.position=[tangx,tangy,0]
    )
)
```

Figure 18.10 *Code adds interface element to* ***Utility Menu****.*

Choose **Utility|MaxScript|Run Script** and choose *plate.ms* in the *Scripts/* directory. The "Mounting Plate" text in Figure 18.10 is loaded into the MaxScript drop–down window and a "Mounting Plate" roll-out is created to hold the "Make Plate" button. When this "Make Plate" button is pressed, the plate creation script is run and a plate is created (Figure 18.11).

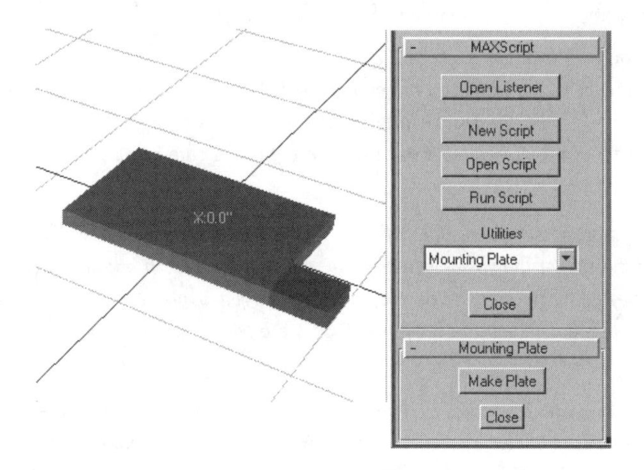

Figure 18.11 *Mounting Plate as option in Utility Menu.*

Tip: Any MaxScript placed in the *Scripts/Startup* directory will be run when 3D Studio is started. In our case, this makes a "Mounting Plate" utility available without manually running the script in the **Utility** roll-out.

MORE MAXSCRIPT

This chapter has given you an overview of the limitless possibilities in extending the capabilities of 3D Studio VIZ and MAX through programming in MaxScript. If you are interested in further study, consider these resources:

Mastering the Art of Production with 3ds max 4. Jason Busby and Michelle Bousquet. Thompson Learning. 2002.

Mastering MaxScript and the Sdk for 3D Studio Max. Alexander Bicalho and Simon Feltman. Sybex. 2000.

Ultimate 3D Links at *http://www.3dlinks.com/books_max.cfm.* Go to 3D Tutorials/3D Studio Max and scroll to 3D Studio Tutorial Links/Max Scripting.

PROBLEMS

Below are verbal descriptions of problems suitable for solution with MaxScript. Read each description carefully and prepare a sketch that establishes geometric and relational conditions before writing the script. Include descriptive comments and be prepared to document your program development through successive screen captures.

Problem 18.1 *Handle MaxScript.. You work for a company that makes Handles similar to that found in Figure 16.1. The diameter of the Knobs on the end of the handle are a function of Handle diameter. Likewise, the longer the Handle, the greater the Handle's diameter. Create a sketch that establishes the parametric relationship of these features. Create a MaxScript that automates the modeling.*

Problem 18.2 *Pressure Vessel MaxScript.. Pressure Vessels like those shown in Figure 10.1 are fabricated in various sizes. As the size of the vessel increases, the opening on the top and the flange also increase, up to a point where they no longer increase as the vessel is enlarged. Support members on the bottom must also increase at a rate greater than the vessel itself. Create a MaxScript that automates this construction.*

Problem 18.3 *Bearing MaxScript. Ball bearings, like those shown in Figure 8.15, are manufactured in various sizes. As the inner diameter of the inside race increases, inner and outer race thickness also increases as does the number of balls, their size, and the overall width of the Bearing. Create a MaxScript that automates this construction. Add this script to the* **Utility** *roll-out and create spinner fields that record determining dimensions.*

Problem 18.4 *Axonometric camera MaxScript. Standard axonometric cameras are not part of the 3D Studio interface. Using the information found on the CD-ROM in tools\axon-cameras, create a* **Utility** *roll-out that allows you to move a selected camera to a representative selection of dimetric and trimetric positions.*

Problem 18.5. *Tank Farm design. You work for a company that manufactures a standard storage tank similar to that in Figure 15.11. Increased capacity is achieved by arraying the tanks into a "farm." Create a MaxScript that determines the number of tanks needed for a specified storage capacity. Have the MaxScript then merge a preexisting tank and create the array.*

Still and Movie Output

CHAPTER OVERVIEW

It isn't long after making your first model that you will want to have tangible evidence of your progress in your hands (or in front of your eyes). The **Smooth+Shaded** viewport display option produces suitable results during the modeling process, but it is still only a rough approximation of the realism that's possible.

Note that VIZ uses the term "design" to describe the workplace while MAX uses "scene." This chapter uses scene to describe the contents of the 3D Studio workplace unless a procedure applies to VIZ only.

Rendering is the process of determining and displaying how surfaces are represented in terms of colors, materials, and lights. Rendering determines where highlights, shades, shadows, and reflections are placed. The result is a raster (bitmap) file. When a succession of these rendered frames are made, each representing a point in time during an animation, the result is called a movie. 3D Studio is capable of rendering still frames and movies to a number of formats. It is in the choice of these formats and their settings that success or failure is often determined.

In many cases, rendering to the screen (what's called the *Virtual Frame Buffer* or **VFB**) is sufficient to check the progress of a design. Rendering to a file (or saving the screen rendering) gives you a record that can be printed later, attached to an email, posted to a Web site, included in multimedia, or included in a technical document.

It is always important to match a rendering or movie to its intended use. Every file format has strengths and weaknesses; each is a compromise between quality, size, resolution, and color depth. It's tempting to create numerous renderings during the course of a design. But as you gain experience, you'll render more selectively. This means you'll spend more time modeling, material mapping, and animating and less time waiting for rendering to finish.

Because digital movies are a succession of rendered frames, rendering a movie will severely tax system resources. Even on workstation class computers major animations require hours, even days, to complete. Final animations can be 50, 100, or even 200 MB in size. So if rendering too often is a poor use of time, animating too often can be a *really* poor use of time.

Note: Several things can be done to make rendering more efficient. First, rendering can be done on separate "rendering stations," freeing modelers and animators to continue their work. A rendering station only needs lots of RAM, a big fast hard disk, and 3D Studio. It can use a cannibalized monitor, keyboard, and mouse. Another solution is *distributed rendering*. A number of computers on a network with a server can form a *rendering farm* where frames of an animation are distributed for simultaneous rendering.

KEY COMMANDS AND TERMS

- **Anti-aliasing**—the process of smoothing jagged transitions in raster data by blending the transition between dissimilar pixels.
- **Clone (VFB)**—the creation of a duplicate display of the VFB so that two or more renderings can be compared.
- **Compressed**—data that has been reduced to a smaller data set.
- **Data Format (File Format)**—the structure of data within a file; the three character extension that identifies the type of data structure contained in the file.
- **Digital Halftone Matrix**—the group of printer dots assigned to represent each halftone dot.
- **Distributed Rendering**—the process of breaking an animation task into portions that can be assigned to individual computers on a network.
- **Encapsulated PostScript (EPS)**—a device independent and resolution independent meta data format (raster and vector) used in print publications.
- **Extracted**—data that has been uncompressed.
- **Frame**—the unit of rendering; the condition of the workplace (design in VIZ, scene in MAX) at a given point in time during an animation.
- **Halftone**—the process of converting continuous tone into a pattern of dots for reproduction.

- **Hidden Edge**—a polygon edge (plane boundary) on a side pointing away from the camera.
- **Hidden Line Renderer**—the option in VIZ that allows geometry to be rendered as a wireframe of 2D vectors in DWG or EPS format.
- **Joint Experts Photographic Group (JPG)**—a 24-bit raster data format employing adjustable lossy compression used in Web publications.
- **Lossy**—the compression technique that removes differences in data to reduce file size. When expanded, missing data is reconstructed.
- **Portable Network Graphic (PNG)**—a flexible color depth raster data format employing lossless compression used in Web publications.
- **Printer Resolution (ppi)**—the capacity of a printing device to image fine detail expressed in printer dots per inch.
- **Quick Render**—the command that renders the active viewport using current rendering settings.
- **Render Design (Scene)**—the command that allows settings to be adjusted before rendering the frame(s).
- **Render Last**—the command that reexecutes the previous rendering although the active viewport may have been changed.
- **Rendering Farm**—the collection of computers used for distributed rendering.
- **Rendering Resolution (dpi)**—the capacity of a display device to image fine detail expressed in dots per inch; the density of dots in a rendered frame.
- **Rendering**—the process of interpreting geometry, materials, and lighting into a visual representation.
- **Silhouette Edge**—the limit of geometry with the background.
- **Supersampling**—the technique that determines the value of a rendered pixel (or group of pixels) by looking at adjacent pixel values.
- **Tag Image File Format (TIF)**—a 24-bit raster data format that includes tags (instructions) to optimize printing on high–quality printing devices.
- **Virtual Frame Buffer (VFB)**—a block of memory used to hold the currently rendered frame.
- **Visible Edge**—a polygon edge (plane boundary) on a side pointing toward the camera.
- **Windows Image File (BMP)**—a 24-bit raster format native to the Windows operating system.

BASIC RENDERING METHOD

All rendering is done to the Virtual Frame Buffer (VFB). This is a block of memory that holds the rendered data until you, or 3D Studio decides what to do with it. The data in the VFB is displayed in a separate display window where colors can be sampled, RGB channels turned on or off, or the data saved, printed, or cloned. When you render single frame or a series of frames in an animation, you see each frame, as it is rendered, in the VFB.

If you have geometry in your scene, you are ready to render. Three icons in the Main Menu provide basic rendering options.

- The **Quick Render** icon renders the active viewport using current rendering settings.
- The **Render Scene** icon provides access to the Render Scene dialogue where settings such as frame size, file name, file format, and more esoteric controls are found.
- The **Render Last** icon repeats the last scene rendered even though you may have switched to another viewport.

STEP 1

Activate the viewport you want to render. You can either left– or right–click anywhere in the desired viewport. If you left–click, you run the risk of moving something if **Select-Move** is the active identification method. Right–click will only activate the viewport.

Note: The display size of the active viewport has nothing to do with the rendering size. The dimensions of the rendering are set in the **Render Scene** dialogue and are not available for adjustment in a **Quick Render**.

STEP 2

Choose **Quick Render**. A **VFB** rendering window is displayed and 3D Studio's scan line renderer begins to render the scene (it calculates and then displays one line of the rendering at a time.) The background color or map used is set in **Rendering|Environment**.

During rendering the **Rendering** dialogue is displayed behind the rendering window and progress is shown in a progress bar. Rendering can be paused or cancelled at any time. Likewise, if the **Rendering** dialogue is closed, rendering is halted. Because 3D Studio knows how many frames are in an animation, rendering can be halted and recommenced later (as long as the animation didn't change and you write to the same file).

Note: Lights are critical in effective rendering. Review The basic lighting method in Chapter 14, Camera and Light Basics.

STEP 3

When the rendering is complete you have several choices.

- Close the rendering window. Rendering continues although you cannot see it. The **VFB** can later be refreshed by choosing **Render Last**.
- Minimize the rendering window. Rendering continues if not completed and the window is stored at the bottom of the interface. When another scene is rendered, this minimized window is expanded to display the new contents of the **VFB**.
- Print the rendering. This sends the rendered frame to the currently selected printer. By choosing **Quick Render** you accept the current rendering frame size and aspect ratio as established in the **Render Scene** dialogue.
- Save the rendering as a raster file. A standard Windows browser is displayed along with a drop–down menu of output formats.

IMPORTANCE OF OUTPUT FORMATS

Data format (file format) is an important consideration when rendering still frames. Several of 3D Studio's file formats have the popularity and flexibility that make them worthy of discussion (Figure 19.1). For a complete discussion of the file formats not discussed here, see Murray and VanRyper's definitive work *Graphics File Formats* (O'Riley and Associates, 1996).

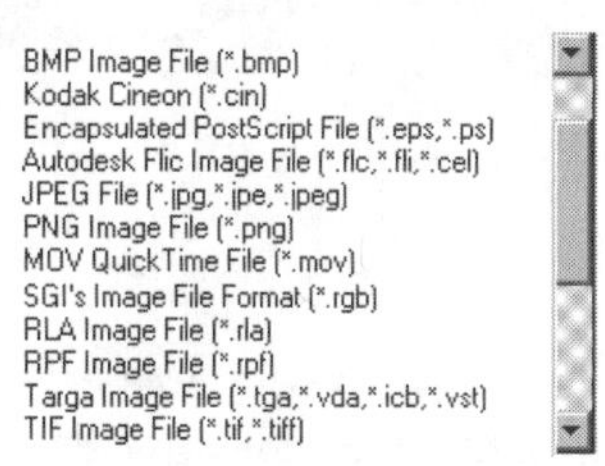

Figure 19.1 *Output formats.*

Windows Image File (.BMP). BMP is the native Windows bitmap file format. Using a form of *lossless Run Length Encoding (RLE)* compression, BMP is appropriate for black and white, grayscale, and color images with either 8- or 24-bits of color depth. Use this file format for still images that will appear in multimedia training and for storing renderings from 3D Studio before editing.

Encapsulated PostScript (EPS). Because 3D Studio renders only to a bitmap, this file format is a little misleading. By choosing this format, you place an EPS header on bitmap data so that it can be used as EPS data. You do not capture resolution-independent PostScript. Use this file format only if the application in which you will use the rendering expects .EPS data.

Joint Experts Photographic Group (JPG). The JPG format uses a *lossy* file compression suitable for distribution of continuous tone images by email or display on a Web page. It's also an efficient way to store a large number of continuous tone renderings. The fewer areas of solid color, the more appropriate is JPG. The higher the compression, the lower the file size but also the lower the quality when the file is *extracted*. Figure 19.2 shows JPG controls. As you move the **Quality** slider toward **Best**, the **File Size** slider automatically moves in the opposite direction.

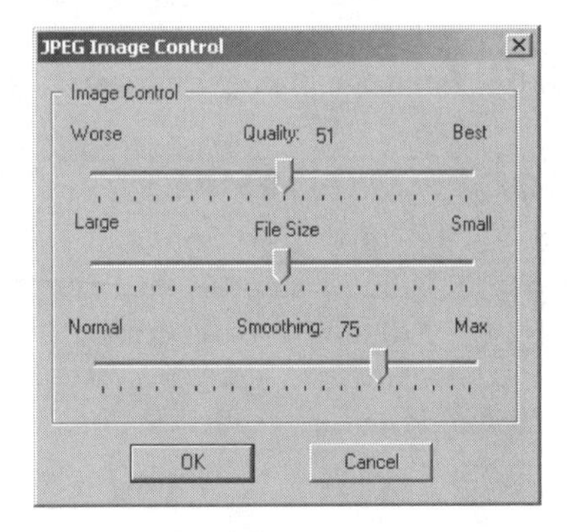

Figure 19.2 *JPG file format controls.*

Note: When a rendering has been stored in JPG format, data has been lost-even with quality set at **Best**. You cannot regain the quality by saving the extracted JPG data in a more robust file format, such as TIF.

Tag Image File Format (TIF). The TIF format as implemented by 3D Studio records the rendering in uncompressed gray scale or color. Because it is uncompressed, TIF is appropriate for saving renderings before they are edited for print publication. TIF contains *tags* that PostScript printers interpret to optimize the printing of continuous tone images. Use this file format for renderings that end up in printed documents.

RENDERING GUIDELINES

* Always render at the smallest practicable size. Quick checks should be rendered at 160 x 120. Output for the Internet can go to 320 x 240. Publication quality renderings should be a minimum of 1024 x 768.

- All renderings (other than **Hidden Line**) are fixed raster resolution. The only way to increase the quality (ability to show fine detail) is to increase pixel dimensions and then reduce the scale of the output in a program such as PhotoShop.
- Rendering time is directly related to geometric complexity. Additionally, it is directly related to the number of lights and materials in the scene, the type of material (ray-traced), and the rendering method (algorithm).
- Under most situations you will want to leave **Anti-aliasing** turned on. This blends the edges of areas that would normally be rough and pixilated.
- Leave **Supersampling** turned off unless materials show signs of odd patterns.
- Use **Force 2-Sided** to render materials on both sides of surfaces. This is independent of setting a viewport to render two-sided.
- Clone the virtual frame buffer to compare one rendering to another.

OUTPUT FOR TECHNICAL DOCUMENTS

Renderings that appear in technical documents can take the form of line renderings or high–quality *halftones*. Line renderings in VIZ (this option isn't available in MAX) make use of a special **Hidden Line Renderer** that must be substituted for the standard scan line renderer.

STEP 1

Choose **Tools|Options|Current Renderer:Production** and click on the **Assign** button. Choose the **Hidden Line** renderer from the list (Figure 19.3).

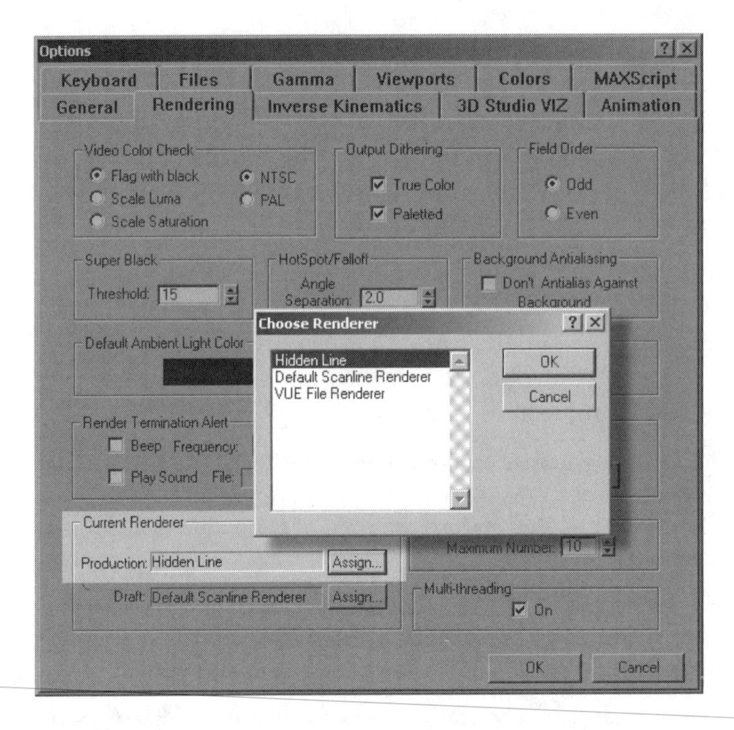

Figure 19.3 *Switching to the Hidden Line Renderer.*

Line renderings make use of a hidden line algorithm that determines which polygon edges are on the front and which are on the rear. It is the only case where VIZ renders a vector file. The **Scan Line Renderer** roll-out is replaced in the **Render Design** dialogue with a **Hidden Line** roll-out (Figure 19.4). This figure contains the settings for optimum line output. Several peculiar things surround this hidden line rendering:

- Always choose the desired file format *before* leaving the roll-out. Although you have the opportunity to switch between DWG and EPS file formats in the **Save** dialogue, this may not work.
- The background of the rendering in the **VFB** will be black. The environment color has no impact on the background of the rendering. The viewport background color has no impact on the background of the rendering.
- The black background means that you need white lines to see the impact of line type and line thickness.
- The vectors generated by this hidden line function will have to be changed to black in the CAD or drawing application to be used. Only the vectors are written to the file, not the background.

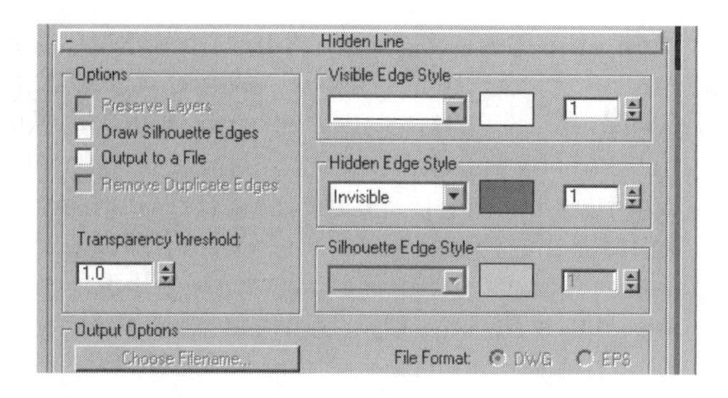

Figure 19.4 ***Hlidden Line*** *roll-out.*

Note: Although intersections of overlapping objects are determined visually, no line of intersection is created when they are rendered as lines. Perform a **Boolean Union** to get a line of intersection between spatially intersecting objects.

STEP 2

Choose Rendering|Render. You are presented with the **Render Design** dialogue. Open the **Hidden Line** roll-out, then choose the following:

- **Output to File.** This activates the silhouette option and provides DFX and EPS output. If you only render to the **VFB**, you can't save vectors, only the typical raster formats.

Tip: As you will see, rendering to the **VFB** and then saving the hidden line raster file is not all that bad. A quick inspection of Figure 19.7 reveals the marginal usability of 2D vectors out of 3D Studio. The alternative is to perform the hidden line rendering, but to a high resolution raster file, 1600 x 1200 for example. This file can then be scaled down for high–quality printing.

- **Visible Edge Style|Line Type** lets you assign a line type to internal plane edges. Choose a solid line or broken line, if surfaces have many edges.
- **Visible Edge Style|Line Thickness** lets you assign a thickness to all internal plane edges. Make this a small number.
- Accept the white color as displayed in the color chip.
- **Hidden Edge Style|Line Type** determines style of lines at the rear. Set this to **Invisible**.

One of the hidden line **Options** is **Draw Silhouette Edges**. This affects the boundary edge of objects in the scene, but again, in a peculiar way. If you set **Silhouette Edge|Line Style** to a solid line and **Silhouette Edge|Line Thickness** to a number twice that of visible edges, you will get a nice bold outline. However, visibility for these silhouette lines is not determined. To get an acceptable representation, the outline must be edited later. Figures 19.5 through 19.7 show the impact of these options on the Pulley Assembly from Figure 8.20. Figure 19.5 shows the **VFB** display of only visible edges. Figure 19.6 shows the same part with **Silhouette Edge** set to a thick line. Note the failure to determine correct visibility. Figure 19.7 shows an enlarged detail of the segmented vectors. You can see that extensive editing is necessary to make this silhouette technique useful.

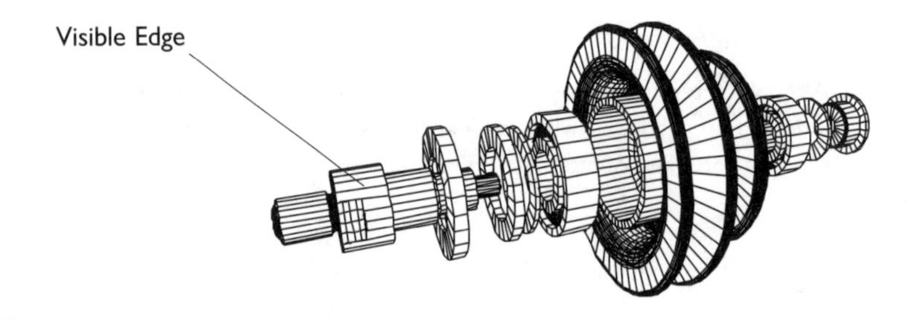

Figure 19.5 *Visible edges displayed.*

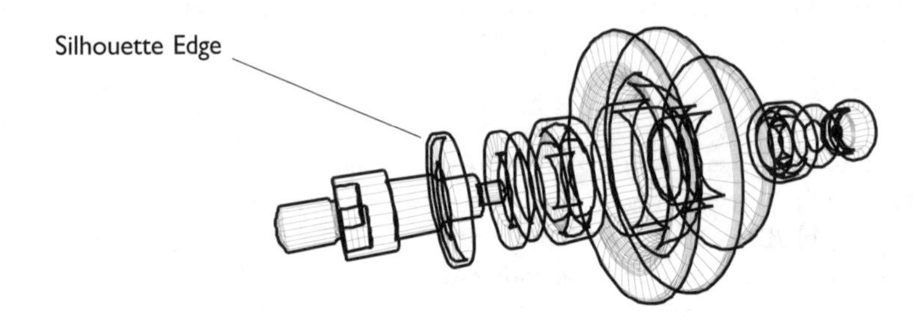

Figure 19.6 *Visible edges and silhouette outline displayed.*

A line rendering has the benefit of being abstracted. You are presented with a lower level of visual information, making part identification and functioning easier to understand. To make the rendering even easier to understand, we omit hidden visible edges all together, much as we omit hidden lines from pictorial drawings.

Note: Depending on the geometry in your design, you may be able to turn the visible edges invisible and only show the silhouette. This may turn the rendering into a cartoon outline but is worth a try.

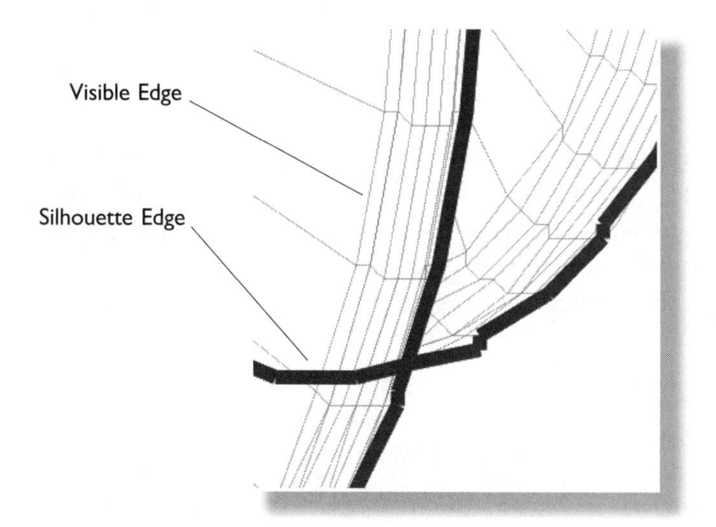

Figure 19.7 *Silhouette outline must be extensively edited.*

The same 3D studio file can be rendered for a high–quality publication. The key to rendering for high–quality print reproduction is to match the resolution of the rendering to the resolution of the printer or printing press. High–quality publications are printed at 1240 ppi and higher. This printer resolution (ppi) should be evenly divisible by the rendering resolution (dpi). This gets a little tricky, but here's the method.

STEP 1

Determine the desired size of the rendering when it's printed. Let's say the rendering will be printed 2" wide.

STEP 2

Determine the dpi of the final rendering: 1240/8=155 dpi. This yields an 8x8 digital halftone matrix capable of displaying 256 separate values for every dot on the rendering.

STEP 3

Determine the overall dimensions of the rendering. At 2" in width the rendering will be 155 x 2=310 pixels wide. With an aspect ratio of .75 (the

800 x 600 frame shape), the height would be 310 x .75=232.5 or 232 pixels high.

STEP 4

Render the scene to a 310 x 232 custom TIF file (Figure 19.8). This file is rendered at a default resolution of 72 dpi so its overall width is approximately 4.3". In a raster editor such as PhotoShop, scale the rendering. This can be done by increasing the resolution to 155 dpi, *without resampling*. Because of our calculations, the rendering ends up being 2" wide.

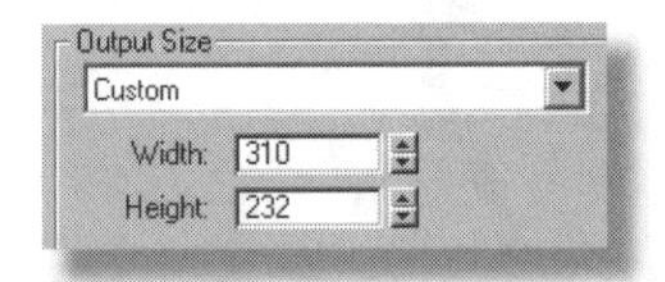

Figure 19.8 *Custom output size based on printer resolution.*

The completed 2"–wide 155–dpi rendering is shown in Figure 19.9.

Figure 19.9 *Rendering scaled to 155 dpi in a raster editor.*

STILL OUTPUT FOR WEB

Two factors impact rendering still frames for display on Web pages. The first is color depth and the second is resolution. Two 3D Studio rendering formats are appropriate for posting images to the Internet: JPG (explained earlier) and *Portable Network Graphic* (.PNG). The PNG format is provided in lieu of the more prevalent *Graphic Interchange Format* (.GIF) and can be successfully used for renderings with large areas of solid colors (like white or black backgrounds). PNG uses lossless compression and has many of the same features as GIF. JPG is more appropriate for renderings with continuous tones and few, if any, solid areas.

Note: How important is it to select the correct format for the Web? For displaying one or two images, probably not very important. But for Web manuals that might have thousands of images, this is critical. For example, the file size of a continuous tone image saved in GIF format may actually be larger than the original file.

All Web pages display graphics at fixed screen resolution. This is usually 72 dpi, but may be as high as 90 dpi. Images of greater resolution must be mapped to the resolution of the screen, something you shouldn't make the Web browser do. A 300 dpi graphic (unless it is scaled in the HTML tag) will end up approximately four times larger as its resolution is mapped to the 72 dpi of the screen. Additionally, you shouldn't make the browser rescale a graphic to make it fit the desired size. Not only does this slow download and display, results can be unpredictable. So use these guidelines:

- Use JPG for renderings having many smooth tones.
- Use PNG for renderings having large areas of solid color.
- Use an output size in 3D Studio that produces a rendering of the correct dimensions at 72 dpi.

Including a 3D Studio rendered frame in a Web page is relatively straightforward. You need an HTML file with an "image source" tag that calls the graphic from its storage location. By using the dimensions of the graphic in the HTML tag, you speed up display because the browser doesn't have to calculate the needed real estate. The line of HTML below brings a JPG file named *rendering.jpg* to the Web page from a directory named *images* that is at the same level as the current HTML file.

**<IMG SRC="images/rendering.jpg" WIDTH=310
HEIGHT=232>**

If the JPG file contains more colors than the browser is capable of displaying (more than 256 including white and black), the image is *dithered*. This means that the missing colors are approximated by placing colors that are within the palette next to one another so that the viewer's eye assembles the correct color. An example of this would be placing a blue pixel next to a yellow pixel so that the effect is green.

Figure 19.10 shows the pulley assembly, rendered to JPG format, displayed in the Web browser.

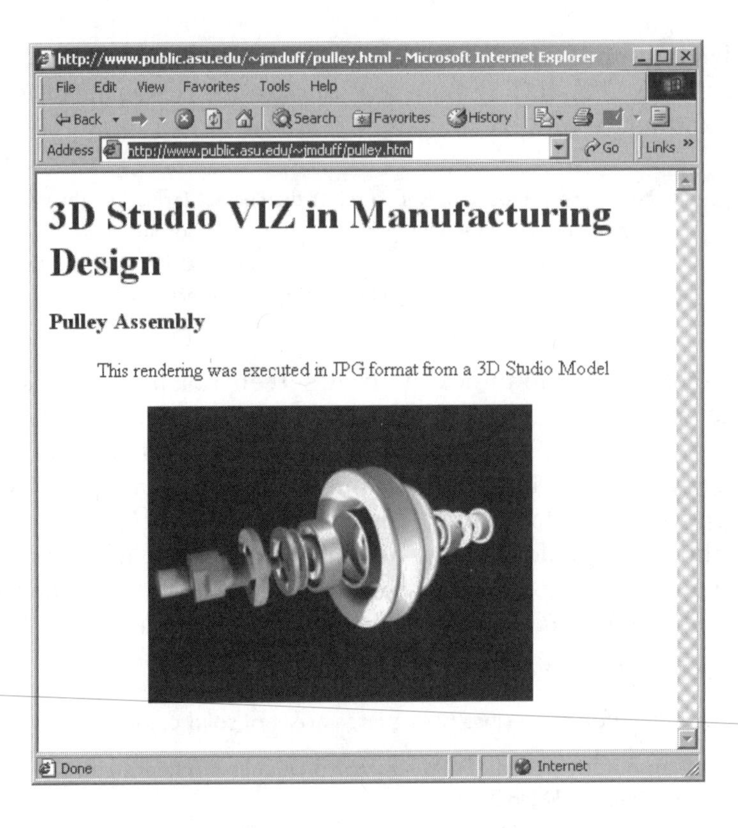

Figure 19.10 *File rendered to JPG format and included on a Web page.*

STILL OUTPUT FOR MULTIMEDIA TRAINING

Today's multimedia training makes extensive use of still and move output from 3D Studio. Renderings from 3D Studio find their way into programs such as Macromedia's Director and Authorware, Asymetrix's Toolbook, and even Microsoft's PowerPoint. Consider the following when preparing renderings for eventual use in multimedia.

- A multimedia authoring application may work from a reduced color palette. 3D Studio works with a 64-bit palette so images rendered from 3D Studio will usually be dithered when displayed in multimedia applications.
- Although many applications accept native files from a number of graphics applications, avoid this. These files contain significant header information unneeded by the authoring application and they increase program size.
- Avoid using TIF files in multimedia programs unless the printing of a document is an important feature (and then a PDF document would be a better choice).

- Even though a few multimedia programs have integral vector and raster editors, you'll almost always do better to use dedicated programs…like 3D Studio.

Figure 19.11 displays a multimedia interface modeled in 3D Studio and rendered to a BMP file. Because of 3D Studio's modeling and rendering prowess, realistic results can be achieved that simply are impossible in 2D drawing and painting applications. As a final touch, our JPG Pulley Assembly is displayed inside the interface as part of an Authorware presentation.

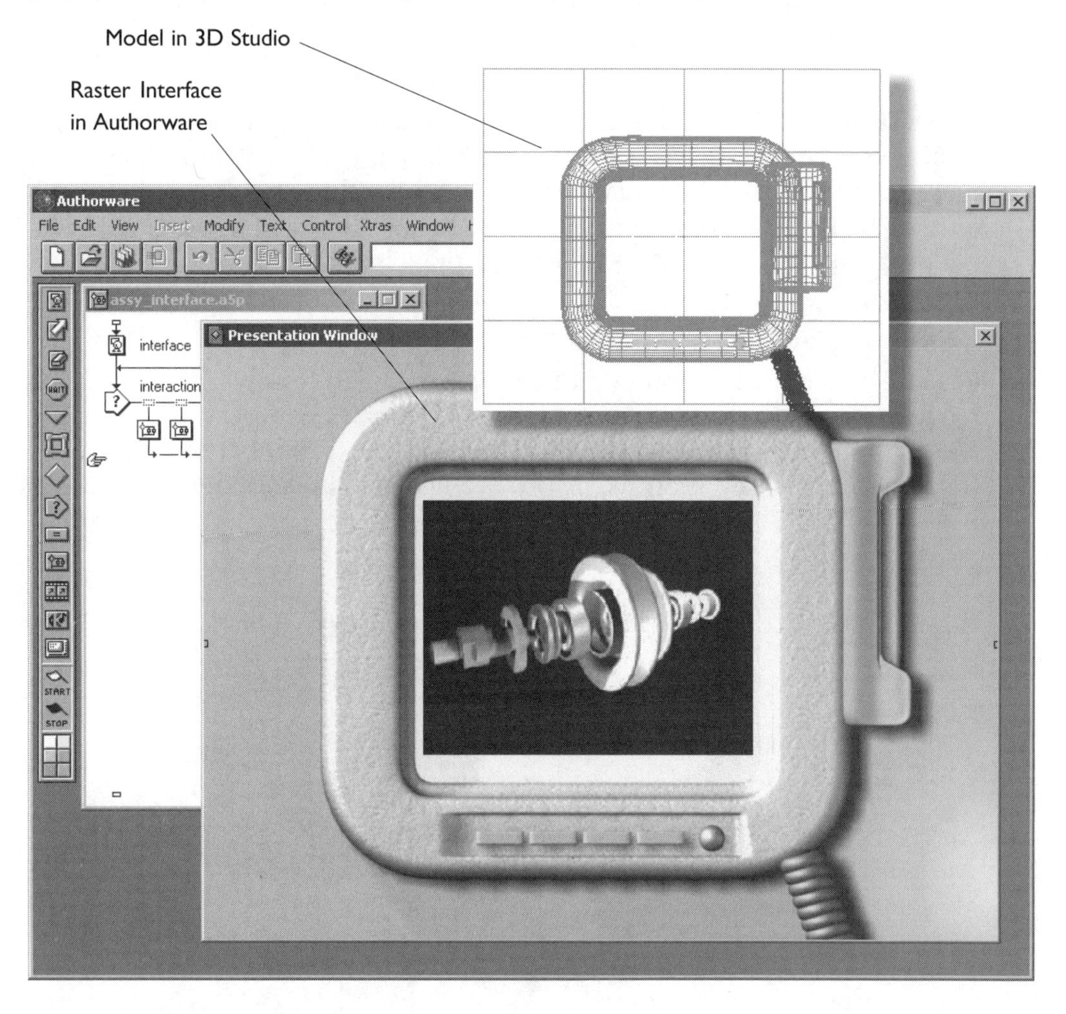

Figure 19.11 *Multimedia interface created in 3D Studio (Macromedia Authorware).*

ANIMATION FORMATS

Animation requires a time-based format; two formats are widely used and discussed here: Microsoft's Audio/Video Interleave (.AVI) and Apple's QuickTime (.MOV).

Audio/Video Interleave (.AVI). This movie format is the ubiquitous choice on Windows platforms and because 3D Studio is native to Windows, this is probably your first choice in output. Other computer animation formats are available such as Real Video and Flash. Both AVI and MOV animations can be reformatted in nonlinear digital editing applications for numerous purposes.

QuickTime (.MOV). This is the industry standard for multichannel time–based data. Though it has a more robust data structure than AVI, the fact that QuickTime for Windows doesn't ship on Windows machines limits this format. However, if your animations are being processed by a commercial video studio, they may well ask for animations in MOV format.

BASIC ANIMATION RENDERING METHOD

Because animation is comprised of a series of rendered still frames, the same rendering considerations come into play for animations that were important for still frames. In fact, you will want to render out a number of still frames, starting with the **Key Frames**, just to see how the animation looks.

STEP 1

Move the animation slider to the frame you want to render (Figure 19.12).

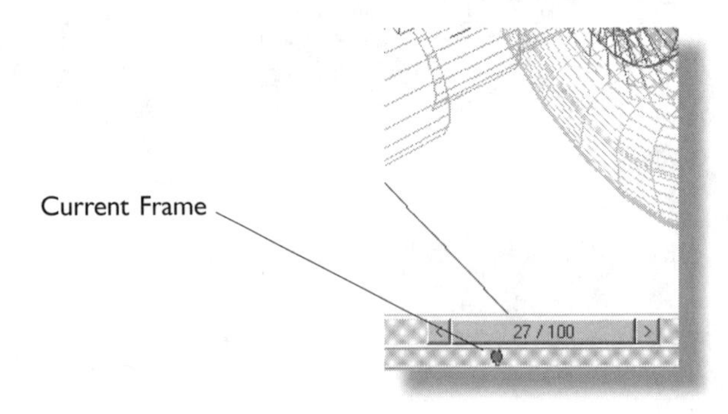

Figure 19.12 *Frame to be test rendered.*

STEP 2

Choose **Rendering|Render** and select **Common Parameters|Time Output|Single** (Figure 19.13). This renders the current frame of the animation. You can print single frames for comparison or **Clone** (twin person icon in the rendering window's main menu bar) the **VFB** each test so you have an on-screen record. It is helpful to make note of the time it takes to render each test frame. If each frame has roughly the same number of lights, materials, and complexity of geometry, you can estimate overall rendering time.

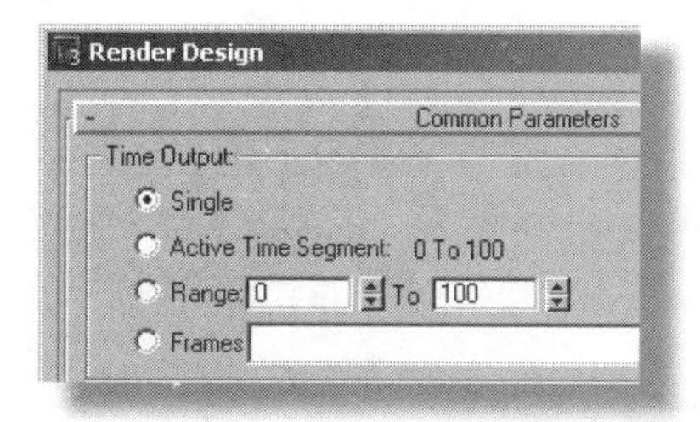

Figure 19.13 *Test render representative frames before rendering all frames.*

STEP 3

Render the entire animation. When you are satisfied with the test frames, you are ready to render the entire animation. Multiply the average time for the test frames by your total frame count to arrive at an estimated elapsed time. This gives you an idea of whether or not you should render overnight, over the weekend, or while on vacation. In the **Render Design** dialogue choose **Active Time Segment** (refer to Figure 19.13). This renders the full animation.

STEP 4

Choose **Render Output|Save File|Files** and browse to the desired storage location (Figure 19.14). Choose **AVI File (*.avi)** from the drop down–menu. Now here's the tricky part: choosing the desired CODEC. If you don't know what CODEC you need, choose **Full Frames (uncompressed)**. If you have a large and lengthy animation you'll need tens or hundreds of megabytes of free disk space. Otherwise, choose the desired CODEC.

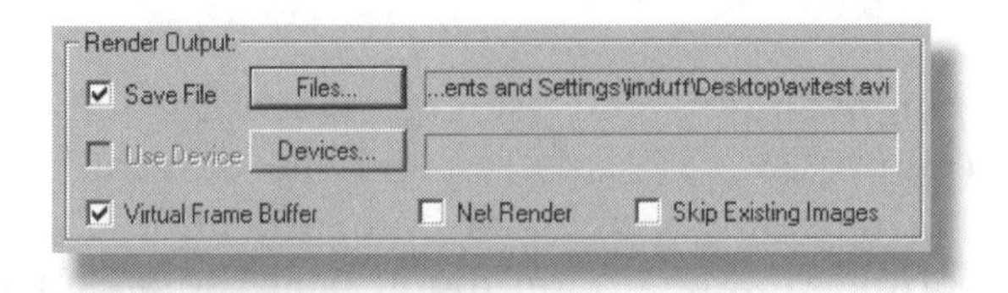

Figure 19.14 *Render the animation to a file.*

When you return from the browse window *you have only established the path for the animation*. Click **Render** at the bottom of the dialogue to actually start the rendering.

Tip: If you are unsure of what CODEC to use, render short time segments of 10 to 20 frames at 160 x 120 using the various CODECs. You can easily determine compatibility as well as compare compression quality.

ANIMATION FOR THE WEB

Both AVI and MOV movies can be displayed on Web pages. However, AVI is recognized natively on Windows machines by both Internet Explorer and Netscape Web browsers. QuickTime movies require the downloading of a free plug-in from http://www.apple.com.

The HTML code for including animations on a Web page is, like that for still images, rather straightforward.

<EMBED SRC=”movies/pulley.avi”>

This HTML tag instructs the browser to look for the file pulley.avi in the movies directory that exists at the same level as the current HTML file. The animation is displayed on the page in a viewer that allows interactive play (Figure 19.15).

PROBLEMS

Practice your still and movie rendering on the following problems:

Problem 19.1 *Reference Problem 16.1 Gear Puller Assembly. Render an 800 x 600 color TIF frame of the assembled Gear Puller in front of a wall. Establish lighting that creates a drop shadow. Animate the assembly in a twenty second AVI including the movement of the Arms as the Spread Nut moves downward.*

Problem 19.2 *Reference Problem 17.1 Double–Acting Cylinder. Create an 800 x 600 color TIF frame of the assembled Air Cylinder with the outside transparent so the inside can be seen. Animate the operation of the Air Cylinder to include colored air representing air on either side of the piston. Finish by rendering a 20–second, 15 FPS AVI file.*

Problem 19.3 *Reference Problem 17.2 Double–Bearing Support. Animate the assembly of the Bearing support in a 15–second, 15 FPS AVI file. Show the assembly of the tubes and fasteners.*

Problem 19.4 *Reference Problem 17.3 Tube Holder. Create a publication–quality rendered frame showing the Tube Holder in favorable pictorial position. Create an environment for the Tube Holder in which several Tube Holders are installed to hold a length of Tube. Animate the installation of the Tube as well as any tools needed to accomplish the task.*

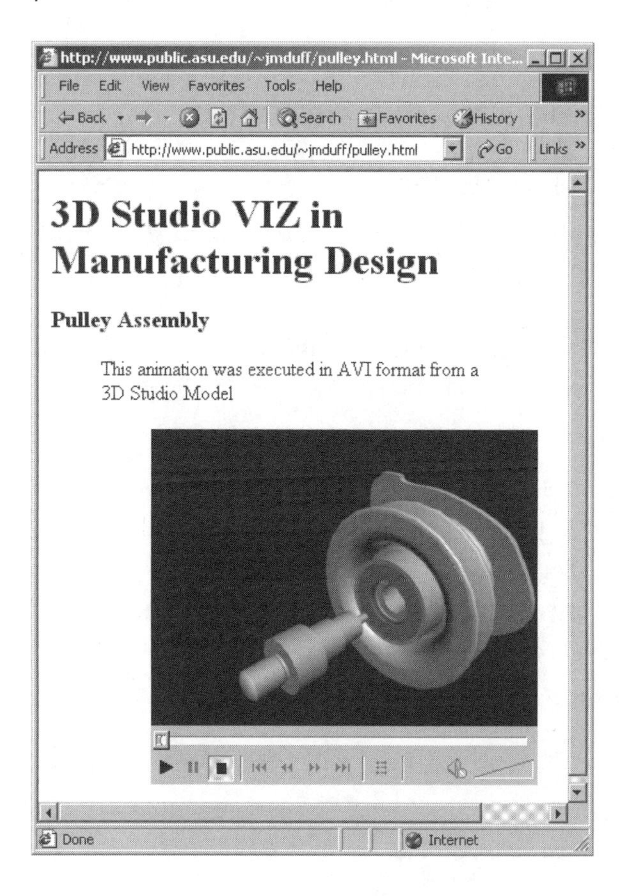

Figure 19.15 *Animation displayed on Web page.*

References for Further Study

BOOKS ON ENGINEERING DRAWING

Bertoline, G. R., Wiebe, E. N., and Miller, C. L. *Fundamentals of Graphics Communication.* WCB McGraw-Hill.

Earl, J. H. *Graphics for Engineers with AutoCAD 2000i.* Addison–Wesley Publishing Company.

BOOKS ON CAD AND MODELING

Leach, J. A. *AutoCAD Companion 2000.* McGraw-Hill Publishing Company.

Frey, David. *AutoCAD2000: No Experience Required*. Sybex Publishers.

Mortenson, M. E. *Geometric Modeling*. John Wiley & Sons.

Stellman, Thomas, and Krishnar. *Harnessing AutoCAD 2000*. AutoDesk Press.

BOOKS ON RENDERING AND ANIMATION

Birn, Jeremy. *Digital Lighting & Rendering*. New Riders.

Kerlow, I. V. *The Art of 3-D: Computer Animation and Imaging,* 2nd Edition. John Wiley & Sons.

WEB RESOURCES

AutoDesk at http://www.autodesk.com/.

Ultimate 3D Links at http://www.3dlinks.com/.

3D Cafe at http://www.3dcafe.com/.

Automated Design Systems at http://www.autodes.com/.

VIDEO RESOURCES

Engineering Drawing & Design Video Sets #1 & 2 (CD). Delmar Publishers.

VMS, Inc. Online at http://www.VMS-online.com/. Source for CAD training videos.

INDEX

C

D

H

I

J

K

L

M

N

O

P

Q

R

S